The Cajun Prairie: A Natural History

To: Sue

The Cajun Prairie: A Natural History

By

Malcolm F. Vidrine, Ph.D.

The Division of Sciences and Mathematics
(Louisiana State University Eunice) and
The Cajun Prairie Habitat Preservation Society and
The Cajun Prairie Gardens
(Eunice, Louisiana)

Malcolm F. Vidrine
Eunice, Louisiana
2010

Front Cover image: *Cajun Prairie* is a 14" x 18" color pencil drawing by Corinne Louise Greenberg. http://thegardenisateacher.com

Cover designed by Van Reed

 malcolmvidrine@yahoo.com
 Published 2010

ISBN (paper): 978-0-615-36813-9

Printed on acid-free paper in Canada.

Dedication

To my wife Gail; she has carried the burden of doing so many things to permit my interests to grow. And to my children; each provided me with a separate adventure. Like so much else, this book will be part of their legacy. I hope it helps to explain our front yard. May this book also explain many more front yards!

Contents

		Page
Preface		ix
Chapter 1	Introduction	1
Chapter 2	The people of the prairie	13
Chapter 3	Pre-settlement to 1870s	23
Chapter 4	1870s-1930s	47
Chapter 5	1940s-1970s	57
Chapter 6	1980s-2010	63
Chapter 7	Abiotic features of the prairie	89
Chapter 8	Biota: Plants	117
Chapter 9	Biota: Animals	141
Chapter 10	The future of the Cajun Prairie	163
Literature Cited		171
Appendix 1	Timeline for the Cajun Prairie	197
Appendix 2	Genealogy of MFV	199
Appendix 3	Spanish land grant	207
Appendix 4	Autobiographical comments	209
Appendix 5	An essay on the Cajun Prairie of the 1600s	213
Appendix 6	Flora of the Cajun Prairie	219
Appendix 7	French/Native American names of plants and animals	229
Appendix 8	Flora of the prairies by Clair Brown	231
Appendix 9	Phenology/blooming times of plants	233
Appendix 10	Factors affecting blooming times	239
Appendix 11	Animals (invertebrates) of the Cajun Prairie	241
Appendix 12	Vertebrates of the Cajun Prairie (Huner's list)	251
Appendix 13	Flight seasons of dragonflies and butterflies	259
Appendix 14	The Cajun Prairie Habitat Preservation Society	261
Appendix 15	The Cajun Prairie Gardens	267
Appendix 16	Creating or recreating a prairie by Charles Allen	269
Appendix 17	Landscaping tips	275
Appendix 18	List of benefits of restoration	277
Appendix 19	Exotics	279
Appendix 20	Obvious blooming plants and landscaping problems	281
Appendix 21	Comparison of native plant species distributions among remnant prairies in the 1980s	283
Appendix 22	Remnant Status: Morbidity and Mortality	287
Appendix 23	Characteristics of the Cajun Prairie	291
Appendix 24	Differences between Cajun Prairie & Midwestern prairie	293
Appendix 25	Books/articles/films regarding the Cajun Prairie	295
Appendix 26	List of volunteers	297
Index		301

Preface

For this effort, I am taking on the role of scribe, but I am using my credentials and personal experiences to annotate the story of the Cajun Prairie—a natural story, i.e. a natural history. I have taken the liberty of making aspects of the story personal, and I hope that by doing this, I have not made the story less valid. I am not an historian, so I have relied on others with those credentials to inform the story. I am not a botanist, so I have relied on others with those credentials to inform the story. I am not a vertebrate zoologist, microbiologist, agronomist and much more, and I have relied on others to inform the story. And I am not the great organizer of people that was needed to do many of the projects described herein; however, I am just one of the workers. The story will make clear, I hope, those people who are the inspired leaders who deserve the respect and reward for their efforts.

This book is really a compilation of the work of several biologists including Charles Allen, Larry Allain, Jim Grace, Bruno Borsari, Bill Fontenot, Peter Loos, Marc Pastorek, Domingo Jariel, Jameel Al-dujaili, Jay Huner, Sheila Renee Hazelton-Robichaux and Avery Williams. Claude Oubre provided guidance in my presentation of the history of the region. Tom Hillman helped with the re-evaluation of the remnant prairies. Myriam Stanford, James Cordes, John Hamlin, and others provided additional guidance. Many others helped in a variety of ways. The Cajun Prairie Habitat Preservation Society and its members have been most instrumental in all aspects of this work. The numerous people who attended the North American Prairie Conferences over the past decade have provided lively discussion and interest in our work. Gerald Patout, LSUE Head of the Library, read the book and provided very useful comment.

My plan of attack was as follows. Chapters 1 and 2 provide an introduction to the Cajun Prairie, an introduction to the people of this prairie and an introduction to the author. Chapters 3 and 4 provide a history of the Cajun Prairie based upon published journals and articles focusing on periods from pre-settlement to the 1930s, when the prairie was considered essentially extinguished. Chapters 5 and 6 provide the first natural history studies on the Cajun Prairie, a discussion of remnant prairies, rediscovery of the prairies in the 1980s and prairie restoration. Chapters 7-9 provide an introduction to the ecology of the Cajun Prairie, including discussions of abiotic factors (climate, fire and hydrology), plants, animals, agriculture and biodiversity. Chapter 10 provides closing remarks. Twenty-six appendices follow the text in order to provide further information on a myriad of topics, including: the author's genealogy, lists of major plants, lists of major animals, blooming times, plant assemblages, The Cajun Prairie Habitat Preservation Society, flight seasons of butterflies and dragonflies, restoration tips, landscaping tips, exotic plants and animals and a status report on the remnant prairies.

Louisiana State University-Eunice provided me with sabbatical leave in the spring of 2008 to draft this book. Opelousas General Hospital Endowed Professorship for Science in Support of Nursing (Cajun Prairie Project) in the 2008-09 year through the LSU Eunice Foundation provided funds to help defray costs of color plates. My own Division of Sciences and Mathematics under the leadership of Renee Robichaux has provided me with not only a reasonable workplace but also with comradeship and ever-worthy intellectual challenge. Although we are heavily overworked and underappreciated, we collectively have published a remarkable number of scientific papers. Only in this atmosphere would I have been able to tackle the challenge of writing this manuscript. So thanks to all my faculty and colleagues at LSUE who in their own way would not let me quit and more insisted that I complete this effort. Gerald Patout and the LSUE library staff provided me with a variety of kinds of assistance, including interlibrary loans.

Appendix 14 introduces The Cajun Prairie Habitat Preservation Society—a non-profit organization focused on preserving and restoring Cajun Prairie. Appendix 26 lists the many people who have helped in various ways with restorations and with this book. If I have forgotten anyone, I am sorry for the omission. Any mistakes or misstatements in this book are truly my own. Works like this are so difficult to do in that they are commonly riddled with minor errors and a couple of large ones to boot. Almost 40 years ago, Nell Causey, my major professor for my master's degree, asked, "Are you afraid to make mistakes in print?" I responded most definitely. She then said, "You only make mistakes in print if you publish—so go out and make a lot of mistakes. For me, it will mean that you have published a lot!" She was right!

I am continuously amazed by the fact that others are interested in the prairie. After so many encounters with those who not only dislike but also despise it, I am so pleased to see changes of heart and interest from young people. The Cajun Prairie was so fortunate to have a savior in Charles Allen—his energy and sheer interest in everything Cajun Prairie always amazed me. He has been a beacon of hope for me and for many of the prairie enthusiasts in Louisiana and surrounding states.

The prairie is a habitat that, like the forest, generates a sense of belonging although remaining somewhat alien for humans. Humans evolved in the mix of forest and grasslands; it is obvious as I watch neighbors and humans in general recreate this habitat around their homes as if this is the only way it should ever be. Prairies however are the best places in the world to grow the massive grain crops and a variety of animals that our human population needs to survive. Unfortunately, population increase and the present use of these resources are unsustainable; alternative futures must be discovered or rediscovered from our past. The collapses of other communities around the globe are harbingers of the future of southern Louisiana—the signs are there. Loss of wetlands, pollution of soil, air and water, failing agriculture and even economic collapse are among those signs. I once said to my friend Bruno, and he often repeats, "A nation only lasts as long as its topsoil." The topsoil in the Cajun Prairie is essentially gone, my own yard—once a fertile field—was covered by an inch of sand and a hard clay-pan. The topsoil is gone. Prairie creates topsoil; prairie is 'embryonic' topsoil.

I want to personally thank all of the prairie enthusiasts who made it possible to have a story to tell. James Wilson, Jr., Carl A. Brasseaux, and Jessica Hornbuckle of the University of Louisiana at Lafayette Press are acknowledged for their efforts in the production of this book.

Chapter 1
Introduction

Could all the vast sheet of prairie and woodland, marsh and water, hill and valley be placed on canvas by a skillful artist, it would give to the admirer of nature's charms "a world of beauty in a nut shell." (Col. Samuel Lockett (1969)).

Scholars have yet to explore the complex relationship between humans and their environment in Acadiana, a subject of tremendous relevance as the area's vulnerable wetlands are literally washed away, threatening the continued existence of regional subcultures that exist in these remote and forbidding habitats. Environmental factors also profoundly shaped the local societies' material culture, yet another subject that awaits comprehensive analysis. (Carl Brasseaux (2005)).

As a youngster, I ran barefoot nearly everywhere most days of the year. I recall wearing shoes for church and school and the occasional trip into town. But the long days working on the farm, hoeing and picking cotton, milking cows and slopping hogs, gardening and caring for the chickens, were barefoot days. Some jobs required shoes, like hauling hay and winter chores, but most did not. So what does all this bare-footedness have to do with anything? I want to write about soil, the very soil that was between my toes, and plants and bugs and people. The prairie of my youth is part of something much larger.

Prairies are the American grasslands (Savage 2004). Grassland is one of the major biomes of the world, sometimes called steppes, savannas, and even marshes. Every continent has grasslands except Antarctica, which is covered in ice. These grasslands are the breadbaskets of the world, and as suggested by their name, grasslands, are now predominately used to grow grasses for food. Major food grasses include corn, rice, wheat, rye, oats, barley, and sorghum/milo. The grasslands as natural areas predominately had grasses but they also had numerous forbs (non-grass flowering plants) and a few trees. They intergrade along their edges with adjacent biomes, e.g., forests, marsh, deserts, etc. One of the largest grasslands in the world is the Midwestern prairie ecosystem of the United States.

The Midwestern prairies developed east of the Rocky Mountains in the rain shadow (lee) of the mountains as they formed (Savage 2004). To the east, the wettest prairies are tallgrass prairies, moderately wet are the midgrass prairies, and driest are the shortgrass prairies that intergrade with the desert biome. The tallgrass prairies intergraded with the eastern forest biome of the Americas. A distinctive tallgrass prairie formed along the Gulf of Mexico extending along the northwestern shores and is called the coastal prairie. The prairie extends from Corpus Christi, Texas, north and east to New Iberia, Breaux Bridge, Lafayette, Opelousas, and Ville Platte, Louisiana—forming a broad northeastern boundary along the Atchafalaya Basin and its associated bottomland hardwood forests. The Louisiana portion of this coastal tallgrass prairie is called the Cajun Prairie.

The Cajun Prairie occupies most of southwestern Louisiana and covers an area of approximately 2.5 million acres (Allen and Vidrine 1989). Like the Midwestern prairies, it was a tree-less grassland with numerous wildflowers (forbs) that produced the magnificent soils that now provide some of the most agriculturally productive lands in the world.

The Cajun Prairie has had many names, but it initially was known by the names of the Indian inhabitants, the Opelousas (Opelousas Prairie) and the Attakapas (Attakapas Prairie) (Kniffen 1965). The Attakapas were mentioned by M. Le Page Du Pratz, who spent 16 years in Louisiana from 1718-1734 (Du Pratz 1758). The prairie was later called the Great Prairie in southwestern Louisiana or the Great Southwestern Louisiana Prairie or the Cajun prairies referencing the named smaller unit prairies making the larger complex of prairies. These prairies form an even more complex ecosystem with distinctive interactive communities inhabiting different soil types as well as different subdivisions that are noticeably based upon the different plant associations.

Maps 1-3 provide views of Louisiana and its prairies and specifically identify the locations of the named prairies of the Cajun Prairie. They also provide a general perspective on the natural vegetation and the courses of the major rivers in and outlining the Cajun Prairie. Map 1 depicts named inland prairies in northern and eastern Louisiana, but more obviously it frames the extent of the Cajun Prairie (labeled The Great Southwest Prairie for obvious geographical reasons). The two major parishes in the open prairies are Acadia and Jefferson Davis parishes. Other parishes however contained some prairie and their histories are intertwined including St. Landry, St. Martin, Iberia, Evangeline, Vermilion, Cameron, Calcasieu, Allen and Lafayette Parishes. The major towns in the prairies were initially tied to ports and bayous, e.g., Opelousas, Washington, St. Martinville, New Iberia, Lake Charles and Mermentau. Later the towns developed along trails at post offices and along railroads, e.g., Crowley, Rayne, Jennings, Eunice, Ville Platte, Kinder, Vinton, Mamou, Welsh, Iowa, Fenton, Duson, Church Point and Vidrine. The towns in the open prairie have decidedly Midwestern names (e.g., Crowley, Welsh, Jennings, Eunice and Iowa) indicating the settlement of this region by Midwesterners as the prairie was changed into a farming economy.

Map 2 details the Cajun Prairie and shows the major cities and the named prairies of the region. One can outline the prairie region by connecting the cities in this order: New Iberia, St. Martinville, Breaux Bridge, Opelousas, Washington, Ville Platte, Pine Prairie, Kinder, Lake Charles, west to Vinton, and east to Gueydan, Abbeville, Jeanerette, to New Iberia. Towns like Eunice, Crowley, Jennings, Welsh, Elton, Mamou, and Church Point are in the Prairie proper. The prairie is divided by streams with gallery forests into named prairies: Cote Gelée, Cypremort, Au Large, Grand Prairie Cherrami (locally called the Gros (or

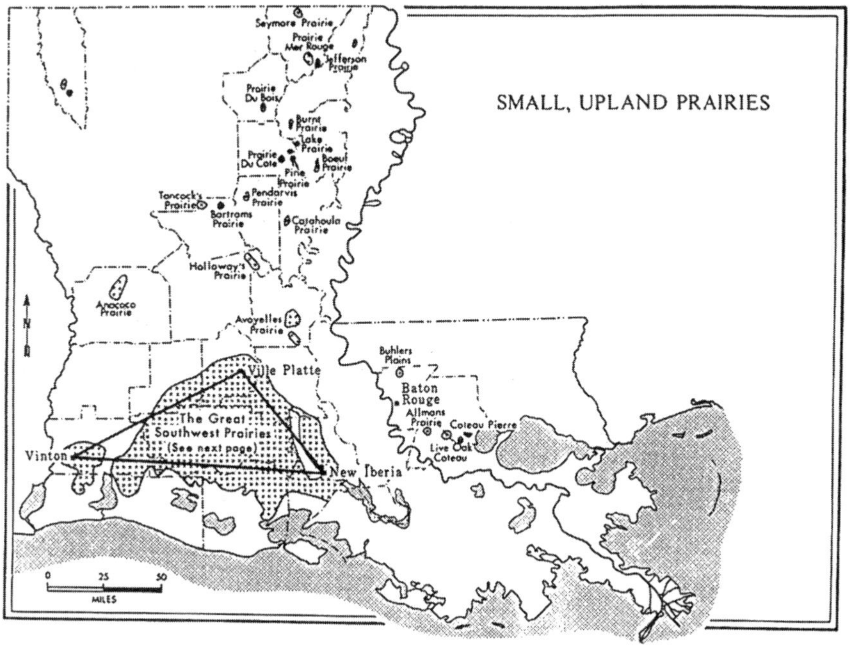

Map 1. Small, upland prairies of Louisiana. Reprinted with permission from Newton, Milton B. 1972. Atlas of Louisiana: A guide for students. School of Geoscience Misc. Publ. 72-1. Louisiana State University, Baton Rouge, Louisiana.

Grand) Chevreuil Prairie (fide Claude Oubre, 2010)), Beau Basin, Grand Prairie, Vermillion, Mermentau, Plaquemine, Hays, Robert's Cove, Prairie Faquetaique, Mamou, Prairie Swallow, D'Arbonne, Hickory Flats, Calcasieu, Little Calcasieu, Sabine, Choupique, and Long Prairie.

The main river drainage of the prairie is the central river known as the Mermentau River. Its entire drainage basin is in the prairie region. The river dissects the prairie into named prairies. Map 3 shows not only the natural vegetation regions of Louisiana but also the fingerlike distributaries of the Mermentau River and their gallery forests as these distributaries dissect the Cajun Prairie. The eastern boundaries of the Cajun Prairie are the Bayou Teche and the Atchafalaya River and Basin, while the western boundaries follow the distributaries of the Calcasieu River and the Sabine River to the far west. The natural vegetation types of the areas surrounding the Cajun Prairie are evident: marshes to the south, Cypress forest and bottomland hardwood to the east, Longleaf pine forests to the north and pine flatwoods (and savannas) to the west.

The streams are generally forested by bottomland hardwoods, but the western

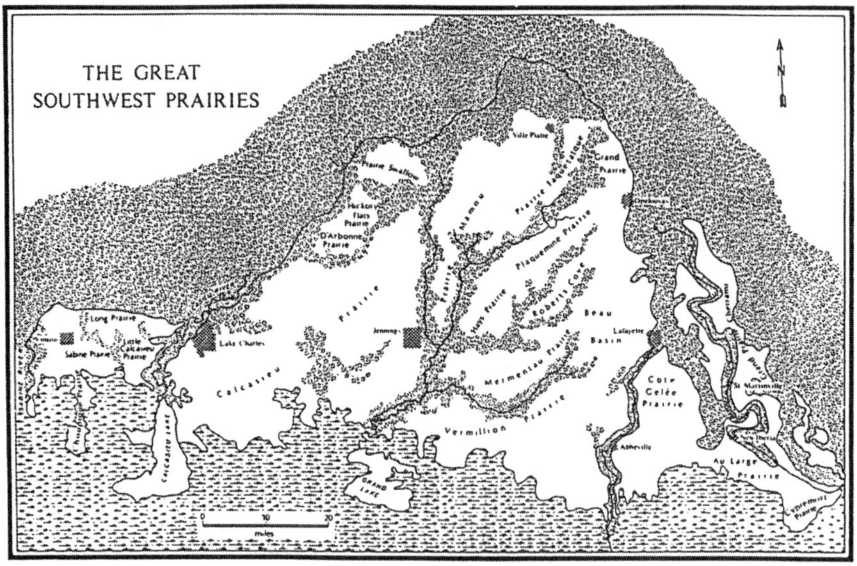

Map 2. The Great Southwest Prairies. Reprinted with permission from Newton, Milton B. 1972. Atlas of Louisiana: A guide for students. School of Geoscience Misc. Publ. 72-1. Louisiana State University, Baton Rouge, Louisiana.

prairies intergrade to the north with flat pine woodland and Longleaf pine forests, often with open pine savannas. The southern prairies intergrade with freshwater marshes along their entire length. The life that lives in and is associated with streams and ponds increases the diversity of the prairie ecosystem. Man-made canals, lakes and ponds also greatly increase diversity. The marshes associated with the prairies are famous for the massive numbers and kinds of birds migrating and residing there. I have spent my life studying the diversity of organisms in the Cajun Prairie.

For the most part, the Louisiana coastal tallgrass prairie region is a rather featureless plain, ranging from approximately 5 feet in elevation nearest to the Gulf Coast to 50 feet at its northern reaches. However, the region is characterized by two rather subtle physiographic features: 1) *occasional* shallow water-filled depressions, referred to as *marais* ("little marsh") or *platins* (nearly circular ponds) by the Cajuns, and 2) occasional series of low, circular mounds, ranging from 6-60 feet in diameter and a few inches to over 5 feet in height. In some areas, as much as 25% of the landscape may be covered by these hillocks (Smeins et al. 1992), which are most often referred to as "pimple or mima mounds." There is ongoing conjecture as to the origin of both of these features. Thus far, the most plausible explanation may center on a "differential erosion" concept, where the lands containing these features were subjected to wind and wave action during

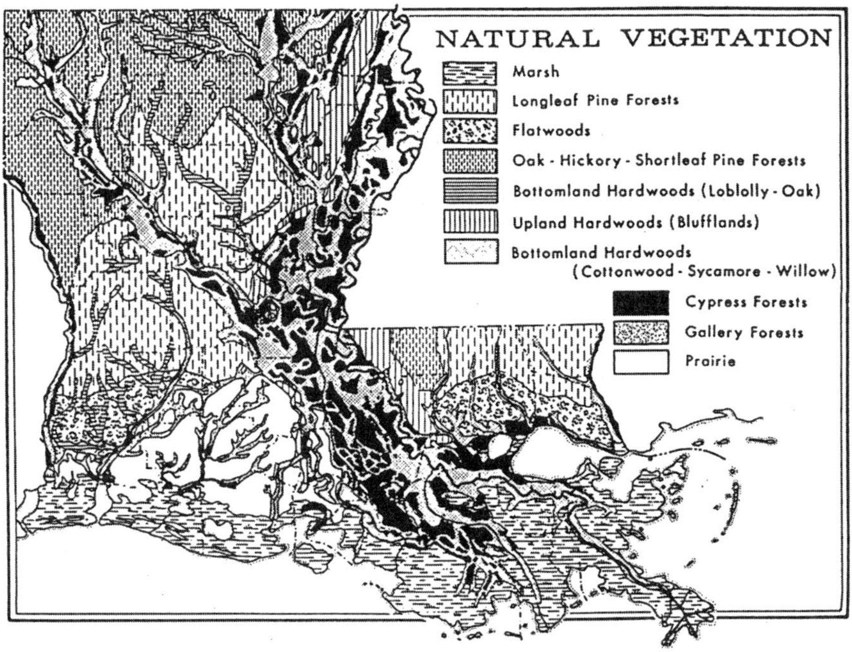

Map 3. Natural vegetation of Louisiana. Reprinted with permission from Newton, Milton B. 1972. Atlas of Louisiana: A guide for students. School of Geoscience Misc. Publ. 72-1. Louisiana State University, Baton Rouge, Louisiana.

historic coastal environments (Smeins et al. 1992).

Physiographic/topographic features allowed early coastal tallgrass prairie travelers to more easily navigate and orient themselves within the region. Several bayous (sluggish streams) dissect it in north-south fashion at various intervals and their riparian hardwood forests developed along the better-drained natural levees of these streams. Ranging in width from 30-3000 feet, these strip-like, gallery forests were inhabited and traveled through by Native American and early Francophone settlers alike. These streams with their associated woodlands served to separate the vast grassland into named prairies replete with various wooded landmarks which could be utilized for more efficient orientation. Several islands, e.g., Avery, Jefferson and Weeks Islands, and lakes, e.g., Grand Lake, Lake Peigneur, Spanish Lake, White Lake, Lake Charles, Lake Arthur, Sweet Lake, Greens Lake, Browns Lake, Black Lake, Willow Lake, Sabine Lake, Lake Misere, and Vermilion Bay provided additional landmarks.

Another physiographic consideration, which contributed greatly to the eventual cultural disposition of southwestern Louisiana's prairie region, involves its degree of physical isolation resulting from the ecology of the lands which surround the region. To the north, the prairie system is bounded by a substantial pine/hardwood "flatwoods" system possessing sandy soils, which are too nutritionally poor and/or wet to clear for crops. Together with eastern Texas, much of this area is considered the Big Thicket (Ajilvsgi 1979) and the West Gulf Coastal Plain (MacRoberts and MacRoberts 2008a and b). To the east lies the vast Atchafalaya Basin, one of North America's largest river swamps; and the entire southern edge of the prairies is bounded by an equally impenetrable mosaic of freshwater, brackish, and salt marshes (Vidrine et al. 2001c).

"Samuel H. Lockett, Colonel C. S. A. and Professor of Engineering, Louisiana State Seminary in Pineville and Louisiana State University in Baton Rouge, conducted a topographical survey of Louisiana between 1869 and 1872 (Lockett 1969 and MacRoberts et al. 1997). His map was used by Clair Brown (1972) in his '*Wildflowers of Louisiana.*' William Darby (1816) also prepared a map, which is noted for showing the Sabine River most accurately (Suarez 1968). Other important maps included Hilgard's (1884) *Agricultural map of Louisiana* and the 1935 map of the Soil Divisions of Louisiana published by the Louisiana Agricultural Extension Service (LSU, Baton Rouge, LA). There also are two in-depth atlases: Newton's (1972) *Atlas of Louisiana* and Goins and Caldwell's (1995) *Historical Atlas of Louisiana*. Collectively, these maps and books provide a wealth of information regarding Louisiana. From historical atlases of railroads to the changing panorama of parishes and political boundaries, the prairie remains obvious and easy to follow. The story might be one best told with maps, but I will limit their use.

The Cajun Prairie is not only a biogeographical and ecological region but also a cultural region—it is obvious in its name. If it were strictly a biological region, maybe a name like coastal prairie would be sufficient. But this is not so, the culture is so distinctive that it necessitates the region taking on its name. The only way that I know to readily integrate this into my story is to link the story of my ancestors to the natural history of the prairie. In the end, as I grow to understand the prairie ecosystem, I will come to understand the role that people, including my family, had to play in the ecosystem. Further, it will allow me to contemplate the future role(s) that I undertake in this ecosystem. The idea of 'sustainability' is never far away from my thinking—so providing for a sustainable future for me and my family hinges on integrating sustainable notions into planning for the future of the ecosystem. This is not just my idea, this is a major movement that rose from Aldo Leopold's view of the 'Land Ethic' (Leopold 1949).

One of the central themes of this book is ecology. Biodiversity is one of the primary attributes of the prairie. Aspects of biodiversity including 'species richness,' 'species evenness,' indicator species, keystone species and lists of species are discussed in chapters 8 and 9. Succession and catastrophe are also discussed. Phenology (blooming sequence of plants) and flight seasons of butterflies and dragonflies are presented. Community ecology topics including plant and animal associations, symbiosis, assemblages and human interests are also discussed. The biosphere is divided into the atmosphere, hydrosphere and rhizosphere in the prairie. I will relate the occurrence of plants and animals in relation to these spheres.

Literally a thousand species of plants are found associated with the Cajun Prairie; more than 600 are considered 'prairie plants' and 253 are 'indicator species' of prairie with Coefficients of Conservatism equal to 5 or greater (Allen et al. 2001, 2006, Allain 2007 and Appendix 6). Thousands of animal species also occur in the Cajun Prairie ecosystem (see Appendices 11 and 12). These prairies were often described as 'seas of grass,' because the general view of the prairie resemble a sea with waves as the winds blew across the prairie. The descriptive names used to describe locations like 'island,' 'cove,' 'point,' etc. exemplify this notion of the prairie as sea. The Cajuns and other explorer settlers were literally men of the sea. Thus the idea of naming places in the prairie as one would name vistas in an ocean or bay is logical.

Culturally, the eastern prairie is distinctively Cajun, the northern prairie is French, and the western prairies are Midwestern; but the flavor of the region is Cajun, thus the name 'Cajun Prairie' sticks (Jones 2007). Each of the Cajun prairies is different biologically (if you evaluate the plant species and varieties) indicating differences in soils. It is striking in the eastern half of the region, which is often labeled 'cotton and corn' region and contrasted to the western region labeled 'rice and cattle' region. All of these regions were cattle and horse ranches and/or open range.

People and their occupations have had dramatic impacts on the Cajun Prairie. These occupations, taken individually, can be treated as having specific impacts, but taken collectively the impact is essentially the loss of the prairie. Major livelihoods included:

1. Early agriculture—the Native Americans'Three Sisters Agricultural System—corn, beans and squash—grown in the region for ca. 10,000 years
2. Fur trapping, fishing, hunting, etc.—the original '*Coureurs du Bois*' made a living off the wildlife of the prairie ecosystem
3. Military—soldiers from France and then Spain were important in settling the posts in Louisiana and along the Mississippi River. Their role was to guarantee the protection of the settlers and ensure the role of government in the region.
4. *PetitS habitants* –farmers with small farms, few animals, common among Cajun and French and Spanish early colonists
5. *Vacher/vacheries*--cattle and horse ranchers/ranches (since the early 1700's—traded with Indians; *Vacheries* refers to the ranch or range while *Vacher* refers to the cow herder himself, not necessarily the owner).
6. Timber—big business in Calcasieu region
7. Oil and natural gas—coastal prairies/marshes in general have been

exploited

8. Cotton and Rice farmers—for 200 years these crops have been grown
9. Soybean and crayfish farmers—more recent crops—since 1950s
10. Sugarcane, milo and other crops
11. Salt mining at salt domes at Avery, Jefferson and Weeks Islands
12. Modernity and urbanization

The Cajuns are introduced as a group of farming people (Fontenot and Freeland 1976 and Ancelet et al. 1991), but other authors point to the fact that most everyone associated with the Cajun Prairie since the Spanish explorations were cattlemen (*vachers*) (Brasseaux 1987, 2005, Brasseaux et al. 1994 and Jones 2007). Evidence is, however, good that they were involved in the many occupations listed above during their colonial years in Louisiana.

The Cajun Prairie (and coastal prairie), unlike the Midwestern prairies, are mild enough during the winters to allow the naturalization of feral cattle, horses, and pigs. This is significant ecologically and of course commercially. It set the course for early settlement by giving significance to settlements like Natchitoches and Opelousas, where Native Americans traded these animals for goods. These posts were soon developed into settlements for colonists once the military set up the posts. The animals changed the plants in the prairie and as a result masked our view of the prairie prior to their existence.

As a result, we can reconstruct an image of the prairie based upon what little eyewitness information that we have and based upon the remnant prairies that we have documented. Our list of plants and animals is restricted by our documentation. We are also able to extrapolate from the surrounding habitats to some extent, but we are hampered by simple rumor. For example, *marais* and *platins* are considered as natural wet areas in the prairie grassland; however, a number of authors contend that these were created and maintained by cattle milling around in an area. A few of these wet areas remain along railroad tracts with the prairie remnants, and as such, we can speculate that the flora in these remaining wet areas is similar to what was in the *marais* and *platins,* whether they are natural or not.

Charles Allen et al. (2001) described the Cajun Prairie as follows:

> *Many authors map the southeastern extent of the true prairie in the United States in east Texas (Weaver 1954). However, there is an area of remnant coastal tallgrass prairie in southwestern Louisiana between the Atchafalaya and Sabine Rivers that has been called Attakapas Country, the Great Southwestern Prairie, or more recently, Cajun Prairie (Allen and Vidrine 1989). This grassland originally stretched across parts of Acadia, Allen, Calcasieu, Cameron, Evangeline, Iberia, Jefferson Davis, Lafayette, St. Landry, St. Martin, and Vermilion parishes. From 1869 to 1872, the prairie occupied an area of 1,000,000 ha (Lockett 1969).*
>
> *Several interrelated factors help explain the presence of prairie in an area which receives an average of 125 cm* (50 inches) *of rainfall per year (National Oceanic and Atmospheric Administration 1987). The terrain is flat, and there*

is a densely-packed, hard clay pan located 20 to 30 cm below the surface. Tree roots usually penetrate much deeper that roots of grasses and other herbaceous plants, but tree roots do not penetrate the clay pan (Brown 1972). Fires caused by lightning and later set by Indians and by European settlers also helped to retard the growth of trees (Allen and Vidrine 1989).

Today, most of the original prairie has been destroyed and replaced by cultivated crops, in particular rice. However, a few remnant prairies currently exist along railroad rights-of-way. Most of these remnants were never tilled or have not been tilled since the railroad acquired the land approximately 150 yrs ago. The remnants are all very narrow, mostly less than 30 m wide and no more than 800 m long. The estimated total area of intact Cajun Prairie today is 200 ha (Allen and Vidrine 1989).

The tallgrass prairie ecosystem is ranked G2 (imperiled globally because of rarity or because of some factor(s) making it vulnerable to extirpation) by the Nature Conservancy (Grossman et al. 1994) and in Louisiana it is ranked S1 (critically imperiled in the state because of extreme rarity or because of some factor(s) making it especially vulnerable to extirpation from the state) by the Natural Heritage Program (Smith 1995).

This book will detail the timeline of these discoveries and activities such that the history of the events is recorded for posterity. News releases, popular articles, and scientific articles are available to detail the story; however, retelling the story in a personal way will provide insight into details not available to the public and serve to knit the published stories together into a 'tale' of sorts.

The primary goal of this book is to create a baseline picture of what the Cajun Prairie might have been like, biologically, in the year 1600 A. D. (or if cattle have been roaming this prairie since the mid-1700's, it may be necessary to push the date to pre-range of feral cattle and horses—cattle and horses released by Spanish in 1540s as they moved across the southern states, and later the Native Americans moved the cattle as they became accustomed to their potential). A second goal is to give a general history of the Cajun Prairie with emphasis on the land and the changes that have occurred. The third goal is to provide a status report on the 'ecology of the Cajun Prairie' and to promote an understanding of the region. With all this in mind, the overall goal is to encourage that group of 'prairie enthusiasts' who are collectively trying to preserve and restore the ecosystem and provide them with information that they may need to continue their mission.

It is necessary to understand/deconstruct the prairie and attempt to reconstruct it. This is exactly what has/is happening with the Cajun Prairie. My family and their peers deconstructed the prairie, and now my peers and I are reconstructing the prairie. This is an effort that involves the entire community of people and other organisms and requires interacting with nature. Success, in my opinion, rests on first understanding what the prairie was like before the unraveling of the past 400 years. Reconstruction can then proceed at a pace that is noteworthy. It is essential to understand that the prairie intergrades with the marsh habitat—we do not know to what extent they interact and depend upon each other, but we should be suspicious. In the long run, I suspect, the restoration of prairie and marsh will be sibling processes; therefore, it is essential for the survival of coastal Louisiana not only as urban centers but also as agricultural lands that we develop proactive and collective approaches to these processes.

The national pastime is gardening. Even the lawn is a garden (Wasowski and Wasowski 2000). However, it is becoming apparent that an element of sustainability is essential in gardening. As such, emphasis is rapidly turning to the 'natural garden' as the future garden. In ecology circles, 'biodiversity garden' is a more descriptive term from the genetic perspective (Nabhan 2009). From hummingbird to butterfly gardening, this change is becoming one of the largest economies in the United States. A more compelling level of gardening is 'reconstruction' of local habitat. This type of gardening is quickly becoming vital as the roles that natural habitats play in preventing global climate change and protecting coastlines and preventing erosion become more commonly understood. Gardening with 'natural prairie' plants by prairie reconstruction was the first type of habitat restoration. Aldo Leopold, professor of wildlife biology and father of the conservation ethic (land ethic), was the first to spearhead 'prairie restoration' at the University of Wisconsin arboretum and on his own property. These projects remain intact today and serve as the model for natural gardening.

The role of fire as a management tool is finally understood. For nearly 100 years, fire was seen as totally destructive. Grasslands are natural burn ecosystems and totally depend upon a regimen of fire management. Even forests are now seen as evolving with fire and literally depending upon fire to achieve maturity (Earley 2004). Fire is one of the primary factors creating biodiversity by adding a type of stress into the mix that causes the release of nutrients to plants in massive quantities. Fire also promotes the development of specific assemblages of species. Among these are the grasses, which provide the major food for livestock and humans.

The varied roles of 'biological diversity' in the ecosystem are just now being re-evaluated. As the world experiences another massive extinction event of flora and fauna, the loss of not only millions of species of unique kinds of organisms but also 'genetic erosion' resulting from the loss of millions of populations with unique qualities are occurring all around us. Our food and medicine industries have come to understand the peril of this sheer loss of variety in the natural world, and 'banking' of genes (seed banks) and bioinformatics are central themes of recent research (Nabhan 2009, E. O. Wilson at http://www.eowilson.org/ and http://en.wikipedia.org/wiki/Bioinformatics).

Other authors (e.g., Brasseaux 1987, 1991 and Ancelet et al. 1991) provided efforts that center on the people of the prairie (90% people: 10% prairie/land); my effort is to be prairie and the people (90% prairie/land: 10% people). In other words, this is more of a historical ecology of the Cajun Prairie—a natural history. The interdependence of agriculture and human settlement cannot be underestimated, but this same interdependence of prairie as habitat and the human settlements has been underestimated. But then again, humans and the natural world often at first appear as adversaries, but after study, they are realized as protagonists and cooperative units in nature. In societies where this relationship is never realized or realized too late, collapse commonly follows for both humans and the ecosystem (Diamond 2005).

Chapter 2 introduces the people of the Cajun Prairie emphasizing their origins and original settlements. My bona fide includes my genealogy, my credentials as a locally-grown person and my scientific credentials.

Chapter 3 presents a general discussion of the time period labeled 'Presettlement – 1870s' and includes writings of early travelers in the Cajun Prairie. The prairie is introduced as cattle country. The French and Spanish settlers are the central players.

Chapter 4 presents a general discussion of the time period labeled '1870s-1930s' and also presents the prairie as rice country (as well as oil and gas country). The German and Midwestern immigrants are the major players on the Cajun Prairie. Major changes occur and the native prairie is essentially destroyed.

Chapter 5 presents a general discussion of the time period labeled '1940s-1970s' and also presents the work of Caroline Dormon and Clair Brown, who provided information on the flora of the Cajun Prairie. Many other scientists ignore the prairie or barely mention it. The prairie is considered extirpated.

Chapter 6 presents a general discussion of the time period labeled '1980s –2010' and also recounts the rediscovery of the prairie. The prairie remnants are studied, and scientific evaluation leads to a modern flora, animal studies and soil studies. Further, these studies lead to efforts at preservation and restoration of the prairie.

Chapter 7 provides an introduction to the geology, soils, and climate of the Cajun Prairie. Special discussions on the role of fire in the ecosystem and on the hydrology of the prairie are engaged. An introduction to sustainable agriculture and the Cajun Prairie model for agriculture in southwestern Louisiana are included.

Chapter 8 provides an introduction to the biota, specifically the plants of the Cajun Prairie. A flora for the Cajun Prairie is included.

Chapter 9 provides an introduction to the biota, specifically the animals of the Cajun Prairie. A faunal list and discussion are included.

Chapter 10 provides discussion of the future of the Cajun Prairie. With the accumulation of knowledge of the prairie as a natural system, preservation and restoration/landscaping of the prairie can then be attempted. The remnants of prairie and varied projects are considered. Integration of prairie with modern culture serves as a closing topic.

Twenty-six appendices are included in order to expand specific components of the chapters. Topics vary from my genealogy to partial lists of animals and plants of the Cajun Prairie.

Thirty-two plates accompany the text. The first plate shows examples of wildflowers of the Cajun Prairie. Plates 2-28 show views of the remnants of Cajun Prairie found along railroad rights-of-way during the 1980s and 1990s. The views depict prairies from different named prairies, wet and dry remnant prairies and seasonal perspectives. Collectively these views provide a potential template upon which a mental image of the Cajun Prairie can begin to be imagined. Plate 29 shows the remnants as they appeared in 2008. Plates 30-31 show seasonal views of the Cajun Prairie Restoration Project and the Cajun Prairie Gardens, both in Eunice, Louisiana. Plate 31 shows an undredged Bayou des Cannes in the heart of the Cajun Prairie, several animals of the Cajun Prairie and a photograph of several of the Cajun Prairie scientists.

Chapter 2
The People of the Prairie and My Bona Fide

Stuart O. Landry Jr. translated C. C. Robin's French travelogue entitled *Voyage to Louisiana*, which described his 1803 trip from New Orleans through Donaldsonville to Opelousas and the Cajun Prairie: "*The first place from which one may reach the Atakapas is Bayou Lafourche, a western branch of the river found about twenty-five leagues above New Orleans. This branch communicates, by various ramifications with the Teche, the first river of the Atakapas.*"...Upon entering the Bayou Teche, the description continued, "*This description of prairies stretching out to either side of the Teche, bordered by a strip of woods which shade its banks, would do as well for all the other streams between the Teche and the Rio Brazos, a stretch of about two hundred leagues. Each of these rivers in the same way is bordered by a strip of woodland beyond which lie the open prairies. The only differences are that the rivers are almost all larger and more navigable, that the prairies are broader and more spacious, stretching up to twenty-five leagues in extent, that the winters, which are mild in the Atakapas, appear to be even more so to the west, and, finally that these countries are as yet uninhabited and are covered with innumerable herds of cattle and wild horses, as well as all kinds of game.*" (C. C. Robin (2000)).

"But before I take up these parishes individually I will give a short general description of the Opelousas and Attakapas countries to which these parishes belong. These countries take their names from two Indian tribes that once inhabited them. Its present inhabitants claim for it the title of "The Garden of Louisiana," with what justice let the following description determine." (Col. Samuel Lockett (1969)).

The general purpose of this chapter is to introduce the people of the prairies of southwestern Louisiana. These prairies resemble those of the Midwest in flora and fauna historically, but there are significant differences. One of the major differences is the process by which the prairie was settled, while another difference is the people who settled this prairie. A timeline of the historical events related to the Cajun Prairie is provided in Appendix 1. Claude Oubre (2001) provided a timeline for the general history of Louisiana. The history of the prairie is knitted to the history of the people, and vice versa. Weaving this story is a tremendous challenge, since the story is all-too-much an oral history, with much of it told in several languages. The local people of this prairie region avoided conventional education and thus left little as written record.

In an earlier paper (Vidrine et al. 2001c), my colleagues and I summarized the history of the Cajun Prairie people as follows:

"*Due to the isolated nature of this prairie ecosystem, few non-native people settled or even surveyed the region prior to the mid to late 18*th *century (Kniffen*

1965). The only human inhabitants of the coastal tallgrass prairie region prior to the mid 18th century were small bands of Attakapas and Opelousas Indians, who seasonally encamped along the gallery forests associated with the various bayous which dissected the region. But by the latter half of the 18th century, this pattern would change. Poste de(s) *Opelousas (near modern day towns of Washington and Opelousas in St. Landry Parish) was the first recorded settlement in the Cajun Prairie region."* Poste des Opelousas in St. Landry Parish was officially established in 1765 (C. Oubre, personal communication 2010). The history of the people of the prairie remains an area of great research interest.

Vidrine et al. (2001c) continued,

"The Cajuns or Acadians, who are the most famous inhabitants of the region, were of French origin. French immigrants, who had settled in Nova Scotia, Canada in an area they called Acadie, were exiled by the British colonial government - ostensibly on the basis of religious differences - in 1755. Some of these exiles went back to France, but most of them spent the next decade wandering the Atlantic coast in search of colonies which might take them in. While some were eventually accepted into various New England colonies, most were not. Eventually, the "Acadians" got word of French colonies, which existed along the north-central coast of the Gulf of Mexico, and began making their way to this region. Between 1764-1788, an estimated 2635 Acadians arrived in the Louisiana colony…These were quickly relocated west of the Mississippi River along the eastern edge of the prairie region (Ancelet et al. 1991, Brasseaux 1987 and 1991). During this period, French and/or Spanish settlement in that region was scanty, and limited to clusters around two military installations: the Poste des Attakapas in present-day St. Martin parish, and the Poste des Opelousas in present-day St. Landry parish."

"The migration of the Acadian settlers was heralded in the epic poem Evangeline by Longfellow (1847). Other non-Cajun myths and stories developed on the prairie and its associated environs. However, it is the Prairie Cajuns that have received the greatest attention. The Jean Lafitte National Historic Parks of Louisiana commissioned a detailed study of the Cajuns and produced a five volume project report (summarized in Ancelet et al. 1991). The park system houses it at the Prairie Cajun visitor center in Eunice, Louisiana."

"Over time, the Cajuns gradually became entrenched within the prairie proper, doubtlessly finding the long growing season (compared with that of France and Canada), plentiful game, and seemingly infinite range of livestock forage well suited for their subsistence lifestyles. But living there was by no means easy, due primarily to the great difficulty experienced in carving out a homestead and/or traveling for even short distances amidst exceedingly thick and tall prairie vegetation. Lifelong Evangeline parish (north-central boundary of the prairie) resident Burkeman Veillon (born: 1921) once recounted to one of the authors' (Bill Fontenot) *stories from his grandfather (ca. 1870-1890) of men traveling on horseback through the "Mamou Prairie" where the only things visible of them were their hats!"*

"Living there in near-total physical and cultural isolation from the rest of colonial America, a strong, resilient, and ultimately fascinating culture would develop over the next 100 years. Assimilation of varied nationalities was one of the most interesting aspects of the development of the Cajun Prairie. Native Americans (local and from the upper Mississippi River), French Creoles, Acadians (Cajuns), Spanish, Germans, varied Midwestern English, African, etc., settled and made significant contributions to the development of the region now called the Cajun Prairie."

"As previously mentioned, it is noteworthy to understand that these settlers lived on the prairie in total obscurity for at least 100 years, employing a demonstrably sustainable lifestyle all the while. It was during this period of extreme obscurity that the folkways underpin what is today referred to as "the prairie Cajun culture." The term "Cajun" incidentally, was originally derived by the Acadians themselves, as a fast spoken French corruption of "Acadian" (pron. Ah-kah-JE-ahn). With the arrival of each new settler family it was necessary for nearby farmsteads to assist the newcomers in erecting their homes, barns, etc. in what was known as a coup de main (="helping hand"). They employed their own special brand of architecture which is still identifiable today as "the Cajun house" or "the Cajun barn" (Newton, 1972 and Ancelet et al., 1991). In the absence of refrigeration, it was necessary to butcher livestock on a relatively frequent basis because only small amounts of meat could be utilized by each family. Thus, the boucherie evolved, which was a weekly butchering attended by several neighboring families. The "host" for the week would provide the hog or steer for slaughter, with the group working as a team to butcher and divide the meat amongst each participating family. Boucherie host families would rotate with each succeeding week (Dorman 1983)."

"Once enough prairie was opened up to allow for such relations amongst neighboring farmsteads, regular entertainment ensued in the form of the bal du maison ("house dance"). Here again, families alternated in hosting these musical events, played by the few folks who had bothered to carry their violins and (occasionally) guitars with them during their long period of exile. It was then that the peculiar brand of U.S. folk music, which is known today as "Cajun music" had its beginnings (Ancelet et al. 1991 and Savoy 1984)."

"Likewise, the language known today as "Cajun French" was allowed numerous additional generations to incubate, again as a direct result of the isolation which the prairie Cajuns experienced in their new surroundings. Of course, this process of differentiation from standard French language had begun long before when these people had first departed France for eventual colonization in Canada and Louisiana. Today, French tourists who hear their first examples of "Cajun French" language upon arriving in south Louisiana can barely restrain the mixture of shock, amusement, wonder, and glee which readily shows upon their faces. To them, Cajun French sounds somewhat like an extremely archaic version of their own contemporary language, with, of course, a substantial number of New World terms thrown in!"

The Spanish had ceded Louisiana to the French in 1800, and the French, namely Napoleon Bonaparte, sold it to Thomas Jefferson and the United States in 1803. Jefferson sent explorers within the decade to the major rivers (Lewis and Clark up the Missouri, Zebulon Pike up the Arkansas, Dunbar and Hunter up the Ouachita (Berry et al. 2006), and Freeman and Custis up the Red Rivers (Freeman and Custis 1821, Flores 1984 and 2008)). Also, surveyors and varied travelers, e.g., C. C. Robin and William Darby, arrived in the Cajun Prairie. They evaluated the people, livestock, pasture quality, water quality, and mineral resources, like salt, and they calculated board feet of timber and the variety of trees as part of an overall look at the resources of the land. A wave of immigration of people from the Midwest, Atlantic colonies, and even Texas followed. Two of the most famous characters were Jean LaFitte, privateer and pirate, and Jim Bowie, knifefighter and speculator; interestingly, these two were contemporaries in both business and adventures (Ramsay 1996 and Thorp 1991). The region became very prosperous until the Civil War, which led to at least a half century of difficulty. Following the Civil War, the Midwesterners would come in ever-increasing numbers as the prairie was developed into Midwestern-style towns and farms. They would however adopt the Cajun lifestyle and language as earlier settlers had done.

Land grants

Two major land granting systems, French long-lots or riverbank lots and the checkerboard or American lots, were employed in order to deed the land of the Cajun Prairie to the people. In many cases, the land was deeded free to settlers as the French and Spanish governments were trying to settle the area and develop it. These governments wanted to not only civilize the area but also to profit from it. Often large corporations were contracted by the controlling government to build and populate communities. While there was a lot of land, few were interested in living in the wilderness. Later the land was bought from the Native Americans. The French had good relations with the Native Americans and carried out many business transactions for land and livestock.

Brasseaux (1987) and Ancelet et al. (1991) provided interesting discussions of land grants of the period. Ancelet et al. (1991) reported, *"As a custom of the area, land was divided into long-lots. These were narrow on the river bank but extended far back into the swamps, with their sides perpendicular to the river bank. Plots of land were measured in "arpents of face," using the French measure established in Quebec in the early seventeenth century. Each Acadian family was granted a plot that most typically varied between three and six arpents in width on the river. (One arpent equals about 192 English feet.) Thus, the typical family concession varied between 576 and 1152 feet in width. These concessions were further subdivided according to the French laws of inheritance, which reserved an equal portion to each child."* While this discussion relates to Iberville Parish and the swamps, a simple replacement of the word 'swamp' with 'prairie' would describe the early French and Spanish as well as the prairie Cajun land grants along the rivers. Drawings of these land grants are readily available (see Fontenot and Freeland 1976 for examples of both the 'long lots' and the 'checkerboard' grants). My parents' land, which was recently settled in their estate (2006), was measured in arpents, and the land had frontage on Bayou des Cannes up onto the Mamou Prairie.

Fontenot and Freeland (1976) described the purchase of land from the Native Americans, *"In 1784 Antoine Blanc purchased a tract of land, one league of front by the depth of 40* (usually 42) *arpents (2,820 86/100 acres) from Nementou, chief of the Attakapas. The deed was signed by Nementou and 13 of his warriors who inhabited the village at the time."* This land was on the Plaquemine Brûlée, a bayou between the Plaquemine and Mermentau Prairies in the heart of the Attakapas country. Also John Lyon purchased land on the Bayou Queue de Tortue from the Native Americans. And likewise, the land on the Bayou Wikoff was purchased by William Gilchrist at the site of an Attakapas village. *"In 1799 the Attakapas sold a village on the west side of the Mermentau River to André Martin for $100. The record of this transaction gives the origin of the name Mermentau, and tells of the Indians who joined the Galvez expedition in 1799."* The river apparently was named for the chief Nementou. The Native Americans continued to reside on the land owned by Martin until at least 1814 as reported by Louis de la Houssaye (Fontenot and Freeland 1976). It is apparent that the land in the Attakapas country was purchased from the Native Americans who lived there in a fairly peaceful way.

Fontenot and Freeland (1976) described land grants in general as follows:

> *"Like the French, the Spanish used the 'riverbank system' for granting of land. Land holdings were long, narrow strips, fronting on a river or bayou. Each grant began at the bank of the stream and extended between parallel lines for an ordinary depth of 40 arpents, an arpent being about 192 feet. The width was also expressed in arpents, so a tract of six arpents by the ordinary depth of 40 arpents measured six by 40 arpents."*

"By contrast is the 'checkerboard pattern' of lands in the prairie areas of the parish; this is the Rectangular System of Surveys introduced by the United States Land Office, based upon townships, sections and quarter sections, in which all holdings seem to be squares or combinations of squares. The open prairies, referred to as au large, remained unclaimed and unsettled until after 1803, the beginning of the American period." Fontenot and Freeland (1976)—they also list the 160 landowner family names as of 1803 of the grants of land. The Native American purchases were sanctioned under the granting system of the Spanish in the late 1700's. These rectangular properties and the roads developed to service them are so apparent on any map of the Cajun Prairie. In fact, many areas are clearly divided into square miles with roads intersecting at each corner. Properties

and fields within the square mile are likewise smaller squares. Having walked hundreds, actually thousands, of miles of this kind of land, I can attest to the layout of the land.

My bona fide

In southwestern Louisiana, the authenticity and veracity of one's story is measured by the initial responses of the storyteller to two questions:

1. Who is your family?: Who's your daddy, your mommy, your people?
2. Where are you from?: What is your heritage, your home, your base?

As a foundation for the story that I want to tell, I will provide the answers to the above questions as well as an introduction to my training and experience as a biologist. History, especially natural history, of a place is extremely difficult as it not only takes into account human activity but also takes into account the climate, geology and the variety of organisms. Each organism has a story to tell. However, the community of organisms has a more dramatic story to tell. My story is part of the story of the Cajun Prairie—interconnected as they are. My genealogy, my life as a prairie local, and my scientific credentials are provided in order to show the necessary authenticity and veracity to the story.

My surname ancestor on my father's line, Jean Baptiste Pierre Vidrine and many of the soldiers, married Native Americans or descendents of Native Americans and earlier French soldiers. Jean Baptiste Pierre retired in the mid-1750s and married Elizabeth Tisserand de Moncharvaux, daughter of the commanding officer and his wife, a quarter-blood Kaskaskian of the Illinois Confederacy of Native Americans (Vidrine 1981). The original land grant placed the family just north of Washington, Louisiana in the Grand Prairie bordering Thistlewaite. Their two sons, Pierre and Etienne, fathered the Vidrine clan in the Americas—many lived their entire lives in the area (J. Vidrine 1981 and per. Comm. 2008). The Vidrines in my family would marry Brignacs, Fontenots, Soileaus, Fuseliers, and Guillorys—all former families of French soldiers in the Evangeline Parish areas. Finally my grandparents, the parents of my father, would represent descendents of the two clans of the original brothers. My family tree is provided in Appendix 2. The Spanish land grant is illustrated in Appendix 3.

With the loss of the French and Indian War to the British, Louisiana was ceded from the French to the Spanish in 1762. The Spanish King awarded land grants to the French soldiers and settlers in an attempt to tame the wilderness and begin making some profit from the colony. About a decade later, he sent soldiers, i.e. men at arms, and their families to the now Spanish colony of Louisiana. My mother's surname ancestor brought his family from the Canary Islands (Villerè 1971) and was stationed in Donaldsonville just south of Baton Rouge at the headwater of

Bayou Lafourche. Josef Hidalgo and Isabel Sambrana Morales traveled from the Canary Islands in 1778 to the colony of Louisiana. They had at least 3 children including Joseph "PePe" Antonio Hidalgo who was born in Donaldsonville on January 6, 1784. His grandson, Joseph Luc, was born in Lafayette in the Cajun Prairie region; later Joseph Luc would marry Louise Martin, a Cajun, and they would have 8 children including my great grandfather, Joseph. His son Felix would marry Angelina Lee, whose mother was Aristella Thibodeaux, another Cajun, from Church Point in the Plaquemine Prairie. Angelina's father, however, was of English, Midwestern and German heritage.

Part of this story is the relationship of the people to the land, and vice-versa. My story directly relates to the land and the people of the Cajun Prairie. My father's family represents 7 generations of French settlers and Native Americans in Evangeline Parish on the Mamou Prairie. My mother's family is entirely European. Her father's family represented the Spanish militia and settled in Donaldsonville. Within a generation or two, they had moved to Lafayette and then to Church Point, the Plaquemine Prairie. And finally her family moved to Opelousas and to the Faquetaique Prairie. But along the way, these Spanish settlers had married into the Martin, Thibodeaux, Matt and Goss families—the Cajuns and Germans/ Midwesterners that settled and developed the rice country of Acadia Parish.

My family genealogy brought together the French militia, Native Americans, the Spanish militia (often called Isleños), Cajuns, Germans, and English in the Cajun Prairie. All of my relatives spoke French in their homes, even my mom and dad had to force themselves to speak English with my brothers and me in fear that we might suffer harm as they had in school for not speaking English.

By the 1940's, the area, formerly known as the Great Southwestern Prairie, was dissected by irrigation canals and developed into rice fields and cotton fields. About three months after I was born in Eunice (1949), my family moved to Duralde. Duralde is a rural community/region of southwestern Evangeline Parish just north of Eunice in the Mamou Prairie. Roughly 20 acres of farmland and 10 acres of forest, measured in arpents, on the Bayou des Cannes were our estate. By the time I can remember, we were growing cotton and cattle, pigs, horses, chickens, etc. We also had a garden and we were using every square foot of the homestead for some venture—a bit of corn, peanuts or watermelons, or a storage area for lumber, tractor, tools, etc. Even corners for plum, peach, and fig trees were allotted. Finally an old barn stood on the edge of the property with two large pecan trees and the customary Chinaberry trees for 'bullets' for our sling-shots. We kept cattle pasturing 'au large' in the woodland, rounded them up for branding in the spring and butchered them with the neighbors on cold Saturday mornings. In Appendix 4, I reminisce about the general activities occurring during a year of my youth.

Endrows and lanes had prairie plants even in my childhood. When we weren't planting, hoeing or picking cotton or some other crop, I had a chance to look at flowers and butterflies. I was amazingly ignorant about them so now I understand why so little attention was paid to the organisms in our prairie. By 1960, mowing and the elimination of the end-rows were the newest fads, and the last of the prairie plants essentially disappeared. Only a few really tough plants with significant evolutionary adaptations up their sleeves survived. The Purple passionflower or Maypop was routinely spread by birds and hung around the fence rows. Annuals, like Partridge pea, with massive seed production capabilities kept reappearing. Unfortunately, exotic plants took advantage of this new situation and simply did not relent until they became seemingly permanent fixtures.

Thus my story is the immediate culmination of the settlement and development of the Cajun Prairie into the vast economic machine that it is currently. The system is however unsustainable. In part, the lack of sustainability is a result of the disconnection between the land and the people that has developed in the last century. Reconnecting the land and the people (the Land Ethic) is one of the goals of this effort. By unveiling the nature of the land, the people have the opportunity to integrate their economics into the economics of nature. The role of climate change and the long-term impacts of many of our actions in both agriculture and the petroleum extraction industries are poorly understood, but people can see the changes that are occurring to the prairie and the adjacent marshes. The impacts of hurricanes are evident and remain the single most obvious threat to the people of the prairie (Dixon 2009 and Vidrine and Venzon 2006).

Even the idea of 'prairie' was lost. As a youngster, I heard about 'prairie,' but it was simply mentioned as a long-disappeared idea. Rather the region was introduced to me as land that I had to help work in order to provide for food and shelter. Whereas, there was pride in family and in the land owned, the connection between the land and the people was very tenuous. The land was literally blowing and washing away, and we were dumbfounded. Even today, I ask my students about the prairie in which they grew up and currently live, and sadly, they don't know what I am talking about. They most commonly respond something about the 'lawn' as though it was 'prairie.' I then relate Charles Allen's quote, "Lawn is anti-prairie." And I am reminded of the Native Americans who were stupefied by the colonists who insisted upon turning the prairie 'upside-down' before planting. It is this veil of ignorance that compels me to continue to strive to awaken as with a spark the view of land inspired by Aldo Leopold's 'Land Ethic.' The land and the future are on loan to us from our children—we cannot lose both and simply contend (usually assert) our ignorance and beg forgiveness or worse, indifference!

My credentials as a biologist/scientist/storyteller complete my bona fide for authorship of this book. I earned my B. S. in Zoology in 1970 from LSU, M. S. in Zoology from LSU in 1974, and Ph.D. from ULL (then USL) in Biology in 1980. I worked for 3 years in New Iberia with Gulf South Research Institute (GSRI) on armadillos and leprosy, 4 years in Jennings with the Jefferson Davis

Parish Mosquito Abatement District (JDPMAD) on mosquito control and biology/ecology of rice fields and 26 years at LSUE as professor/scientist. I have published more than 100 scientific articles dealing with an array of topics as well as 9 books and 10 popular articles. I was with Charles Allen when he rediscovered remnant prairies in southwest Louisiana, and I have visited, photographed, and studied each of the remnants that we found along railroad rights-of-way in southwestern Louisiana. I have also directly participated in the reconstruction of 3 prairies (the Cajun Prairie Restoration Project in Eunice, the LSUE Prairie Project and my own Cajun Prairie Gardens). My formal training was in field biology with an emphasis on aquatic ecology and evolution.

I had the opportunity to reside in several different parts of the Cajun Prairie. While I grew up running through the Mamou Prairie just north of the Bayou des Cannes and Eunice, I now live just south of Eunice in the Prairie Faquetaique. When I was completing my master's degree from LSU, I moved to New Iberia on the Au Large Prairie or rather next to the Bayou Teche in the gallery forest. I then went to ULL and lived in Lafayette in the Beau Basin Prairie. When I completed my doctorate, I move to Jennings and literally walked the Calcasieu Prairie in search of mosquitoes. For the first decade at LSUE, I lived in the gallery forest along Bayou des Cannes just west of Eunice separating the Mamou Prairie and the Prairie Faquetaique. Each location was different and provided a unique perspective on the prairie in general. The loveliest pastoral scenes that make me think of landscapes observed by C. C. Robin, William Darby and Samuel Lockett are the views from the prairie escarpment looking over the 50+ foot drop along Interstate 49 just north and south of Opelousas. The elevation permits viewing nearly a mile down the slope, where cattle may be grazing in green pastures. Absent the wildflowers and native fauna, the imagination has to fill in some gaps, but the view is pure history. But most of the prairie must be viewed close-up. I am constantly reminding people that the prairie from afar is usually mistaken for a pasture or worse an abandoned field. Walking through the prairie with most of the flowers either at your feet or in your face gives you a personal and intimate view not available in the forest lest you walk through the treetops as in the tropical walks of Costa Rica. For the student of nature, each walk is an opportunity to discover some new or remembered plant or animal or some ecological theatre live and functioning.

Thus, I serendipitously stumbled into the nature of the Cajun Prairie again and again in my life and career. First I grew up in this prairie, and then I studied mussels and their parasites in the streams of the prairie. Later I studied mosquitoes and other organisms in rice fields and canals. Finally I had the opportunity to study the flying insects, including butterflies and dragonflies, and the plants of the prairie. I had eaten cornbread twice a day throughout my childhood and I didn't understand why I had tired so of this diet until I studied the history of the Cajun Prairie. This diet of cornbread was vital for more than 200 years for the people of the Cajun Prairie. This cornbread was just another connection for me to the generations that preceded me. The prairie seemed all but lost just as we rediscovered it, and certainly few were really mourning the loss. Now I have come full circle. The Cajun Prairie is my home, and I now think quite differently. I recall two distinctive conversations with my mentors, Samuel L. H. Fuller and Daniel J. Bereza. In two separate incidences in the 1970s, I was driving Sam or Dan through the prairie on the way to a stream to search for mussels. I took the liberty to share with these Philadelphia scientists, who thought that Louisiana was all marshes with people getting around by boat, that the state was mostly dry. I told them what I knew about the prairie. Both were fascinated, and each strongly suggested that I get to know my place in this prairie and the prairie itself. I took this advice under advisement.

Chapter 3
Pre-settlement -1870s

"Later in 1817, Darby described the livestock he found in the region. He understood that the cattle, horses, and methods of managing such livestock were brought to Louisiana from Texas, that former Spanish province. Those Texas cattle then so prevalent in Opelousas and Attakapas were "high, clean limbed and elegant in appearance," while the horses, of Andalusian or Nubian descent, were, "like their ancestors, small, compactly built, and inconceivably durable."" (Jones (2007) discussing William Darby's comments on the livestock of the Cajun Prairie).

"Crossing the wide prairie, strewn with flowers, whose stems raise them to the height of the horse on which the traveler is riding, surprise follows surprise in this varied vegetation. One rides suddenly upon herds of cattle, who raise their haughty heads above the grass as one rapidly approaches." (Robin (2000)).

Earlier Native American settlements were common along the bayous (Gibson 1976a and b), but little evidence exists of their activities. The oldest Native American settlements may date back to 10,000 B. C. or roughly 12,000 years and are known from Avery Island (Gibson 1975). The Younger-Dryas cooling ca. 10,000 BC probably had a dramatic impact on people in Louisiana, but no evidence remains. These Paleo-Indian cultures may be linked to later cultures (Saunders et al. 1997 and 2005). The history of the Native Americans of the Cajun Prairie must be told elsewhere.

The first explorers of Louisiana territory were Spanish (Francisco Coronado and Hernando de Soto in 1539-42) (Smith 1996 and Oubre 2001); however, neither visited the Cajun Prairie. They traveled across country and left livestock, including cattle, horses, sheep, and/or pigs, possibly escapes or possibly purposely; in either case, these animals became the primary livestock for Native Americans and early European settlers. These animals became major trade items, and they were later used to forge new alliances. European settlement in Louisiana territory began in the early 1700s, 150 years after the Spanish explorers traversed the country. Little is known of the Native Americans in the Cajun Prairie until the time of European settlement, when the Attakapas and the Opelousas lived in the Cajun Prairie.

Louisiana was the name of an administrative district of New France. Originally the territory that included most of the drainage basin of the Mississippi River and stretched from the Great Lakes to the Gulf of Mexico and from the Appalachian Mountains to the Rocky Mountains. The present-day U.S. state of Louisiana is named for the historical region, although it occupies only a small portion of the territory claimed by the French. On May 17, 1673, Louis Jolliet and Jacques Marquette began the exploration of the Mississippi River, which they called the

Sioux Tongo (the large river) or *Michissipi*. In 1675, Marquette founded a mission in the village of Kaskaskias (on the Illinois River), which became permanent in 1690 (Wikipedia, 2008: *http://en.wikipedia.org/wiki/Kaskaskia* and *http://en.wikipedia.org/wiki/Kaskaskia,_Illinois*).

The first European settlement in the Louisiana territory was Natchitoches in 1713; New Orleans followed in 1718, and the Poste des Opelousas (currently Washington and Opelousas) was established in 1765. These trading posts traded goods for cattle to ship to New Orleans, Natchez and other towns along the Mississippi River. French militia, made up of able bodied local male settlers aged 18 to 45, served these posts until the end of the French and Indian War, in which the British defeated the French and Illinois Nation. In 1762, the French ceded the Louisiana territory to the Spanish. An interim French government served the colony until 1766 when Spanish Governor Antonio de Ulloa finally arrived. Dissatisfied with his commercial regulations the colonists ejected Ulloa in 1768. He was replaced by Governor Alexandro O'Reilly who firmly established Spanish control of the colony. Spain would govern and settle the area until November 30, 1803 (C. Oubre 2001 and personal communication 2010).

The Poste des Opelousas et Attakapas was the first European settlement in the Cajun Prairie, then known as the Opelousas and Attakapas district. In 1769, the Spanish commandante turned a Native American trading post near what is now St. Martinville into the Poste des Attakapas. The town of New Iberia was established by 1778, when emigrants from Malaga, some 500 of them, arrived (Villeré 1971). In 1765, 1766, and again in 1785, the Acadian migration brought as many as 2000 immigrants (C. Oubre 2001 and personal communication 2010).

The Poste des Opelousas, although not officially established until 1765, was originally a trading post established by Blanpain and LeKintrek who held an exclusive trading franchise from the Superior Council. In 1740, LeKintrek arrived and began trading with the Opelousas Indians at the junction of the Boeuf and Cocodrie where they formed the body of water he called the Riviere des Opelousas (now Bayou Courtableu). Other settlers came to the area in the interim between 1740 and 1765 when the first group of Acadians arrived. Control of the area was turned over to the Spanish Government in 1766 when Governor Ulloa confirmed the French Commandante Pellerin as the official Spanish Commandante. In 1801-3, the Spanish returned it to France, and France quickly sold it to the United States as part of the Louisiana Purchase in 1803. The exact western border of the Louisiana territory was not decided until 1819 (Adams-Onis Treaty of 1819, ratified in 1821, settled the western border of Louisiana at the west bank of the Sabine River). In the census of the Opelousas Poste taken in 1771, there were ten Acadian families listed as living in Opelousas. St. Landry Parish was later established out of Opelousas County, shortly after the Louisiana Purchase transfer in December 1803. St. Martin Parish was later established from the area known as the Attakapas County (C. Oubre 2001 and personal communication 2010).

Together the Opelousas County and the Attakapas County comprised southwestern Louisiana—then a literal wilderness. Much of the land was claimed by both the French and the Spanish, and later the Spanish and the United States. It included the "Neutral Strip" that was often referred to as 'badlands' or 'no man's land' as neither authority enforced laws to any extent in the region making it excellent habitat for a variety of criminals. The eastern boundary of the strip was the Calcasieu and the Arroyo Hondo while the western boundary was the Sabine River (C. Oubre 2001 and personal communication 2010). Both the French and the Spanish were more than willing to give free land grants, 40 or 42 arpents, to anyone wishing to settle and hopefully tame the land. Members of the French and Spanish militia were among those deeded land in the hopes of bringing civility to the region.

Carl Brasseaux's (1987) book entitled, '*The Founding of New Acadia*' provided a good introduction to the early Cajuns, their lives and livelihoods. Several of his comments are insightful about areas other than the major posts:

> *Few Acadians sought homes along Bayou Plaquemine Brulé, and fewer ventured into the wilderness beyond this prominent plains landmark. The hardy souls who, with a handful of Creole and Anglo settlers, did brave the rigors of life on the northwestern prairies, established themselves at the Faquetaique district of the Opelousas post (the area southeast of present-day Eunice) around 1790. ... The occupation of Prairie Faquetaique marked the western limit of Acadian expansion in the eighteenth century.*
>
> *The Acadian migrations typically involved family units....These migrants consistently ventured only as far as the nearest prairie stream, or coulee, establishing themselves along the wooded banks where firewood and building materials were plentiful.*

Brasseaux (1987) also pointed out that in Acadie many of these very same Acadians had been ranchers since 1707. In 1765, the earliest Acadians to be brought into Louisiana were from regions in Acadie with ranches. "*The Acadians' skill as animal husbandryman, the fecundity of their cattle, the lushness of verdant grasslands in the Attakapas district, and the region's mild climate (which extended the life span of weak and aged cattle), interacted to produce a remarkable growth rate among the Acadian herds. By 1771, only six years after their arrival as destitute immigrants, the Attakapas ranchers owned an average of 22 cattle and 6 horses – approximately twice the corresponding livestock holdings in the last extant predispersal Chignecto* (Acadie, i.e. Nova Scotia) *census...In fact, by 1803, Acadian livestock production had increased by at least 500 percent, and in the Quartier de Vermilion alone, the average Acadian Vacherie included 125 cattle and 23 horses.*" Notwithstanding Brasseaux's excellent introduction to the ranching skills and livelihood, the story can be expanded in order to include some additional ecology.

Chapter Three

The best way, in my judgment, to revisit the Cajun Prairie during the 1800's is to read the journals (or travelogues) of the visitors, travelers and surveyors of the period. They contained some bias as all journals do, but they were revealing. Three major travelers are presented, namely C. C. Robin in 1803, William Darby in 1805 and Col. Samuel Lockett in 1869-72. Collectively, they gave us a good view of the region; however, they provided limited natural history aside from a rather thorough listing of trees. The prairies had few trees; thus, we limited our use of their discussion of the trees. Bill Jones (2007) collected a number of facts about the early cattle business and the Cajun Prairie. His comments helped to move the story along. Otis Hebert (1968) provided a good introduction to the history of the Calcasieu Prairie—a region that lacked the detail study of eastern named prairies. Finally, Sidney Louis Villeré (1971), Maurine Bergerie (1962), Lauren Post (1962) and Malcolm Comeaux (1983) related some general historical and ecological aspects of the early years that help round out my discussion of this period.

C. C. Robin

Voyage to Louisiana, Stuart O. Landry's translation of C. C. Robin's travelogue related his travels through Louisiana at the time of the Louisiana Purchase during 1803-05 (Robin 2000). After Landry's essay on the identity of C. C. Robin, the translation provided a historical view of the eastern part of the Cajun Prairie. The travelogue reported an 1803, circuitous journey traveling up the Mississippi River from New Orleans through Donaldsonville on the Bayou Lafourche, across the Atchafalaya Basin, up the Bayou Teche in the St. Martinville area (Poste des Attakapas), and finally up the Bayou Courtableu to the Washington/Opelousas area (Poste des Opelousas). His arrival in the Opelousas area is described as follows:

> "*Toward the upper end, the banks of the river* (Teche) *are wooded, and behind this curtain of forest, about two hundred paces wide, spacious prairies stretch out, as far as the eye can see, broken here and there by patches of woods, forming a most agreeable landscape. Patches of swamps, little lakes and winding creeks are also met with from time to time. Almost all of the houses here are found between the strip of woods and the prairie, not along the river, and this gives them a more airy and cheery aspect. No region on earth offers to civilized man more opportunity for happy hours. Herds* (of cattle) *wandering at large multiply without requiring any other trouble to their owners beyond that of reassembling them in the spring to mark* (brand) *them so that they can be recognized. The prairie land is already cleared and awaits only the plow or the spade. The landowner may easily house himself. It takes only a few days to build a cabin. It takes only a few mornings' work to place this prodigious land into production sufficiently to support a family. These prairies are well stocked with*

game and, especially during the winter, they are covered with ducks and geese, so the inhabitant has his choice of birds as if they were in his own poultry yard."

"As one ascends further up the Teche, towards the Opelousas district contiguous with the Atakapas on the north, the terrain is higher and becomes slightly hilly and is cut by deep ravines, and, in addition, the patches of woods among the prairies are thicker and more numerous, giving to the country a more varied and picturesque aspect."

"Northwest of Opelousas, the terrain becomes higher little by little. No longer does one find prairies, but sandy hillocks covered with pines and oaks. This area was not deposited by the river but rather laid down by the sea."

"The beautiful regions of the Atakapas and Opelousas, even though they are close to New Orleans, and lying along the Atchafalaya, and easily reached through the waterway of the river, are the most recent settlements of the colony. Only the vigorous Canadians have penetrated here. Doubtless because only they are unafraid to risk the trackless windings of the waterways which lead here and are likewise unintimidated by the ferocity of the native inhabitants, who are called 'Atakapas,' which means 'man-eater.'"

"What has most contributed to the importance of this post was the establishment here of the Acadians at great expense by the Spanish government. These unfortunate victims of their patriotism had been taken to San Domingo, where their population might have saved that island from France at the time, if so many of them had not perished of neglect. They have found a more favorable climate on the soil of Louisiana, where their originally small numbers have considerably multiplied, both on the River and here at the Atakapas. Among them are some who have become extremely rich, who have amassed herds of several thousand head of cattle. A great number of these people it is true simply vegetate in these beautiful regions. The difficulty of navigation, subject to expensive portages across land, at certain times of the year, and the all too common obstacles to commerce under the Spanish regime, which prevented the settlers from exporting their produce and importing necessary good, caused some of them to decline into indolence. Rich and poor (if one can speak of poor, where one has the means to satisfy all of his ordinary wants), they have, both here and along the River retained their customs, these being much like those of our farmers in France. Through observation, I cannot repeat too often, to point out how important it is, in founding colonies, to provide good stock as colonists."

"These Acadians work their land themselves. The women and children go into the fields to pick corn and cotton, they take care of the barnyard, milk the cows, and spin the cotton into thread of which they make coarse muslin shirts, fine cloth, mosquito nets, and that multi-colored striped cotton cloth so agreeable to the eye, resembling very much our siamoises, out of which they make shirts and blouses, and for the men, pants and jackets."

"The families descended from French officers or merchants, live quite differently. They live in indolent ease, even those with little money. They use a portion of their slaves as indoor servants, in an attempt to recapture a sense of the easy and sumptuous life. Several have fallen into this decadence. The Acadians, simpler and more economical, are prospering and will become, in consequence, more useful to the colony and the mother country."

"The Acadians like to live to themselves and they have the good sense to have little to do with the families who are pretentious. These latter, however, seek out the acquaintance of the Acadians, and like very much to be a part of the joyful celebration of their balls."

"All of the inhabitants have received free grants of land from the Spanish government, which their descendants have subdivided. These grants are always of forty arpents deep by a varying width, often thirty arpents. Thus, as in the rest of Louisiana, when a reference is made to one arpent of frontage, it means a total area of forty arpents, ten arpents of frontage means four hundred of total area, and so forth."

"All of the domestic animals of Europe do marvelously in this country. The climate is much better for poultry than that of France. The inhabitants raise chickens almost all winter long. From February or March on, numerous clutches of eggs are set. The sustained warm weather of the country, the vast area of ground for scratching, many little insects, seeds and soft grasses, all of these make the chicks thrive."

"Pigs multiply here in an astonishing manner. The settlers do not bother to shelter or feed them. The sows farrow in the woods and undergrowth. In order to get them used to coming to houses, and to tame them a few ears of corn are put out for them from time to time. Wolves and tigers which no one takes the trouble to hunt down, destroy many pigs, both here and in other districts of Louisiana. Sheep do equally well all year round."

"Crossing the wide prairie, strewn with flowers, whose stems raise them to the height of the horse on which the traveler is riding, surprise follows surprise in this varied vegetation. One rides suddenly upon herds of cattle, who raise their haughty heads above the grass as one rapidly approaches."

"The prairies of the Atakapas and of Opelousas, the true sources of provisions for New Orleans, provide the city the year round with fresh meat. This district also provides salt meat for maritime commerce. These two districts furnish to the colony a large part of its beef and milk produce. There can hardly be fewer than 200 thousand horned cattle."

"In the distance, the occasional houses scattered here and there along the brownish curtain of woods that border the windings of the rivers and bayous provide relief for this rich tableau of nature, so new to the Europeans."

"Near those houses where the herds pass most frequently, the prairies change. The plants are no longer the same. In these trampled places grow those species most used by the animals for food. There, for instance, particularly flourishes a little clover; the couchgrass and several other grasses. Other plants, so common away from the house, disappear in these places. This observation, which I verified in so many places, proves again that each animal tends to favor the growth of those plants which are necessary to it in those places where it lives."

"The use of corn is universal both among the poor and the rich. Corn is prepared here in an infinite number of ways. Usually it is ground with wooden mortar and pestles. The white corn is made into flour, and from this can be made a very good mush with either water or milk. The latter being thicker, is eaten with that type of soup called gumbo. …Bread is also made of corn, as in Europe."

In my experience growing up in the 1950' and 1960's, cornbread was served at two meals each day. It could be had in a number of ways—dry cornbread, cornbread with milk, cornbread with syrup, warm or cold, or in stuffing, etc. Breakfast of cornbread and milk was common. Cornbread with eggs and or syrup was also very common. Warm cornbread was eaten with almost every meal. So much that I seldom wanted cornbread, and seldom do I eat it today. A version of the Native American 'Three Sisters' agriculture system (growing corn, beans and squash in a polyculture) (Harris 2003) was mentioned by Robin:

> *"Corn has this advantage over wheat, that it produces more abundantly. One can obtain two harvests a year and do the harvesting in one's spare time. If time is pressing, one can, when the corn is ripe, content himself with breaking the stalk of the ear and letting it hang down. Hanging thus, protected by its shuck, the rain runs off without penetrating it, and the birds cannot get at the kernels. In their corn fields the inhabitants plant watermelons and other species of melon that they call French melons. But they plant especially an immense quantity of gourds or pumpkins. Their corn fields produce thousands of them. I have seen settlers harvest 8 to 10 thousands of them."*

Beans are not mentioned in this quote, but we planted corn and beans and squash together in our fields in the 1950's and made good crops of each. We had a row of melons, both watermelons and canteloupes. We grew cucumbers in the field. We also grew potatoes (sweet and Irish) and peanuts. These were grown away from the house on the eastern edge of the property. Near the house a more kept garden grew tomatoes, peppers and cucumbers. Beets, turnips, radishes, carrots and a variety of greens (lettuce, cabbage and mustard) were grown in good numbers.

Robin wrote at length about cotton growing. It had not changed that much from his commentary in the 1950's—the 60's brought many new chemicals including 'per-emergent herbicides' and insecticides. Robin also discussed butterfly larvae—caterpillars—as pests in cotton fields. He clearly described 'host-specificity' of caterpillars and noted that they die if they no longer can find the proper host plants. He discussed the ability of caterpillars to devastate cotton fields during their flowering. He commented on the nature of these caterpillars to increase in number each year and as more and more people planted the same crops. His conclusions clearly predicted what is known today about agricultural pest ecology. He detailed the use of cotton in cloth for clothes, furniture, blankets and more. He wrote of 'fire,' which was not understood as part of the natural ecology of the prairie until the 1950s. The Native Americans and *vachers* used fire routinely as part of their management regime in the prairie (Jones 2007). Robin commented:

> *"One abuse, that of the inhabitants setting fire to the prairie and the pine*

woods, is extremely damaging to the herds and to the countryside in general. The tall dry grasses would be at least food of some sort for the herds. It would serve to shelter them from the cold. It would serve as bedding to keep the animals from lying on the damp ground. The new grass which sprouts early in the year would itself be protected by the old from the effects of the first frost and this would afford the animals a source of green food. This dry grass would serve as a manure for the new plants and would, in the long run, tend to raise the ground and contribute to drying the soil."

"The Atakapas, the most beautiful region in Louisiana, fortunate land where Nature has done everything to make men happy; where riches multiply with so little labor and expense; where the earth holds back nothing from man, where everywhere man is unequal to the earth; where there is so much need of industriousness; this land, at its very inception, has already become a prey to fatal divisions."

Robin continued for pages to list crimes including rustling of cattle, murder, etc.

"The first inhabitants who established themselves in the Atakapas on those beautiful prairies to the west of Louisiana, found the region healthy everywhere. They placed their houses near, or among, the woods which stretch in strips along the rivers and obtained the benefit of a cooling shade for themselves and for their cattle. The waters of the bayou which wound around them never lost their limpidness or their purity. Contagious diseases were unknown then."

"The symptoms of contagion only appear in the places where civilization begins, in measure as careless man uncovers drowned land, abandoning them to the action of the burning sun. this what is happening today in the Atakapas, a region so healthy that stricken New Orleanians come there during their convalescence. The colonies are hastening to cut down those great trees near their settlements leaving near them uncovered stagnant water, which they neglect to drain off, and this will increase the unhealthiness of this district in proportion as the number of settlements increase. The evil is increasing, and during the first summer that I was there (1803-5), many putrid fevers occurred there for the first time, because the summer, drier that ordinary, caused a higher degree of impregnation of the air by these swampy inhalations."

He went on to discuss tetanus, lock-jaw, diarrhea, worms, and other interesting conditions. Landry annotates the translation with "diphtheria, of course."

A small farm with a polyculture (*petit habitant*) was the general lifestyle of the colonists of the Cajun Prairie, but the rapid development of ranches for cattle (*vacheries*) has been apparently understated. Vidrine et al. (2001c) described the early culture as follows:

"In addition to these very modest beginnings within the realm of crop cultivation, the Acadians also encountered remnant semi-wild groups of cattle

and horses, which had escaped from Spanish colonial operations further west. Within a relatively short time the Acadians evolved into adept cattlemen, running their herds "free range" (au large) style across the prairie proper during the summer months, and driving them southward into the marsh region during the winter months - a practice which persists in some parts of the region today. Even for those settlers who did not choose to enter into larger livestock operations, the husbandry of hogs and poultry was considered essential to their individual subsistence efforts. Thus, to varying degrees, all settlers throughout both the eastern and western zones of the prairie region necessarily entered into both crop cultivation and livestock operations in a dual system which is still employed by the majority of contemporary prairie region inhabitants."

Bill Jones and William Darby

A recent book by Bill Jones (Jones 2007) provided an excellent introduction to the cattle industry in southwestern Louisiana as well as to the Sabine Prairie and associated marshes to the Gulf coast (often referred to as 'No Man's Land" or simply 'Badlands'). He integrated William Darby's (1816) comments into his introduction. I will use the two books together in this discussion. Jones reported that cattle were grazed in the prairie in the summer and in the marsh in the winter. Trail drives were the primary way to herd the cattle from one pasturing place to another up until 1972. New government regulations changed all that. But first Jones recounted:

"In the beginning, for the first 125 years of the state's settlement, it was grass – grass and the cattle and horses that grazed it – that was the dominant scene of southwest Louisiana."

"The Acadians were not the first ranchers in Louisiana. The Creoles and European settlers here before them had ranched in the prairie region west of New Orleans even before the posts at Opelousas and Attakapas were established. The first settlers onto the prairie region not only discovered a sea of grass, they found horses and cattle."

"Where did the horses and cattle come from? Neither species is native to North America. They were brought over on ships by the explorers. The Spanish entradas out of Mexico moved into Texas bringing large herds of horses, cattle, and sheep with them. They intentionally stocked the land with horses and cattle as they traveled, turning out one stallion and one mare, and one bull and one cow, at each major river crossing."

"Since the 1500s, the Native Americans, seeing the advantage these domestic animals gave the Spanish, have stolen and traded for them, bringing them north, even to Louisiana. When Tonti, looking for the lost explorer La Salle, made it to the Caddo Indian village at Natchitoches in 1690, he found they had many horses. In fact, the Natchitoches Post (founded 1713), which predates New Orleans by five years, was started to facilitate livestock trade with the Native Americans."

In 1765, the Cajuns began arriving from Nova Scotia. The Dauterive Compact (Rees 1976) as described by Jones discussed the primary roles of early Cajuns as ranchers:

> *"Governor Ulloa, however, was not the only one to greet the new settlers and to make them a proposition. Antoine Bernard Dauterive, a vacher (rancher), owned more than five thousand head of branded cattle near the Opelousas Post west of the Mississippi. He must have known something about Acadians and their history in Nova Scotia, for his proposition dealt with cattle. To the Acadians who landed in New Orleans without money and hardly any possessions, Dauterive agreed to give each family five cows and one bull each year for six consecutive years. At the end of six years, the Acadians had to return to Dauterive an equal number of cattle of like kind and quality. In addition, he would divide the offspring, with the Acadians keeping half. Dauterive and his partner, Edouard Masse, also offered the Acadians land along the Bayou Teche from Opelousas south to Attakapas Post, today's city of St. Martinville."*
>
> *"The Acadians—with a long history as graziers, trappers, hunters, and fishermen in Nova Scotia—liked Dauterive's offer better than Governor Ulloa's idea of farming (sic, The Dauterive contract was established in 1765 with the first group of Acadians to arrive. Ulloa does not arrive until 1766, so there could not be a comparison such as is made here (C. Oubre, personal communication 2010)). The names on the original Dauterive Compact read like today's Acadian Who's Who. There were four Broussards, a Guilbeau, a Duga(s), a Tibaudau (Thibodeaux) and an Arcenaud (Arceneaux), all of whom were referred to as "Acadian chiefs." These names appear again and again, even into the twentieth century, as important players in Louisiana's cattle industry."*

They chose to ranch rather than farm.

> *"The very first Acadians to arrive in Louisiana became cattlemen… and herded cattle on trail drives to New Orleans…In 1783, early ranchers taxed themselves to construct a better road on which to drive their cattle to the New Orleans market…"*

Wild cattle presented both a problem and an opportunity. Problem: the wild cattle mix with tame animals, lead them away from the herd and often spread diseases. Opportunity: if you catch and brand them, they are yours. Jones (2007) continued:

> *"Despite the great mix of races and backgrounds involved during the infancy of southwest Louisiana's ranching, in the end, it was the Acadians who proved to be the dominant ethnic and cultural force on the prairies. This was so true that the land they lived on goes by their name—the Cajun Prairie."*

As a point of natural history of the prairie, Jones made clear that *"they burned the prairie every winter, and boy, in the spring, it would come out pretty – you talk about pretty grass."* The cattle business is central to understanding the future of the Cajun Prairie. As the land loses topsoil and potential for modern crops—a return to grazing and native prairie grasses is likely.

William Darby's comments provided some insight. In 1805, two years after the Louisiana Purchase, American surveyor, William Darby, set out from New Orleans and described the region in his report (Darby 1816):

> *"Continuing westward, a new and astonishing scene would open: the wide, green plains of Attakapas and Opelousas, varied by the irregular chains of woods, narrow and indented."*

Darby was immensely impressed by what he saw.

> *"The face of the earth exhibits an expanse of grass, interrupted only by an occasional clump of oak or pine trees...The winds breathe over the pathless waste of savanna. The wild fowl is seen flitting or the deer skimming over the plain."*

In describing the Attakapas prairie between Bayous Teche and Vermilion and the Opelousas prairie between the Teche and the Mermentau River, Darby noted:

> *"Here you behold those vast herds of cattle which afford sustenance to the natives, and the inhabitants of New Orleans. It is certainly one of the most agreeable views in nature, to behold from a point of elevation, thousands of horses and cows, of all sizes, scattered over the interminable mead, intermingled in wild confusion. The mind feels a glow of corresponding innocent enjoyment, with those useful and inoffensive animals grazing in a sea of plenty. If the active horsemen who guard them, would keep their distance, fancy would transport us backwards into the pastoral ages. When we estimate the extent of ground that must forever remain covered with grass, it is no extravagant declaration to call this one of the meadows of America."*

When Darby got to Prairie Mamou, between Cane Bayou (Bayou des Cannes) and Bayou Nezpique, he found a prairie devoted to the rearing of cattle, with the larger ranches (*vacheries*) holding several thousand head each. It was clear that by the time of Darby's visit in 1805, ranching was the biggest business in southwest Louisiana. Darby (1816) described:

> *"The prairie Mamou is devoted by the present inhabitants to the rearing of cattle, some of the largest herds in Opelousas are within its precincts. Three rich stockholders have, as if by consent, settled their vacheries in three distinct prairies. Mr. Wikoff, in the Calcasu prairie, west of the Nezpique, Mr. Fontenot in prairie Mamou; and Mr. Andrus in Opelousas prairie. These three gentlemen must have collectively, at the moment this article is written, fifteen or twenty*

> *thousand head of neat cattle, with several thousand horses and mules. It may be presumed that Mr. Wikoff is at this time the greatest pastoral farmer in the United States."*
>
> *"In prairie Mamou, you encounter, in great numbers, mounds of earth, ten or twelve feet wide, twelve or eighteen inches high, scattered in immense variety over the whole plain. The origin(s) of those hillocks have given birth to many speculations; all perhaps wide of the truth; the most reasonable hypothesis ascribes them to a kind of mole. These elevations are much more fertile than the other parts of the prairie, the high growth and deep green colour of the herbage, give to the prairie where they are abound a singular and diversified appearance."*

In the Mamou, Calcasieu and Sabine prairies—Darby considered the soils very poor as compared to the more eastern prairies—modern writers divide these into so-called 'cotton and corn' rich soils verses 'rice and cattle' poor soils. It appears that Darby noticed what now seems a given. The Sabine Prairie, located between the Sabine and Calcasieu River, owes its importance to *'its position, rather than from its intrinsic value"* for either soil or timber. This is where cattle or horses often entered Louisiana, smuggled at first and then later bought in openly on well-defined trails to New Orleans. Because the soil was not as good for farming as the soil farther east, the Sabine Prairie was used mostly for cattle even today (Jones 2007).

Darby specifically noted that he found cattle both 'under herd' and 'open range' (free range or *au large*). He also noted the year-round grazing available in the region and the rancher's inattention to providing winter feed for the periods when no grass was available. Jones indicated Darby's attribution of livestock in Louisiana:

> *"Later in 1817, Darby described the livestock he found in the region. He understood that the cattle, horses, and methods of managing such livestock were brought to Louisiana from Texas, that former Spanish province. Those Texas cattle then so prevalent in Opelousas and Attakapas were "high, clean limbed and elegant in appearance," while the horses, of Andalusian or Nubian descent, were, "like their ancestors, small, compactly built, and inconceivably durable.""*

Ecologically, Darby spent a great deal of time listing trees in varied areas—no doubt identifying areas for future logging operations. The timber industry would explode in the next 50 years.

Jones ended his book with a series of anecdotal stories relating to anthrax and tick fever problems in cattle. By 1972, Federal laws required that all the wild cattle be round up. This ended the trail rides between summer and winter pastures in the western prairies. Today, cattle are trucked to pastures—a common event even in my neighborhood. All day long, trucks move cattle (and provide rich aroma for me as I walk in my prairie garden) back and forth to the fenced pastures. A large fenced pasture was immediately to my south for the first 12 years at my home. Bahia grass (*Paspalum notatum*) is the pride and joy for baling hay and pasturing the cattle. It is still a major weed in my garden, but it does not like fire or competition from prairie plants. It is also important to be aware that some winters were so bad in south Louisiana that large numbers of feral cattle died, but the cattle naturalized and grew to large numbers as has been recounted. Further, it is important that the comparison with the Midwestern prairies be evident. Those prairies with snow on the ground for months and severe, drawn-out cold could be pastured by bison but not cattle—feral cattle were not common at all. The impact of this on the vegetation is not well documented, but several of the comments above indicate that changes in the vegetation in association with cattle were evident. Cattle also tend to stay in one area and linger around or in wet areas, thus creating mud-holes. Bison and cattle are very different types of grazers, and Bison and Big bluestem are capstone (keystone) species of the prairie (Manning 1995).

Otis Hebert

Otis Hebert (1968) provided insight into the history of the Calcasieu Prairie and its prominent settlements.

> *"Prior to the Civil War, commerce on the Calcasieu River was negligible. The first important trade was lumber. In 1872, the channel at the head of Calcasieu Pass was deepened to 5 feet wide and 8 feet deep. Channeling had begun and continues today."*
>
> *"It was in 1818 that the first map appeared listing the area as "Calcasieu Country," the name given to it by the Spanish conquistadores in the middle of the 1700's. Martin LeBleu settled on the banks of English Bayou about 1770. He and his wife were from Bordeaux, France (same as the Vidrines). About 1800, Charles Sallier, with a Spanish land grant, built the first home within what is the present city of Lake Charles. "Charles Lake" bordered his land grant."*
>
> *"In 1840, Calcasieu Parish was carved out of St. Landry Parish. The old Spanish Trail ran from Lake Charles to Vermilionville (Lafayette). Midwesterners flooded into the Calcasieu Prairie in the 1880's and gave it a decidedly different flavor–non-Cajun. Part of it remains as Indian Country with Native Americans, albeit non-native as they had moved in from Alabama (nearly two centuries ago)–Koasati Indians, members of the Coushatta Tribe who live in Allen and Jefferson Davis Parishes."*
>
> *"Sulphur began to be produced in 1894, and the town of Sulphur was laid out. The first settlers were Acadians. In the 1920's, the supply of sulphur was exhausted, but now the company started developing oil fields."*

In the same book, Crisler (1968) treated the Bayou Teche history. He provided some details regarding the Civil War in the southern and eastern parts of the Cajun Prairie. Maurine Bergerie (1962 and 2000) provided histories of the Civil War in

New Iberia and the Attakapas region in general. Robert Gahn (1972) provided an overview of the Civil War in the northern parts of the Cajun Prairie. Bill Jones (2007) summarized the impacts of the war on the western parts of the Cajun Prairie. The impacts of the war would last for at least half a century and probably up until World War II. My interests do not presently center on the Civil War, except to note here and later the prairie-wide disruption in the varied economies including the loss of most of the cattle herds and the timbering industry in the war effort.

Following the war, conditions would slowly change for the better. Railroads would bring great change to the prairie; first they lead to its eventual decimation, then their remnant rights-of-way prairies provided the opportunity to learn much more of the prairie flora and eventually provided the seeds and propagules to restore the prairie. The Southern Pacific Railroad was built along the Old Spanish Trail and exists there today. Prairie remnants exist along this railroad both in Acadia and Jefferson Davis Parishes in the Mermentau Prairie (between Crowley and Mermentau) and the Calcasieu Prairie (west of Welsh, east of Elton and between Iowa and Fenton along U. S. 165). Prairie remnants also exist along U. S. 165 and the old Missouri-Pacific Railroad north and south of Kinder in the Hickory Flats and D'Arbonne Prairies. Very few remnants exist along other railroads, with the exception of the old line between Eunice and Mermentau, where the Frey community (part of the Plaquemine Prairie) still has a bit of privately owned prairie remnants. But I am ahead of myself in the story, so let us look at the Cajun Prairie just after the war as the railroad is being planned and built.

With Colonel Lockett's journal entries, the history of the prairies during the period after the war and before the wholesale development of commercial rice production was elucidated. His journal was published in parts and has been cited based upon its recent republishing by Lauren Post in two formats (Lockett 1969 and Post 1969; here I will try to maintain consistency and use the Lockett 1969 citation throughout the book; however, both citations will appear in the Literature Cited).

Colonel Samuel Lockett

> *"Samuel H. Lockett, Colonel C. S. A. and Professor of Engineering, Louisiana State Seminary in Pineville and Louisiana State University in Baton Rouge, conducted a topographical survey of Louisiana between 1869 and 1872. Only parts of his survey were published in his lifetime. The entire manuscript was edited and published a century after Lockett began his survey. Lockett's accompanying map went through two editions, 1876 and 1891 (Lockett 1969)"* (MacRoberts et al. 1997).

Lockett traveled through many parishes, but his most notable trip was on July 25 from Lafayette to Opelousas and then across the Mamou Prairie to Chaumont's

Ferry on the Bayou Nezpique. He related the tale as follows:

"From St. Martinville, an afternoon's ride took me through a beautiful prairie country to the village of Breaux Bridge."

"At this point I left the Teche and, turning sharply westward, made my way across the Prairies and the alluvial lands of Vermilion River to the parish of Lafayette."

"From this parish I next turned my attention to the Empire Parish of the state, St. Landry. To reach its courthouse town, Opelousas, I traveled through Beau Basin Country, along the Carencro Hills, and across the bottom lands of the Carencro Bayou to the Grand Coteau."

"From Grand Coteau my ride was continued to Opelousas through a country composed of prairies and woodlands and cultivated fields."

"The next morning at 7 A.M. I set out for a trip across the great Prairie region. The open plains I had already traversed along the Teche and between Vermilionville (Lafayette) and Opelousas were as little lakes to the great ocean-like expanses before me. Mr. Thompson rode with me a dozen miles to put me on the right road, or rather trail, across the Prairies. After being left to my own guidance, I felt a good deal like a mariner on an unknown sea. But I had my compass along and knew the direction I had to take, so I determined to keep my bearings and religiously take that of the two roads which held nearest to the course I wished to pursue."

"I made Mr. David Courville's by dinner time and was much pleased to find I had followed the path marked out for me. Thence my route was to be through the great Prairie Mamou to Chaumont's Ferry on Bayou Nez Piqué. My instructions were to inquire for Joe Chaumont's, as every one in the Prairie knew his ferry and all the roads leading to it. This I did at several houses that I passed at long intervals on the Prairie...."

"About an hour after I left Mr. Courville's, when in the very midst of the broad prairie, with not a tree or a house in sight, a dense, black, angry-looking cloud began to rise on the northwestern horizon. It came up with tremendous rapidity, and soon a fearful storm of wind and rain came sweeping across the open plain towards me."

"It was the twenty-fifth day of July, and had there been a thermometer on the Prairie, previous to the storm, the mercury would have been in all probability making desperate efforts to get out of the tube at the hermetically sealed end; on the contrary, the wind and rain of the storm were cold and penetrating, and I had to ride through it for an hour or more and keep my wet clothes on during the rest of the day. The Prairies became a vast sheet of water and made the bon chemin (literally 'good road'—he provided a long discussion about the roads and how poor they were and that this was a major concern of both travelers and locals) *all the more difficult to find. But by frequently consulting my trusty compass I kept on my course, and finally reached M. Joe Chaumont's just before dark, having made a day's journey of thirty-two miles."*

"The whole Prairie, from one end to the other, was filled with herds of cattle and droves of horses, but all of this prairie stock is of a very inferior quality. The

horses are mere ragged little ponies, with some hardiness, but very little strength or capacity for enduring a long trip."

Lockett then traveled south and west to Lake Charles. Later in the book, he described the topography of Louisiana—the main thrust of his interest:

"The Prairies"

"From this comparatively uninteresting part of Louisiana, we will now turn to that which to me, at least, is the most pleasing part of the state. I mean the region of the Great Prairies. This region lies almost entirely west of the Bayou Teche and south of Bayou Cocodrie, making up the old Opelousas and Attakapas countries. On the south it is limited by the impassable sea marsh into which it passes often by imperceptible gradation. On the west Calcasieu River and the Sabine River form the boundary lines."

"All of this extensive area, thus broadly defined, is not one unbroken, treeless expanse. Coulees and bayous course through it, generally in a north and south direction, on the borders of which grow fine forests of timber. From these principal belts of timber, spurs run out into the open prairies like headlands into the sea, thus dividing the whole region into separate tracts each having its own name. Faquetaique, Mamou, Calcasieu, Sabine, Vermilion, Mermentau, Plaquemine, Opelousas, and Grand prairies are the largest. There are many other local names that it is needless to mention. The surface of the Prairies though generally level is yet not perfectly so. It is gently rolling like the billows of a deep sea. In fact, one cannot ride through the Prairies without having their striking resemblance to large bodies of water constantly recurring to his mind. The grass that grows upon their surface waves in the wind and looks like ripples on the bosom of the ocean; the dark blue borders of woods are like distant shores, the projecting spurs, like capes and promontories, the "coves," like bays and gulfs, and the occasional clumps of detached trees, like islands in the sea."

Cajuns referred to the projecting spurs as Points and the coves were referred to as Anse (C. Oubre, personal communication 2010).

"Soil. The soil of the Prairies is either of a grayish-yellow or of a cold grey color, but it is much better than is generally supposed and improves wonderfully by being properly cultivated. The subsoil is a good, tenacious clay. The eastern part of the Prairies has a better soil than that farther west, yet even the latter amply repays the laborer for his toil. By manuring, tramping, draining, and deep plowing, the Prairie soil gets better every year that it is cultivated, and may be counted on for making from forty to sixty bushels of corn to the acre. Cotton, cane, and rice may also be raised with profit, and very probably an excellent quality of tobacco. Hay in any desired quantity could be undoubtedly be made by enclosing parts of the Prairie and mowing the grass when fresh and juicy."

"Products. The chief products of the Prairies are beef cattle and horses,

neither of very good quality from the negligent manner in which they are raised. I saw few flocks of sheep and can think of no reason why there should not be hundreds more. Poultry of all kinds can be raised with the greatest ease; vegetables and melons, figs, peaches, and fine strawberries can all be grown with success."

"Population. Most of the population of the Prairies is of Acadian origin, and, with but few notable exceptions, they are a rather thriftless people. They are kind, hospitable, and sociable amongst themselves, but shy and suspicious of a stranger, especially if, like the writer, he speaks no French. The most of them are mere squatters on the Prairies. Their houses, half framed and half built of mud, are located sometimes on the open prairie, sometimes on the skirts of a belt of timber, and often without even a yard or garden enclosed. A neighboring marais will be surrounded by a rude pieux fence and a small crop of rice raised. Their horses and cattle run at all times on the common prairie."

"Climate. The climate of the Prairies is admirable, as good as a climate can possibly be. It is breezy and cool in the summer, mild in the winter, dry and healthy at all times. The Creoles of the Prairies are proverbially healthy and long lived. Altogether, I look upon the Prairie region as naturally the loveliest part of Louisiana. It occupies 3,880 square miles."

In a latter section of the book, Lockett dealt with the prairie parishes. First, the western prairie parishes, then the eastern prairie parishes, and finally the 'coast marsh' prairie parishes were treated. First, the western parishes:

In Vernon parish, he addressed several prairies, *"From Huddleston northward there are scattered tracts of this same land until we reach Anacoco Prairie. This very similar to Pendarvis' Prairie in Catahoula Parish, and, like it, is on the outskirts of the Hog Wallow Lands. The soil is black in color, rich in lime, and produces very fine crops of corn. The prairie is nearly surrounded by hills, and, with its nearly level surface covered by dark green crops and waving grass, presents very much the appearance of an inland lake. The isolated prairies are quite different in many respects from the great Prairies farther south. The latter undoubtedly have an intimate connection with the Coast Marsh farther south and were probably once great marshes themselves. The detached prairies seem to be rather of lacustrine origin. They are with but rare exception very fertile, as much so as the best Alluvial Lands, while the larger prairies are not. The small prairies are covered with partly decomposed shells. I saw no shells on the larger ones. The small prairies of Louisiana are like the "cane-brake" lands of Alabama and Mississippi. They are nearly impassible in winter, become friable in the spring and summer when cultivated, while the large prairies are not peculiar in these respects."*

"*Calcasieu* (Parish, which includes what are now Beauregard, Calcasieu, Allen and Jefferson Davis parishes).

This parish is bounded on the north by Rapides, on the east by St. Landry, on the south by Cameron, and on the west by the state of Texas; Bayou Nez Piqué forms its eastern boundary line and the Sabine River, its western. The Calcasieu River courses through the parish from north to south and has the following principle tributaries.... North of Bayou Blue and between the Nez Piqué and Calcasieu, the country is a succession of small prairies and timbered flats, the prairies predominating as we go south, and the timber as we go north. The principal of the lesser prairies are Prairie Swallow, Godfrey's Prairie, and Prairie D'Arbonne."

"The timber of all this region, with the exception of the creek bottoms, is mostly longleaf pine with scattered blackjack, post oak, and hickory.... South of Bayou Blue begins the great Calcasieu Prairie, which is the most extensive uninterrupted prairie in the state. It embraces all of the country between the Nez Piqué and Calcasieu River on the east and west, and between Bayou Blue and the sea marsh on the north and south. Its length, from northeast to southwest, is over fifty miles, and its breadth is frequently as much as twenty-five miles. Crossing the Calcasieu River between Lake Charles and Lake Calcasieu, the Prairies continue their westward sweep to the Sabine River, a distance in a direct line of about thirty miles. The last prairie is bounded on the north by a line of timber mainly parallel to the west fork of Calcasieu and averaging three miles from that stream. Points of timber extending southward from this belt of woodland and numberless islands of trees separate this prairie into a number of subdivisions which have local names. The two largest of these subdivisions are Little Calcasieu and Sabine Prairies."

"Both branches of the Calcasieu...are quite navigable to the largest boats and very picturesque, with their banks fringed by a luxuriant forest of red and white cypress and live oaks."

"All around Lake Charles and for several miles up both branches of the Calcasieu, the sawmills are so thickly located as to give the whole country the appearance of an immense lumber yard. There are seventeen steam sawmills within a space of ten miles in diameter. These saw and ship millions of feet of the best pine and red and white cypress lumber yearly, giving constant employment to over sixty sailing vessels."

"The lumber trade is at present the principal source of wealth of the parish, but it has other resources yet undeveloped which will one day be of greater value than even its inexhaustible forests. Large areas of sea marsh are admirably adapted to the cultivation of rice, and all of the southern portion of the parish will produce oranges, figs, and bananas in profusion."

"Like the other parishes containing prairies, its stock-raising interests are but little better than in their infancy. And even for the production of the great southern staples, cotton and sugar, much of the parish is far better adapted than is generally supposed."

"All of the pine country of the Calcasieu abounds in game; deer bound across your path as you travel the long, lonely roads; bears, panthers, wolves, and wild cats from the swamps, prey upon the farmer's sheep and hogs, and of wild turkeys, ducks, and smaller game there is no limit to the supply."

"I will now notice the most remarkable region in Calcasieu, or probably in the state, namely, the oil and sulphur region fifteen miles west of Lake Charles.....

Then is encountered an open, wet prairie or half-dry marsh, dotted with little mounds and islets of trees. This is the oil and sulphur region. The evidences of oil consist in a number of black banks of hardened bitumen on the northern border of the marsh prairie and on its surface; quite a number of bubbling springs, emitting an inflammable gas; and crude petroleum, which may be found by walking over the marsh. So abundant is this natural discharge of crude oil that the log-haulers for miles around obtain their only supply of lubricating material from these springs."

The eastern parishes

"But before I take up these parishes individually I will give a short general description of the Opelousas and Attakapas countries to which these parishes belong. These countries take their names from two Indian tribes that once inhabited them. Its present inhabitants claim for it the title of "The Garden of Louisiana," with what justice let the following description determine."

"Six parishes now occupy the region we are about to consider. St. Landry, St. Martin, Iberia, Lafayette, Vermilion, and St. Mary." (Acadia and Evangeline parishes were once part of St. Landry Parish.) "They lie between the Atchafalaya River on the east and Bayou Nez Piqué and Mermentau River on the west."

A list of trees of the area was provided.

"These prairies are all vast, treeless expanses, covered with a luxuriant growth of grass. Generally they are quite level, but occasionally huge swells will cross them from side to side like the mighty billows of a deep sea. The view from one of these swells is very enchanting. The green carpet spreads out all around you, bounded by the distant horizon or terminated by the dark lines of forest that project into the prairies like the headlands of a lake. Scattered here and there are the little farms and homesteads of the few inhabitants, and roaming everywhere are their immense herds of cattle and horses."

"Through these prairies numberless tracks are made by their cattle in their journeys to and from their watering places which, as previously described, are natural ponds, marais, and coulees. The inhabitants know which of these trails are their bon chemins, but a stranger would do well to rely only on his compass and look sharply to his bearings in traversing these broad spaces so destitute of striking landmarks, as every trail will seem equally bon to him."

Lockett later commented on Avery Island and in my opinion sums up the Cajun prairies with these quotes.

"Could all the vast sheet of prairie and woodland, marsh and water, hill and valley be placed on canvas by a skillful artist, it would give to the admirer of nature's charms "a world of beauty in a nut shell." But the great beauty of this island, with its expansive views, is not its most notable feature. The wonderful

deposit of pure rock salt, which was discovered during the war, lying in one of its deep valleys, has given the island a very just celebrity throughout the United States."

And finally, Lockett provided a detailed mention of Spanish Lake, near the southeastern terminus of the prairie region:

"Lake Tasse is sometimes called Spanish Lake. Its lower limits come within two miles of New Iberia. Its length is nearly five miles, its shape, nearly round. One-third of the surface is water and two-thirds, grass, usually called "floating prairie." It has generally been supposed that nearly the whole of this grass surface consists of matted roots a foot or two in thickness with water underneath. This proves to be an error. Some of the grass next to the water surface is afloat, and fishermen run on this surface with skiffs, work holes through with their paddles, and catch immense quantities of fine fish."

"The greatest depth of this lake is about twenty feet. Its margin is mostly fringed with water grass and lilies. The bank on the west side is high and dry, and on the east side there are fine building sites."

"These warm waters were filled with fish: trout, perch, gar, sacalait, rock, bar, choupique, gaspargo, buffalo, cat, blowing fish, softshell turtle, and sardines. Some of the trout are two and a half feet in length, and sacalait, eighteen inches. This lake is fed by springs of excellent water that break out around its margin. There is a large, boiling spring in the middle of the lake. The nearest approach of the Teche to this lake is about 700 yards. Its surface is about six or eight feet above the level of the bayou. There is water power here sufficient to drive a sugar mill."

The "coast marsh' prairie parishes:

"Cameron. Cameron has been recently created out of the southern part of Calcasieu and the western part of Vermilion; the Gulf of Mexico bounds it on the south, and Sabine Lake and River, on the west."

"It contains but a small habitable surface and a sparse population. Its surface presents but two grand features, namely: a vast, impracticable sea marsh on the south and a part of the great Calcasieu Prairie on the north.

"Vermilion. This is a large parish in actual extent, but the following distribution of its topographical features will show that it possesses only a small portion of habitable lands. It contains about 100 square miles of timbered lands, about 500 square miles of dry prairie, and over 1,000 miles of salt marsh, lakes, and bays....The Vermilion River runs through its eastern portion, and the Mermentau River and Lake form its western boundary."

"The lands along the streams are of good quality and produce fair crops. The prairie lands are more level than those of the Prairies already noticed. They are susceptible of cultivation, and those who have given them a fair trial say that they are actually improved by cultivation for a number of years. By turning the stock upon the fields every two or three years, they may be kept in fine condition

for an indefinite period. However, all of these prairies are such excellent natural pastures their inhabitants pay but little attention to agriculture and devote their time and energies to watching their immense herds of cattle and horses. They make no effort at improving their stock, and the cattle are not as valuable for beef as they might be and are nearly worthless for milking purposes. Although their horses are a breed of very small ponies, hardy, it is true, they are too light for heavy work."

Lockett (1969) concluded:

"The most interesting special features of Vermilion Parish are Pecan Island and the Grand Chenière. Both of these are long, narrow bodies of land near the waters of the Gulf, surrounded by the sea marsh...As their names imply, the one is covered by a dense growth of pecan trees and the other is almost an unbroken line of oak grove. They both contain settlers and, but for the swarms of mosquitoes which infest them, would be most desirable localities of residence."

Other Authors on ecology of Cajun Prairie in the early settlement period

Several other authors have made comments on the settlement of the Cajun Prairie. These include Sidney Louis Villeré, Lauren Post and Malcolm Comeaux. I expect more authors to be added to this list.

Sidney Louis Villeré (1971) provided the manifests for the ships carrying militia from the Canary Islands in the late 1770's and provided the following introduction to the creation of New Iberia:

"Nueva Iberia. On January 15, 1779, Don Bernardo de Galvez, the Spanish Governor of Louisiana, mentions in a dispatch, the arrival of families from Malaga, Spain. Most of these recruits settled in the Attakapas country of Louisiana at a place called "Nueva Iberia." Here, they were under the Captain-Commandant Don Francisco Bouligny. They cultivated at first flax and hemp, but without success. Soon, they were raising cattle on the vast prairies of the Attakapas. There were at least 500 people involved in the migration, but all were not from Malaga. A few scattered Canary Islanders joined their Spanish-speaking fellow nationals of "Nueva Iberia." The vanguard of Acadian exiles had previously been conducted to the land of Evangeline by the Militia Captain Louis Antoine Andry as early as 1765. But by 1778 and later, the Acadians arrived from different points of the compass. This new influx added to the natural complexities of the Spanish administration in Louisiana."

Maurine Bergerie (1962) provided a rather detailed historical introduction to the Attakapas Prairie people and their history. She discussed their agricultural travails and successes. Bergerie also reported harvesting honey from domesticated bees in the 1890s in the region. The sugarcane crops would turn the area into a profitable region. However, bouts of yellow fever, cholera and other diseases were common during the early years. Michael Wade (1995) detailed the history of a sugarcane family and the history of sugarcane in the region. He also described the region known as Grand Marais as a giant swamp just south of New Iberia.

Lauren Post (1962) reported that 'carpetgrass' was the major grass used by cattle in the prairie—it was introduced from Central America and the West Indies before 1832. Cajuns called it *gazon*. "Whatever the native grasses were, many of them were crowded out of vast areas by carpetgrass. He also listed many cattle brands and reports thousands of them. Carpetgrass is any of several coarse, sod-forming grasses of the chiefly New World genus *Axonopus*, especially *A. fissifolius*, cultivated for turf and pasture in warm humid regions. This is a native species to Louisiana. In many parts of the Cajun Prairie, wholesale planting of Bahia grass (*Paspalum notatum*), an exotic introduction, is practiced (Thomas and Allen 1993).

Malcolm Comeaux (1983) related some general ecological aspects of the early years:

> *"Some Acadians also settled to the west, along the natural levee of Bayou Teche, in the area of present-day town of St. Martinville. Land for these farmers was subdivided into long, narrow strips, usually 40 arpents deep and 5 wide. The early Acadians invariably settled on the high, easily worked lands near the river and avoided the heavy soils to the rear"*
>
> *"The prairie of Southwest Louisiana was an open, relatively flat area when the Acadians first arrived. It was not one unbroken grassy area, but many grassland areas separated by strips of woods which skirted the bayous of the region. Each of these open areas, called a prairie or cove, had a name, but together they formed one distinct region."*
>
> *"Water for livestock came from small circular ponds (platins) that were once very common."*
>
> *"The prairie can be divided into two cultural zones, the corn-and-cotton section and the rice-and-cattle section. The corn-and-cotton country was to the east, on the richest and most productive lands of the prairie." A farmer planted 40 percent in corn, 40 percent in cotton, and used 20% for pasture, a garden, and for farm and animal buildings."*
>
> *"The situation in the rice-and-cattle country to the west was quite different. The soils there were not so rich, and a hard claypan lay beneath the topsoil. The area was thinly settled by Cajuns, who lived scattered along the edges of the woods, raising cattle and planting small gardens."*
>
> *"The first Acadian occupation on the prairie was cattle raising, but Cajuns were not the first cattle raisers there. ...By 1769, just a few years after the first Acadians arrived, there were nearly 4,000 head of cattle on the prairie. At that time, the Spanish government considered the cattle industry important and required stock fencing and branding. A grantee of land had to have at least 100*

> *head of cattle."*
>
> *"Large herds, for example, were shipped by steamboat from Washington. The development of rice farming greatly changed the cattle industry, for cattle could no longer be raised "at large.""*
>
> *"The growing of rice is an old tradition in Louisiana. It was grown by subsistence farmers along the levee lands as early as 1719. It was planted in the backswamp and given little attention. On the prairie, it was called "providence" rice as its success depended upon watering by Providence."*
>
> *"Two Cajun life styles evolved on the prairie. To the east was a densely settled area of petits habitants that were definitely Cajun. These small largely self-sufficient farmers grew cotton as a cash crop and corn for themselves and their animals. ...On the western end of the prairie a different life style developed. These Cajuns maintained large herds of cattle, and had little agriculture except for gardens. In time, however, this area developed into a wide open environment producing rice and cattle, with much of the land owned by Anglo-Americans. This entire prairie, however, was strongly Cajun in character, and many of the early Anglo immigrants were absorbed into Cajun culture."*
>
> *"Cajuns have inhabited four environments, and, as a result, have exhibited four distinct life styles."*
>
> *"The four environments, marsh, swamp, prairie and levee land, still influence the lives of Cajuns as they have for 200 years."*

The roles of the early settlers in the ecosystem fell into general categories: fishermen, shrimpers, oystermen, farmers (*petit habitant* or rice/soybean farmers), and ranchers. Later the oil exploration led to a whole new set of roles. The first oil field was developed in 1901 in the Mamou Prairie near the junction of the Bayou des Cannes and the Bayou Nezpique (at Evangeline, northeast of Jennings, Louisiana). This changed the Louisiana prairies and marshes. Oil fields have a patchy distribution in the prairie. When driving around, it is not uncommon to spot a group of grasshopper wells and tanks in the open prairie or at the edge of a gallery forest. Usually these are relatively small areas, but a very large area exists near Evangeline, the original oil field, and near Welsh. There is even a small section of remnant prairie along the railroad and adjacent to a large oil field area just west of Welsh, Louisiana. The role of the oil industry in the prairie is not that well documented, but the role is extremely well documented for the marsh area.

As far as the early years, cattle apparently had the most important role in the history of the Cajun Prairie (Brasseaux 1987). Cattle were introduced by the Spanish explorers. Those cattle and horses left behind multiplied and became numerous enough for the Native Americans to round-up and trade with the French at varied trading posts. The majority of the cattle that would become livestock on the prairie would be brought from Alabama and grown on the ranches (C. Oubre, personal communication 2010). The Cajuns and early settlers along the bayous pastured their cattle on the prairie as a free-range. The cattle were identified by branding. We did the same but ranged our beef cattle in the forests along the bayou, while we were using the prairie to grow cotton and pasture a few cattle, especially the milk cows.

The livestock (cattle, horses, sheep and pigs) changed the vegetation and possibly created *platins* by milling around in wet areas. Overgrazing may have been very significant, since cattle range locally and do not typically migrate like the other great herbivore of the prairies, the Bison. Little is recorded on these matters. The cattle and horses played an extremely important role in early trade between Native Americans and Europeans and later between colonists. The initial settlers were not interested in farming, but rather in ranching. Both land and livestock were free. Spanish officials gave free land grants of 42 arpents—grants were for elongated rectangles that fronted the bayou and accessed part of the prairie. The homes were usually built near the bayou, the area between the home and the prairie was used for growing vegetables, and the prairie was used to range cattle. With free livestock and land and a market, the settlers had an opportunity.

The early settlements were created at trading posts, but later they developed along major trails like the Old Spanish Trail. They routinely located near navigable waterways for commercial trade and travel. Traveling in the prairie was difficult during most of the year. With the building of railroads and the move to farming, settlements would develop in the open prairie along railroads and in the middle of the prairie. While cattle were king in the early years, rice farming and oil exploration would radically change the landscape after Lockett's visit. By the 1880's, rice was commercially grown, and canals and lakes for irrigation of rice were developed throughout the prairies.

Native American names for people, places, animals, plants, food and medicine are commonly retained today. For example, the town Mamou and the plant (Coralbean, in USDA Plants database) with the same name are named for a Native American chief (Reese and Allen 2004). Also, the Bayou Nezpique is named for a Native American chief (with a pocked nose as the French translates), the Mermentau River is named for Nementou, another Native American chief, and the Bayou Queue de Tortue is named after Celestin la Tortue, another chief. The list is extensive (see Appendix 7).

Chapter 4
1870s-1930s

"In 1906 the prairie was beautiful, with its cattle, its wild flowers, surrounded by pine trees. The people traveled in high-top buggies." This quote is taken from *The History of the Village and Surroundings of Pine Prairie, Louisiana* by Mrs. R. W. Helmer (Robert Gahn (1972)).

"Not a single tree was found in what would become Eunice in the first decade of the 1800s." Based upon two long days of pouring over the land surveyor's notes from 1803-5 for historic St. Landry Parish in the Opelousas Courthouse, no trees were noted in open prairie areas as the surveyors staked out the land. Without much stretch of the imagination, not a single tree probably occurred in Crowley, Jennings and Welsh in the early 1800s. (Charles Allen, personal communication 1990 and Allen and Vidrine (1990b)).

Following Lockett's visit to the Cajun Prairie in the early 1870's, life in the Cajun Prairie changed dramatically from a ranching economy to a farming economy. Whereas cattle had ranged freely on the open prairie for more than 100 years, the prairie would now become the farming center of Louisiana. For 100 years, rice had been cultivated in wet areas on a small scale in order to provide additional sustenance to the *petit habitants* of the region. With the onset of commercial, indeed industrial, rice farming, the entire western prairie would be set into crop rotation. Cotton farms were scaled down and later disappeared. Corn fell from its former importance. Crayfish culture developed into an industry as farmers began harvesting crayfish in the ponds that were used for rice cultivation (Huner 1994). As a side note, Bullfrogs (*Lithobates catesbeianus* in French the *ouaouaron* (the name mimics the male's call) (Fontenot 2004)) also developed as a 'crop' harvested from rice fields.

The cattle industry remained important, but it changed following the Civil War. Bill Jones (2007) described the events very well as follows:

"The Civil War wrecked the cattle industry in southwest Louisiana, just as it wrecked the entire South."

"In 1860, there had been almost seventy thousand cattle in domestic herds on the Cajun Prairie, but it would take more than a decade to regain those numbers."

"And in what might be seen as divine retribution, four years of terrible weather followed the war, causing massive crop failures across

> *the South. Banks in New Orleans failed. Louisiana, including the Cajun Prairie and the Neutral Strip* (or No Man's Land), *was bankrupt. By the 1870s, the grass prairies of southwest Louisiana, like the entire South, had a Third-World society structure, with a very few rich families, very few middle-class farmers and cattlemen, and many poor people. The rebuilding process took a long time. The end of the Civil War marked the beginning of an eighty-year economic depression in the South that lasted until World War II."*
>
> *"The biggest port used by the ranchers was Washington Landing on Bayou Courtableu. In 1877, fifteen thousand cattle were shipped to New Orleans out of Washington on 'round boats." The steamboats out of Washington Landing received this name because they went a roundabout way to reach that port; up the bayou to the Atchafalaya River and Red River, then downstream to the Mississippi, and eventually to New Orleans."*
>
> *"Cattle were moved by land as well. By 1877, large herds were reported going through Lake Charles east to New Orleans, where they were readily sold. The trail was traveled so heavily that it had a distinct drop of approximately three feet, which could be seen for many years after Lake Charles became a growing, bustling town."*
>
> *"East of the Calcasieu River, the large-scale ranching became concentrated in the marshes."*

The open prairies quickly became the place for farming of rice and soybeans.

Rice and later soybeans became major cash crops—like a 'gold rush,' and the land was tilled. Some land remained fenced in and was used for planted pasture (usually with Bahia grass or other exotic varieties) and haying. The prairie disappeared within the lifetime of those born at the turn of the century. However, Lockett had also provided a clue to the mineral industries that would become a major set of industries in the prairie region. He and other authors had referenced the salt domes and readily mined salt of Avery, Jefferson and Week's Islands. But his brief description of sulphur, oil and natural gas in the Sabine Prairie and their potential was visionary. The first oil well (Jennings Oil Company Number 1, Jules Clement, was completed at Evangeline, 5 miles northeast of Jennings, September 21, 1901, by W. Scott Heywood and Associates) struck oil. This was the start of the vital Louisiana oil and gas industry. That first oil literally started another 'gold rush' for the 'black gold.'

Robert Gahn (1972) recounted the story of agriculture in Evangeline Parish, the northern most extension of the Cajun Prairie and the heart of the Mamou Prairie. Cotton was a major crop on the prairie, but with the building of Miller's Lake into a literal irrigation system, rice became the primary crop in the central and southern parts of the parish. His cotton and rice history included:

"While cotton was introduced to Louisiana as early as 1728, the exact date when it was first grown in Evangeline (Parish) is not known. By 1817, 6,848 bales were grown in this territory which is considered an excellent yield." (from the St. Landry Democrat, April 23, 1887). By 1939, 23,203 bales..."

"One of the prime investments that John LaHaye and Gus Miller made jointly was a large holding of land adjoining the Gus Miller Lake. They both foresaw, by practical experience and observation, the possibility of damming and controlling Bayou Nezpiquez into the now existing Miller's Lake. This represented one of the great early developments of Evangeline Parish. A pro rata share on the lake was agreed upon by the builders and is still in existence today. This lake was enlarged in 1931 and annually floods, by gravity irrigation, from 30 to 40,000 barrels of rice. Its 35,000 acres provide some of the best hunting and fishing in Louisiana today."

Additional statistics for the area and for agriculture were included in his book. My main thrust is not agriculture, but rather to show that agriculture developed rapidly and broadly across the landscape rendering the landscape into a quiltwork of annually plowed fields. Gahn's (1972) point dealing with the construction of a dam on Bayou Nezpique providing for the formation of Miller's Lake was indicative of many such lakes that have been formed since. These lakes provided water for rice, fishing and recreation. Miller's Lake remains a major land feature just north of the community of Vidrine, but dozens of lakes dot the Cajun Prairie.

German immigrants from New Orleans and other areas settled the towns of Crowley, Fabacher (Ritchie), Robert's Cove and Jennings in the heart of the Cajun Prairie. Their rice cultivation techniques changed the face of the open prairie (Merrill 2005). Mary Alice Fontenot and Rev. Paul P. Freeland (1976) provided the story of Acadia Parish and unveiled the initiation of rice as a commercial crop in the Cajun Prairie:

"Joseph Fabacher, born 1830 in Bavaria, emigrated alone to America when he was a small boy. By the time the Civil War started he had amassed a fortune in operating a distillery in New Orleans, later founded the Jackson Brewing Company. The war interrupted his distillery business; afterwards he went into the real estate and persuaded a friend, Zeno Huber, to assist him in founding a German colony in southwest Louisiana."

They "came to Prairie Faquetaique in 1870 and by January of 1871 the first group of colonists from Germany had arrived. A few months

> *later there were about 60 persons in the colony." "...each immigrant had planted about 40 acres of rice, in addition to corn, Irish and sweet potatoes, sugar cane, and oats."*

The settlement was called Fabacher—later Ritchie. Another small community named Frey, named for its originator John Frey, was also established just south of Fabacher across the Bayou Mallet. The Midwestern style of settlement became obvious in the open prairie. Fontenot and Freeland (1976) continued:

> *"The German immigrants were the first farmers in southwest Louisiana to raise rice for market and to grow rice on high land."*
>
> *"The history of rice cultivation in south Louisiana goes back to the days of the first settlers, who planted the grain for their own use. This was the so-called "providence rice," planted in shallow ponds which once dotted the prairies. In times of drought the ponds dried up; the result was crop failure. Given sufficient rainfall, the planter had an ample supply of rice to last his family until the following season. Hence the name, 'providence rice.'"*
>
> *"The German farmers soon learned how to aid providence. They constructed levees around their rice fields to hold the rain water. When the grain began ripening, the levee was cut and the water allowed to drain off."*
>
> *"The German farmers demonstrated that commercial rice farming could be profitable; their example was soon followed by numerous other planters. Improvements in irrigation, cultivation, harvesting and milling were to follow."*

The year 1880 was the turning point—4600 barrels of rice worth $16,000 proved that the Cajun Prairie land was valuable for rice cultivation. Railroads were being built to move rice and cattle to market. Cornelius (C. C. or Curly) Duson married Eunice Pharr (the town of Eunice was named for her) (Reed and Reed 2009). Duson became the local lawman and was famous for always 'getting his man.' He was often described as a local hero (Fontenot and Freeland 1976). Duson (with his younger brother William W. Duson), actually Scotch-Irish with his father's last name being McNaughton, were responsible for building the towns of Crowley, Eunice and Mamou on railroad extensions in the open prairie. Crowley was established in the heart of the open prairie in 1887, Eunice in 1894 and Mamou in 1907. C. C. Duson laid out these towns in a Midwestern design—a central street with all of the rest of the town laid out in square blocks extending out from the central street (towns that developed along rivers developed by extending up and down the river and out from the river). These became the central towns on the eastern open prairie. On the western prairies, towns like Jennings, Iowa,

Elton, Welsh and Fenton were also established on the Old Spanish Trail or on new railroads, and as such were referred to as railroad towns.

In contrast to the earlier French and Spanish land grants that ran from the bayou out onto the prairie in linear/long rectangular or triangular patterns, the land on the open prairie was deeded out in square units. When roads were built, they too formed squares roughly one mile per side. This remains evident in all the prairie parishes today.

Within the decade (1881-1890), warehousing and milling operations were developed in the Crowley region for rice farmers. Canal systems were dug and built to control water from bayous and other large bodies of water. Joseph Fabacher, John Frey and the Duson brothers pumped water to rice from large reservoirs. By 1900, rice irrigation became a business on its own (Fontenot and Freeland 1976), and rice farming became the predominant activity on the prairie. The major canal systems in the Cajun Prairie on a modern map are: Duson, Mamou, Louisiana, Abbeville, Kaplan, Hunter, Welsh, Sturdivant, Southwestern, Indian Bayou, Farmer's Land and Canal Co., Sabine, Bunker Hill, Acadia, Grand, Little Platte, and Point Claire. Many other smaller canals not on the state map belong to individual farmers or groups of farmers and can be readily seen as you drive across the prairie. These canals are fed from lakes (e.g. Miller's Lake), from streams and from deep water wells. The entire prairie is now subject to irrigation and drainage in one way or another as it is now planted in rice, soybeans and crayfish culture ponds. Lakes, canals and gulleys are everywhere for the irrigation and drainage of fields and residential areas. The entire area is a quiltwork mosaic of fields, sculptured to flood and drain. In the latest incarnation, farmers have recently installed underground watering systems with large deep water-wells and are in the process of abandoning the canal system.

These large water wells use nearly 80% of the water annually removed from the Chicot Aquifer, the major underground deep water source for the Cajun Prairie. The entire area is drained by gulleys, some natural, some man-made, which have effectively changed the entire hydrology of the region from 'runon' to 'runoff' (Vidrine et al. 2006 and Vidrine 2008). The resultant erosion has changed the sandy-bottom or clay-bottom streams, bayous and rivers, into mud-bottom streams with muddy runoff nearly all year round. The loss of topsoil is in epic proportions, and many farmers have little topsoil remaining. This companioned with land-leveling has completely depleted the organic matter in the topsoil, leaving sandy fields with thick clay subsurfaces. Some wells are now drawing salt water in years of extreme drought. Other areas have water contaminated with petroleum products, which may become a major problem with 'fracking' of old oil fields and new natural gas fields (Fox 2010).

The French, Cajuns, Spanish and other early settlers lived in the gallery forests along the bayous or on the very edge of the prairie. They mainly grew corn, cotton, beans, vegetables and a bit of 'providence rice.' Row or mound crops of corn, beans and squash (Three Sisters) were as common as they had been for Native Americans. The Germans developed 'rice' cultivation and the cultivation of grains in open 'non-row' crops. Once Eunice and Crowley were settled, the settlers quickly moved into the prairie and began mimicking the Germans. Many young Cajun girls ended up marrying the German boys as they were the available women of the region. The early families practiced Catholicism; thus many of these families raised their children as Catholics. Most families spoke French—it was the primary language. I grew up in a Catholic family where French was routinely spoken. Schools, however, insisted on English!

The Cajun Prairie was strewn with ponds and marshes (*platins* and *marais*) such that the west seasons (winter and early summer) made the whole of the prairie a massive marshy habitat. This was especially true from Eunice to Crowley and east to Lafayette and Opelousas—an area with considerable glacial till (loess soil) over the clay pan. This region often contained so much water that traveling from one place to another in a wagon or on horseback was very difficult and time consuming. Companion the wet areas pocketing the entire area and the tall grass literally hiding the trial ahead and you get an idea of the perils that lay ahead of any traveler. Any trail would soon turn into a muddy quagmire with ruts from wheels and hoofs that were a foot deep. Samuel Lockett wrote of this in his travelogue. In dry weather, the trip was rather harmless especially in the open prairie after a burn. The coming of the railroads in 1875-1895 completely changed this—a train ride was far preferable to any other mode of travel in the wet season (Georgie Manuel, Eunice, Louisiana, personal communication 2008).

Vidrine et al. (2001c) summarized the changes in the prairie as follows:

> *"Almost immediately after the completion of the railroad, outside speculators began purchasing the southwestern Louisiana prairie lands adjacent to it, and laying out towns which include present-day Iowa (Calcasieu parish), Jennings, Welsh, Roanoke (Jefferson Davis Parish), Estherwood, Crowley, and Rayne (Acadia Parish). The fact that most of these towns have Anglo-derived names is no small coincidence, since the vast majority of land purchases and subsequent development were being conducted by non-Cajun investors who were flocking into the region by the hour. Soon, well-organized advertising campaigns were launched, enticing Midwestern farmers to move their operations into the area, which boasted annual rainfall totals and a growing season which were far beyond that of their interior U.S. environs. The advertising campaign worked to perfection, and train loads of Anglo and German settlers began pouring in. Evidence of this invasion can be readily appreciated today as one travels through the above-mentioned towns to note that not only the surnames, but also the architecture of the homes, barns, and older grain driers are all stamped with a distinctive Midwestern U.S. style."*

"Newly-arrived German settlers in particular brought with them two inventions which would drastically and immediately alter both the culture and the ecology of the Cajun Prairie: the diatonic accordion, and artificial irrigation. Cajun musicians immediately took to the accordion, and instrument whose force and volume necessitated the move of Cajun folk music out of the farm house and into the dancehall, thus exposing it to ever greater numbers of listeners. Likewise, the Germans' concept of pumping water out of local bayous in order to flood artificially diked prairie lands (whose dense clay subsoils proved perfect in their water-holding capacities) signaled "the beginning of the end" to the ecological prairie, which would be hastily cleared in wholesale fashion over the next four decades. Thus, by 1920, over-grazing (the cattle industry had attained similarly epic proportions upon the completion of the railroad) and large-scale rice production, with its accompanying drastic land-clearing operations and alterations in hydrology had reduced the ecological prairie to a mere fraction of its former domain."

"It should be noted that for their part, the Cajuns heartily embraced the Germans and their accompanying technologies, for the Germans did not linger within the pre-fabricated railroad towns, but instead dived right into the prairie proper alongside the Cajuns. In fact, within the space of only one or two generations, the majority of German settlers had become virtually indistinguishable from their Cajun brethren! Like the Spanish prairie settlers before them, the Germans were overpoweringly and rapidly absorbed into the Cajun lifestyle in an ongoing cultural phenomenon, which remains in force today. It is obvious that southwestern Louisiana prairie dwellers of Spanish, German, African, and even Native American extractions all contributed shares of folkways which entered the mix of what is known today as "prairie Cajun culture". However, it is equally apparent that the overwhelming majority of descendants of these various ethnic contributors have taken on the foods, customs, religious customs, music, and language of the Cajuns. In part, the underpinnings of this phenomenon can be explained by the simple fact that Cajun women substantially outnumbered the women who would arrive later with the non-Cajun settler groups, and as a result a greater number of non-Cajun men would necessarily marry into Cajun families. And as is usually the case with most ethnically mixed marriages, the children are more apt to take on the customs, beliefs, etc. of the mother (Ancelet et al. 1991, Brasseaux 1991, and Brasseaux et al. 1994)."

The shrink-swell nature of the soils was inimical to tree growth and the growth of many other plants. However, the same soils and the micro-elevations evident all around would later be understood to be responsible for the tremendous diversity of prairie plant species (Grace et al. 2000a and b). It also limited the types of plants that could be grown in agriculture. Rice of course excelled in this habitat. Cotton, corn and melons had to be grown in dryer areas. The dry season and the prairie grasses combined to create an incendiary habitat that readily burned. This remains evident in the remnant strips and in restoration sites.

Drainage became a central focus for development of the prairies. The rice farmers wanted canals to distribute water. Canals, gulleys, coulees, and bayous were made to be efficient drainage. This continues today as streams are dredged, dammed, desnagged, straightened and otherwise changed to improve drainage. Streams once were springfed as the rain slowly drained along the roots of the prairie plants and seeped into the streams—I remember sticking my entire arm into a cool (68-73^0 F—our mean annual temperature (Smeins et al. 1992)) spring in Bayou des Cannes before it was dredged and straightened. The streams of the western Cajun Prairie were sandy bottomed and ran clear until the prairie was destroyed and water was runoff from the land. Erosion became intensive and streams turned brown with dissolved silts and organics and slowly choked off. The bottoms of these streams are now deep silt and mud or clay that has the silt washed by current downstream. The Gulf of Mexico and the marshes receive this silt with all of the residues of herbicides, insecticides, urban wastes, and more. Bayous became drainage features—quite a difference from the original use of the bayou by the early Cajuns.

Later neighborhood 'disputes' developed over dredging bayous. The Bischoff family and others worked for years and succeeded in preventing the dredging of the Bayou Mallet. The Fournerat family and others prevented the dredging of a seven mile section of Bayou des Cannes. Both areas remain undredged today and a tribute to these families. Literally all the other streams in the Mermentau River drainage have been dredged and straightened. Large turtles and catfish inhabited these undredged streams. Pat Mire studied handfishing in these waterways and the folklore of the bayou (Mire 1992; Pat is shown handfishing in Bayou des Cannes in Plate 32). Damage to the surrounding forests and grasslands as a result of stream alteration is evident especially in the hydrology (see chapter 7).

Never knowing of Aldo Leopold's land ethic as it was developing in the mid-1900s, the people unequivocally destroyed the prairies and the streams of southwestern Louisiana. The area was so profoundly changed that biologists—Featherman, Bick and Lambremont—admittedly didn't bother studying the area's plants, dragonflies, and butterflies, respectively. Neither did I like musseling in the area—access to the rivers was difficult, and often little was found. It was more productive in non-dredged, non-muddy streams in ease of both access and collecting, and diversity of mussels was much greater. The Calcasieu River drainage remains high in diversity with its natural sandy bottom. The once sandy streams of the prairie are now mud-bottomed drainage ditches.

Whereas feral cattle, horses and pigs had a dramatic and long-term impact on the vegetation of the prairies, the impacts of farming and petroleum production were much more dramatic and damaging. Loss of soil structure, loss of landscapes (*marais* and *platin* and mima mounds), use of herbicides and insecticides, exotic weed introduction, wholesale changes in hydrology, erosion, loss of pollinator species and cessation of fire were major changes caused by the farming of rice, cotton, corn, soybeans, sugarcane, wheat, milo and crayfish. Both row crops and grasses, although different in their methodology, have similar impacts. Petroleum production involved many of the same changes.

One of the most difficult problems facing the prairie and the entire southern United States is the introduction of the Chinese tallow tree. It is a vicious weed that is difficult and expensive to control. Jim Grace (1998) placed the problem into context—the very future of wildscapes depends on the public sector understanding the nature of this exotic species. The problem is exacerbated by the loss of fire as a woody vegetation control strategy. Many other exotics have been introduced, and the impacts of these species are mounting.

The 1920s were difficult in south Louisiana; great poverty companioned with a general sense of hopelessness were common (Vidrine 1924, Euzebe Vidrine, author of this book/confession, was the last man hung in Evangeline Parish—his crime was murder, serial murder). The late 1920s and the entire 1930s brought the Great Depression to the Cajun Prairie. It also brought the first recorded music, thus Cajun music was recorded for posterity. As the music told, times were very hard, and the people of the prairie struggled to survive. Many wealthy landowners lost their land; others even less fortunate entered extreme poverty. The depression that began with the end of the Civil War deepened in the prairie region, and the people pressed closer to the land. World War II and the post-war economic boom ended the depression and provided many opportunities for the people of the prairie. This story however is not the focus of this book. Literally nothing is recorded of the biota of the prairie—the Cajuns relied upon the spoken language and music to retain information (see Brasseaux 1987, Ancelet et al. 1991 and Jones 2007). We learn much about the people but only the mere basics, essentially trivia, about the prairie species. The people of the prairie however took on the name 'Prairie Cajuns.' Their activities in the prairie ecosystem not only changed the nature of the ecosystem but also changed the interactions between the people and the ecosystem. This will be the essence of the story of the prairie that I plan to reveal.

Chapter 5
1940s-1970s

"Chinese tallow tree probably arrived in North America in the 1700s. The plant has a number of traits that make it a natural for invasiveness. Its seeds can be spread in any number of ways, such as by birds, especially pileated woodpeckers and grackles (which are themselves non-natives), running water, packing material, or mud on soles of globetrotting horticulturalists. In addition, the seed remain viable for many years. In 1772 Benjamin Franklin sent some seeds to a friend in the Georgia Colony, noting in a letter, "I send also a few seeds of the Chinese Tallow Tree which will, I believe, grow and thrive with you. Tis a most useful plant." (Pilkey and Pilkey-Jarvis (2007)). Today, this tree may well be the most common tree in the Cajun Prairie—what a sad commentary! In the 1940s, it began to be used as a shade tree, with some ancient specimens at old homesteads.

"One of the penalties of an ecological education is that one lives alone in a world of wounds." (Aldo Leopold (1949)).

By the 1940's, the area, formerly known as the Great Southwest Prairie (see Map 1), was dissected by irrigation canals and developed into rice fields and cotton fields. The use of fire for management was greatly reduced. The Cajun Prairie, during the Great Depression, was a place of extended depression. The depression following the Civil War was extensive, and the people of the prairie were either doing well or very poor. My family, both mom and dad, spoke in detail of great poverty. With the Second World War, my father served in the Army Air Force in the European Theatre. He returned and married my mom in 1946. They settled in Eunice until I was born. Then they moved to Duralde and the Mamou Prairie, where I grew up. My prairie home would become fertile ground for my early exploration of nature.

It was surprising to later find that literally nothing was known about this prairie. Before it was obliterated, the prairie region came quite close to receiving a proper botanical survey in 1870 by Louisiana State University botanist, A. Featherman (1871). Alas, Professor Featherman had arrived at the prairie region during the summer season, and due to the lack of blooming specimens, felt that his collecting time would be better served elsewhere (Vidrine et al. 2001c). Many early naturalists had simply avoided the Cajun Prairie for one reason or another. George Bick and Ed Lambremont found so few dragonflies and butterflies, respectively, and they saw so much disturbance that they thought their time would be better spent north or east in less disturbed regions (Bick 1957 and Lambremont

1954).

Clair Brown, prominent botanist at LSU for nearly 5 decades, studied varied habitats in Louisiana including prairie habitat. His (1997) paper with a re-evaluation by MacRoberts et al. (1997) primarily discussed the "isolated" prairies of central and northern Louisiana. Several references were made to the Cajun or coastal prairies, as though he had thorough knowledge of them. According to MacRoberts et al (1997), "*Although Brown intensively studied these prairies, he left only a scant record of his work...*" Not much regarding prairie flora and the condition of remnant prairies is mentioned, although his (1997) paper, written in the 1940s perhaps, provided some clues. Of much importance to us is his view of fire. MacRoberts et al. (1997) report, "*Fire as a factor in prairie dynamics was not appreciated in Brown's time; he lived in the era when fire was ecological anathema. Today, fire is considered essential to normal prairie dynamics. Note how Brown tries to explain the existence of prairies entirely from edaphic factors (Collins and Wallace 1990).*" As a result, he had great difficulty appreciating that the prairies that he noted as having trees and shrubs had not been burned in a long time.

Brown (1997) noted that the southwestern portion of Louisiana is usually called the prairie section of the State. Further, he noted that, "*the region is made up of numerous prairies, each more or less enclosed by a rim of trees which occur along the streams, and thus separate these various prairies.*" "*According to our usual interpretation of climatic conditions, Louisiana should have been entirely forested. It is apparent that to find an explanation for these non-forested areas one must examine the edaphic conditions, in particular the chemical and physical features. There is present in the soil an unusually compact layer which is from 6-9 inches below the surface, occasionally as deep as 18 inches or three feet. This layer is impervious to the downward percolation of water, and likewise prevents the upward movement of water from the watertable. This influences the vegetation. In early spring the soil is well-saturated and the vegetation has an abundance of what are usually thought of as marsh inhabiting plants such as rushes and sedges.*" Annual cycles of wet and dry seasons severely stress the vegetation, especially trees. Again, he specifically doesn't mention fire; fire was not considered as part of the natural cycle. In the 1940s, when this was written (his grant for field work to study these prairies was provided by the Louisiana Academy of Sciences in 1938 to the tune of $50), the 'Smokey the Bear' campaign was in full swing and 'preventing forest (and so-called wildfires) fires at all costs' was deemed essential and the natural thing. The result is however devastating as many prairies were simply lost to woody invasive plants in the absence of fire. The concept of a 'control burn' came later. He gave accounts of several prairie remnants and notes, "*it has been exceedingly difficult to find any virgin prairie spots.*" He did note that there are different kinds of prairie and specifically that many of the inland, small prairies were distinctively different from the coastal prairie in southwestern

Louisiana in soils and plants. During Brown's lifetime, the forests of Louisiana were literally cut down, twice, and silviculture of pines in the southeastern United States was developed. Likewise, the prairies were literally plowed under for cropland and rangeland. Haying and mowing became the mainstay for the maintenance of grasslands.

Brown (1997) also pointed to "*natural ponds and marais that form so striking a feature of the prairies of St. Landry and Calcasieu.*" Brown (1972) described the prairie and marais as, "*If you fly over this prairie, you will notice a southward converging drainage pattern with trees along the streams forming a band of woods only a few hundred feet wide. There are marais ponds with such trees as the American elm, swamp red maple, green ash, and other hardwoods if the ponds dry out in May. The marais ponds with cypress and swamp tupelo usually do not dry out until October. There is a more or less sandy ridge, a few feet higher than the prairie proper, which extends from Kinder to Iowa, Louisiana, and which was covered originally with longleaf pine. No matter where you stand in the prairie, you can see a rim of trees in the distant background. Thus our southwestern prairie area is not one big prairie but is divided into many by a network of streams.*" He further related, "*The presence of indigo, sundew, butterwort, blazing-star, and many others is the same as in the pine flatwoods. This singularity of vegetation suggests that the prairie would have been pine flatwoods, as in eastern Louisiana, had not the development of the clay pan effectively prevented the establishment of the pine. The remnant of prairie vegetation, aside from the grasses, sedges, and rushes, has quantities of milkweed, blue-star, compass-plant, clover, vetch, spiderwort, obedient plant, sneezeweed, blazing-star, and indigo.*"

Most of the early authors made a big deal of wet prairie features and some made a big deal of the pimple or mima mounds (both Darby and Lockett). Thus the prairie contained some very specific 'microrelief' of a few feet. Stories of walking through a marais during the wet season mention walking a half mile with water rising to chest-deep in the deepest part. Such areas were simply not traversable in the wet season but by boat.

However, there are a couple of works that gave us a clue as to the prairie flora in southwestern Louisiana. Caroline Dormon, who is responsible for the development of the Kisatchie National Forest system in Louisiana, wrote two very useful books. Her earlier book, *Flowers Native to the Deep South* (1958), listed several plants as prairie plants in southern Louisiana—this provided us with a short list of plants for Cajun Prairie. Her listings include: *Iris nelsoni*—then called Abbeville irises and *Silene subciliata* (Prairie Fire Pink)—"*blooms in August on prairies and sunny woodlands in Texas and Western Louisiana.*" While the two plants above are less than the best prairie plants in my experience, they do occur either in the marsh south of Abbeville or along the Calcasieu River in deep sand near Kinder. Other plants on her list are more informative: *Gaura lindheimeri* and *Sabbatia angularis* in prairies in south central Louisiana, she wrote "*they form sheets of bright bloom.*" *Physostegia sp.* (Finger false dragonhead and/or Obedient plant)) were reported to occur near Crowley in the center of the Cajun Prairie. She also reported *Lobelia loudoviciana* and *Liatris halei*—neither species is currently listed for the Cajun Prairie. However, the genera listed included *Gaura, Sabbatia, Physostegia, Lobelia*, and *Liatris* as having prairie as habitat. She lists dozens of other prairie plants but she does not allocate them specifically to that habitat.

Caroline Dormon provided a second major contribution to the Cajun Prairie. In her book *Natives Preferred* (1965), she initiated natural landscaping in the south, especially Louisiana. One chapter, *Wild and Sunny*, opened with a sentence on prairies and indicated landscaping with prairie plants.

Clair A. Brown in his *Wildflowers of Louisiana and adjoining states* (1972) delineated the Cajun Prairie very well and listed 53 plant species as 'prairie' species (Appendix 8). This provided a working list and actually gave us a working flora for the Cajun Prairie. He however limited his prairie plants to those found on the elevated soils of prairie remnants. He did mention railroad rights-of-way between Lafayette and Lake Charles and between Opelousas and Kinder as having small patches which were not scooped out to form the railroad berm (bed). "*Here we find many of the attractive prairie species.*" He added, "*Most of the prairie area has been under intensive cultivation or grazing so that it is doubtful that any virgin prairie sod exists.*" However, this list is eye-opening and gives good insight into the flora of the Cajun Prairie. This list is however very short compared to our current knowledge and it gives little information on populations and communities of plants in the prairie.

Geyata Ajilvsgi (1979) in her *Wild flowers of the Big Thicket: East Texas, and Western Louisiana* listed 109 plant species from mixed grass prairies. Her plant list included both native and introduced (exotic) species. Her total list of plants included many species not labeled for prairie that are prairie plants in the Cajun Prairie. However, her prairie plant list closely resembled that of the Anacoco Prairie and other inland prairies rather than the Cajun Prairie (Allen et al. 2006). Her plant list was however very useful and like that of Brown (1972), it provided a starting point for a prairie flora for the Cajun Prairie.

Vidrine et al. (2001c) provided the following introduction: "*Historical documentation of southwestern Louisiana prairie fauna prior to the turn of the 20th century is similarly lacking. Fortunately, Louisiana State University zoologist George Lowery, Jr. amassed as much pre-twentieth century bird and mammal data as he could, including it in volumes which he later authored. Of the historical bird life on the prairie, most noteworthy from Lowery (1974a) were several references characterizing the Whooping Crane (Grus americana) and Attwater's Prairie Chicken (Tympanuchus cupido attwateri) - both presently hovering on the brink of extinction - as "common" and "abundant," respectively, in the region. Similarly, Lowery (1974b) recounts the details surrounding the discovery of both the red wolf (Canis rufus) and the Louisiana prairie vole (Microtus ochrogaster ludovicianus) - both thought to be presently extinct - within the prairie region of southwestern Louisiana. Much like that of the region's plant life, it is generally believed that historical bird, amphibian, reptile and mammal life in this region was extremely diverse (Lowery 1974a and b, Johnson 1999 and Dundee and Rossman 1989). Both the diversity and abundance of contemporary mammal life is substantially high. During the winter months in particular, notably large numbers of raptors, waterfowl, shorebirds, sparrows, blackbirds, and other birds still migrate into the prairie region, now almost totally supplanted with rice, sugar cane, cattle, and crayfish farms (Vidrine et al. 1995 and Huner 2000).*"

Fish and other aquatic animals received some attention (Douglas 1974 and Hoese and Moore 1977). George Bick (1957) spent some time in the Cajun Prairie looking for dragonflies, but he soon left for he felt the area was too intensely disturbed. Edward Lambremont (1954) and Gary Ross and Lambremont (1963) studied butterflies in Louisiana and spent little time in the Cajun Prairie. Few papers on the biota of the Cajun Prairie exist from the first half of this century. The late 1960s and early 1970s saw the development of a legion of biologists, who would enter the prairie biome and literally census the varied biota. Charles Allen and I are members of this group.

By the end of the 1970s, the Cajun Prairie was considered 'lost.' It was certainly disappearing from the collective memories of not only the present generation but also the children of my generation. After questioning literally a hundred of my parents' contemporaries, their collective view of the prairie is summarized as '*paille jaune*' or 'yellow straw' recalling the autumn appearance of the prairie. In March, 2010, I drove by several pastures in clear view of I-49 south of Opelousas that presented a similar view of acres of 'yellow straw.' Pictures of the prairie are black and white and have little detail or lost detail. I was not fortunate enough to see any pictures that even resemble the image of the Cajun Prairie that resemble the pictures on the Plates in this book. Sadly, many thought of the prairie as 'good riddance' as the idea of mowing grass into a carpet of lawn was inculcated into them nearly one and all. The view of the prairie plants as weeds remains so prevalent as to cause any land with these plants to be referred to as 'wasteland' and subjected to ridicule. I can't tell you how many people didn't mind giving me a piece of their mind on this issue as I worked on varied restoration projects. I was surprised that my parents were easily won over, and my mom gave me tremendous support in my efforts. I recall her eyes going wide as she remembered flowers of her childhood upon seeing some of my wildflowers in bloom. Her enthusiasm inspired me then, and my memory of her compels me now to make her memories come alive for my children.

Chapter 6
Rediscovery of the Prairie, 1980s -2010

"*The chemical and munitions industries, freed from their wartime responsibilities, turned to manufacturing and promoting lawn chemicals and fertilizers. And did they have a ready market! In 1945, millions of servicemen and women came home, got discharged, got married, and got mortgages. There was an immediate need in the country for 5 million new homes, and builders responded, led by William Levitt, the inventor of the mass-produced subdivision. ...By 1955, three out of four new homes in America were being built on his mass-produced subdivision concept – rows upon rows of look-alike but affordable homes that sported yard after yard of look-alike lawn-centered landscapes.*" (Andy and Sally Wasowski (2000)).

"*Today, the prairie is gone. Millions of acres have been plowed under for farm and grazing land. Gone forever is the wide expanse of waving grasses and many of the birds associated with it.*" (C. C. Lockwood (1986)).

The Cajun Prairie had essentially disappeared as a result of urbanization, oil and gas production and agriculture. The land was cut into lots for homes and lawns, the soil was turned-over/plowed-under in order to plant crops, or the land was grazed so heavily that most of the plant species were extinguished. And it is possible that all three insults were placed upon the same piece of land during the previous century. The hydrology of the prairie was entirely changed by the digging of canals, dredging of rivers, land-leveling, and straightening of streams. The prairie was pronounced 'extinct' by the Louisiana Natural Heritage Program. Then Charles Allen and I discovered 10 remnant prairies along railroad rights-of-way in 1986. This discovery became common knowledge within the next decade (French 1987, Allen and Vidrine 1989, Anderson 1997 and Mire 1989). We decided to restore/recreate this prairie in a landscape project in Eunice, Louisiana, under the initial leadership of Mayor Curtis Joubert. A non-profit society, The Cajun Prairie Habitat Preservation Society (Appendix 14), was created to serve as the governing/managing group for the preservation and restoration of Louisiana prairies in the Eunice area. In 2000, we published a series of papers and presentations both to the North American Prairie Conference in Mason City, Iowa, and elsewhere, and thus introduced our findings regarding the Cajun Prairie to national and international prairie audiences of scholars and enthusiasts. This is the short of the story, but I would like to lengthen the story so as to avail the reader to more of the details of the story.

By 1984, interest in both the cultural and ecological aspects of the Cajun Prairie ignited both within the Jean Lafitte National Historical Park system and within the staff of the Lafayette Natural History Museum in Lafayette, Louisiana and culminated in the opening of Prairie Cajun/Cajun prairie exhibits at those facilities in 1986. William R. (Bill) Fontenot, Lafayette Natural History Museum, and his team had searched for prairie remnants. They did not have much luck finding any remnant prairies. Charles Allen and I were searching for mosquitoes and studying mosquito-plant associations in rice fields, and we serendipitously discovered several remnants of Cajun Prairie.

Charles and I found remnants of Cajun Prairie along railroad rights-of-way that were mentioned by earlier authors (Brown 1972). We were sharing a small grant/contract ($500 and a truck with gas) from the Jefferson Davis Parish Mosquito Abatement District (JDPMAD) in Jennings and the Mosquito Research Laboratory (USDA) in Lake Charles in the summer of 1986. Our goal was to compare the mosquito communities in rice fields with the plant communities in search of some obvious relationship—a relationship that might allow us to predict from the plant community the potential mosquito community. While driving around looking for rice fields and checking mosquito light traps in Acadia Parish near Midland, Charles spotted several prairie plants along the railroad right-of-way just east of Midland. He had me stop as I was driving, and he took his ever-present plant-press and began collecting and pressing plants. I simply asked, 'What are you doing?' He replied, 'This is a little bit of prairie!' I remember Button snakeroot (*Eryngium yuccifolium*) and Ashy sunflower (*Helianthus mollis*) blooming—and I remembered long lost tales of prairie from home and from school. We naturally began looking along railroad rights-of-way all through our search area---Acadia and Jefferson Davis Parishes. We began to find these narrow strips of remnant prairie of varying lengths along railroads. Many of the remnants were very highly degraded with varied kinds of disturbance, but in mid-summer prairie plants really stick out. Summer ended quickly, and we were back in school. Our brief encounter with prairies had yielded a stack of plants for the herbarium and a tell-tale sign of what Cajun Prairie might have been like.

In the winter of 1986, we visited Garrie Landry and William (Bill) Reese at the University of Louisiana-Lafayette (then University of Southwestern Louisiana) Herbarium. Bill Reese, then a professor emeritus of botany, still came in to school and worked in the herbarium. After we told him, almost in passing about our find, he sternly looked at us over his glasses and said, 'This is something you should follow up!' Charles was a different person after that. He began planning biweekly visits to each of the prairie remnants and searches for more remnants. The Natural Heritage Program with the Louisiana Department of Wildlife and Fisheries and the Louisiana Native Plant Society both provided provided small grants for $500 for fuel for our sampling efforts. For two years, Charles and I revisited the prairie remnants every two weeks. He was possessed! I followed him around for a month or two, trying to help with my red cones to spot red flowers that Charles invariably could not see against a green background. Then after noting that he needed an hour at each location to search and collect and 'press' plants for museum (herbarium) collections. A field botanist always carries his map, pencil and field press—a simple device that allows for the collection and drying of plants literally while the collector is on the run. Plants are collected (usually with flowers and roots attached), flattened neatly between old newspaper sheets, labeled with location and date and other notes, and finally pressed tightly between cardboards and wooden boards forming a neat (usually 14" x 24") strapped bundle. While Charles was doing all of this, I needed something to do. I promptly bought a 35mm camera (manual Pentax) and started taking pictures of scenes and individual plants (ca. 10,000 slides) and I started netting butterflies and dragonflies—an interest I had developed while working in rice fields at the JDPMAD for 4 years. So the team was set—Charles doing plants and me doing photos, flying insects, etc. A more detailed discussion of this project follows under the heading of 'The Cajun Prairie Remnants.'

As we began our biweekly visits in early spring of 1987 (with the help of a $500 travel grant to pay for fuel by the Louisiana Natural Heritage Program), the burning of the remnants by local farmers was evident. My first thoughts were filled with doom for the plants, but within a week and certainly by the end of March, the prairie turned green and came to life. Burned prairie remnants were splendid in bloom; non-burned remnants were difficult to access (walk) and assess and as such were quickly abandoned in search of burned areas. About mid-way through the bloom season in 1987, Charles decided to contact the press; he began using the expression 'Cajun Prairie' and made sure to use it continuously—finally replacing the use of expressions like 'Louisiana Coastal prairie' and 'The Great Southwest Prairie' and 'Louisiana prairie' and many more. Having had the opportunity to read much of the natural history of the area, it occurred to me that Charles was the first to use the expression 'Cajun Prairie' in order to describe this biological region—I am not sure about the historians' use of the term nor about the common use of the term as I had not heard it called this before. I also discovered that he was absolutely second-to-none in his ability to garner press coverage. It started with a piece in The Daily World (French 1987) and exploded in numerous statewide newspapers. He then displayed a new talent—he started writing popular articles and sent one in to Louisiana Conservationist—*Wildflowers of the Cajun Prairie.* We reprinted copies of the article, made T-shirts, collected seeds to package and sell, and began going to weekend festivals and classrooms/meetings (anywhere 3 people gathered—we were there, talking or touring or selling). Two additional articles appeared in the Louisiana Conservationist (*Butterflies of the Cajun Prairie* and *Dragonflies of the Cajun Prairie*) (Allen and Vidrine 1990a and Vidrine and

Allen 1993).

Charles and I presented slide shows/talks at weekend festivals and to as many schools and venues that he could line-up. Exhausted, I met Pat Mire—when he volunteered to make a video of our talk—I seized the opportunity. I did not have to work hard to convince Charles (Mire 1989). It was evident that we were in the field working on the video during the summer of 1989 from my field notes that read: "*July 29, 1989 – visit west Midland Cajun Prairie remnant with Pat Mire. August 9 – Jerry Devillier videos Charles and me for movie under Pat's direction at Midland. (64 degrees in morning – very unusual weather). September 8 – film in South Fenton prairies. September 16 – Eunice Prairie.*" The video changed our lives. Pat also was able to convince the Lafayette Open Channel to incorporate the video as part of their programming; thus, the video was shown numerous times to local and cable customers in the Cajun Prairie region.

In 1988, Charles had the foresight to begin organizing a non-profit, tax-exempt society—The Cajun Prairie Habitat Preservation Society—in order to focus our efforts and maximize our opportunities for grants and for volunteers. At the same time, we approached the Mayor of Eunice, Curtis Joubert, and asked him to help us start a prairie restoration—it was popular in the Midwest. He said that he wanted that 'wildflower plot.' He succeeded in convincing Union Pacific to allow us to plant 10 acres in Eunice in 1988—and to collect plants from the railroad rights-of-way to transplant into the site. We began planting in the winter of 1988-89—we had numerous school groups and wildflower enthusiasts (many from the Louisiana Native Plant Society and even some from the Eunice Bulb and Blossom Society) collect seeds from the remnant prairies in late summer and fall—bags full of seed. The December 9, 1998, morning was the coldest so far that winter, but the Mayor led the tossing of the seeds to start the process of restoring prairie in Eunice. Each student had his name on the bag and was allowed to attend the planting day and to throw out their seed onto the site. Nearly 100 students were present and excited to participate (and to miss class). Both The Eunice News and The Daily World newspapers led Sunday morning (December 11, 1998) with stories of the new 'wildflower project.' Sporadic seeding and transplanting over the next three years by a wide variety of volunteers completed the initial effort to restore prairie. A more detailed discussion of this project follows under the heading of 'Restoration Projects.'

The year 1991 was an important year for the Cajun Prairie. In January, Joel Hilburn from Carencro High School wanted us to help his students make a 'postage stamp prairie' restoration. This gave us a great opportunity to begin another project and refine our thinking about prairie restorations and our advocacy of them. In the same month, Bruno Borsari arrived to teach at LSU Eunice and to work with me on the prairie project in Eunice, just as Charles planned a move to Monroe to continue his career as a botany professor for the University of Louisiana-Monroe (ULM) (then Northeast Louisiana University).

Bruno and I continued to work on the restoration projects, and we did some of the outreach work, but we scaled it down. I focused more intently on the insects of the prairies. I continued driving to prairie remnants and surrounding environs in search of dragonflies and butterflies. I also continued searching for new plants—thus I found the Snowy orchid (*Plantanthera nivea*) in the North Iowa prairie remnant. I also searched for the various colors of flowers of Downy phlox (*Phlox pilosa*), Fewflower milkweed (*Asclepia lanceolata*) and other plants to add to the Eunice Prairie. I also found several different phenotypes (possibly genotypes) of other plants (4 Big bluestems, 3 Indian grasses, 3 Little bluestems, 3 Eastern gama grasses, and more) and added them to the Eunice prairie or my collection in containers. One of our greatest failures happened in 1992 and signaled the end of the initial prairie restoration in Eunice. I had grown nearly 200 seedlings of Fewflower milkweed (*Asclepias lanceolata*) (formerly called 'Red milkweed') for butterflies and for the prairie in every color from yellow to scarlet-flowered. These were my favorite prairie plant species, although it had a predilection for wetter spots, it grew well in garden soil as I had discovered at home. A number of special memories were also attached to this plant relating to Charles' red-green color-blindness and to Monarch and Queen butterflies. He could not easily see red on a green background, so plants like the Fewflower milkweed and the Woods poppymallow (*Callirhoe papaver*) would be invisible in the green prairie. I served as his 'red cones' in the event of their presence, and he often asked me, "Is there anything red out there?" I had developed a great fascination with butterfly gardening and prairie plants—the Monarch and their milkweeds were the symbols of the concept. On one cold Saturday morning in 1992, Bruno and I transplanted the Fewflower milkweed into the Eunice prairie restoration site (yes, all 200 of them)—it was invigorating to think of the May-August blooms of this species in such numbers. As luck would have it, it stopped raining and dried to dust. By summer, we could find only 3 plants, and they were greatly weakened. During 1992, complaints about the prairie restoration and our 'weeds' by varied members of the community reached a crescendo, but they paled to the loss of this genetic diversity of the Fewflower milkweed.

The next few years were spent studying the restoration site as the plants developed. We continued to give talks and write papers including a series that we published on butterflies and dragonflies in the Louisiana Environmental Professional (Vidrine et al. 1992a, b and c). Charles revised his *Grasses of Louisiana* with all the new prairie data (Allen 1992). I took time to write my book on mussels of Louisiana (Vidrine 1993). Bruno left for Africa in 1994, and I continued to monitor the restoration project in Eunice. We wrote a book in 1995 after 6 years of restoration and a summary article that chronicled our efforts (Vidrine et al. 1995 and 2001a).

In 1994, Harland Guillory obtained a grant from the Louisiana Educational Quality Support Fund (LEQSF) for LSUE to build a greenhouse for an outdoor classroom and for a place to grow prairie plants for our restoration projects. Bruno used the greenhouse for the latter purpose and grew plants for a prairie restoration on the LSUE campus as part of his horticulture class.

In 1996, my wife Gail and I bought a 3 acre piece of worn-out rice field. We built our home and started our restoration of prairie—the Cajun Prairie Gardens (see Appendix 15). A more detailed discussion of this project follows under the heading of 'Restoration Projects.' We have chronicled this effort in several articles (Vidrine et al. 2003 and 2008). Shortly thereafter, Bruno returned from Africa, and he began a prairie plot in his front yard. He would face a variety of legal actions and other kinds of problems with his effort. In 2001, he would terminate his effort upon receiving his doctorate and leave Louisiana for Slippery Rock, Pennsylvania.

During the 1990s, Jim Grace, Larry Allain, and Stephen Johnson (USGS National Wetlands Center) began to monitor the remnant prairies of Texas and Louisiana. Popular articles on the Cajun Prairie by Allain and Johnson (1997) and Johnson (1996, 1999 and 2000) appeared in *North American Native Orchid Journal*, *Wildflower*, *Reptile and Amphibian Hobbyist* and *The Prairie Reader*. Bill Fontenot wrote a book (1992) and an article for *Louisiana Gardener* (1999) relating to wildflowers and the Cajun Prairie ecosystem. Amy Althans (2002) also wrote an article on Cajun Prairie for *Louisiana Gardener* magazine. Jeanne Frois (2003 and 2004) wrote short articles in *Louisiana Life* magazine on the Cajun Prairie. Also in 2003, the Louisiana Natural Heritage Program celebrated Earth Day with a poster on the Cajun Prairie (Allain et al. 2003)—a beautiful celebration of the diversity of life on the prairie in Louisiana. The remnant prairies were measured for future reference in that same time period (Allain et al. 2004). Five hundred or so acres based upon our conservative estimates in the 1980s had quickly dwindled down to less than 100 acres by 2000.

In the mid-1990s, Charles had worked with many organizations. A major cooperative venture developed with the Lacassine National Wildlife Refuge and Vicki Grafe. The Duralde prairie restoration began—Charles and his students from ULM began restoring prairies in this 200 acre location north of Eunice. Larry Allain organized the preparation of a brochure on the prairie (coastal prairies of Texas and Louisiana) (Allain et al. 1999 and 2000). This brochure opened many new avenues. Larry also spearheaded the development of several restoration projects, including LouAna, Dugas landfill and Gueydan restorations. Scientifically, Jim Grace, Charles and Larry put together the first flora for any of the Cajun Prairies based upon their work at the Florence Hunting Club remnant (Grace et al. 2000a and b). The work also analyzed the primary factors governing diversity in the remnant prairie. It was apparent from the data that micro-relief and moisture regime were most critical. Detailed studies on mima mounds and *marais* (marsh-like) habitats clearly showed striking vegetational differences. Much more work needs to be done, but the remnant prairies are in decreasing condition.

During Charles' tenure at ULM, he also encouraged a legion of students to join in the study of the Cajun Prairie. He also made strategic contacts with prairie enthusiasts in other states, including Peter Loos and Marc Pastorek. Charles retired from Monroe in 2001, and he reinvigorated the society. It had been inactive for several years. The newsletters of the society began in 1998 as did the rejuvenation of the activities. Two annual meetings with speakers and tours provided opportunities for new members and an excellent learning opportunity. Things began rolling. Land was donated to the society in mitigation efforts following destruction of prairie remnants. On March 28, 2003, The Daily World featured a story entitled, 'Prairie society obtains deed to tract' by William Johnson and featuring a picture of Peter Loos, president of the society, and Sara Thames, vice president, and me, treasurer (my last action as treasurer), signing the contract of purchase of the Cajun Prairie Restoration Project in Eunice. The society purchased the restoration site/park in Eunice—10 acres for $50,000. After a major effort involving an auction led by society members, the society began selling seed. At $50/lb, the land was paid off, and money was available for other projects.

In 2000, I decided to go to the North American Prairie Conference; Charles and Larry had decided the same. In Mason City, Iowa, we introduced the Cajun Prairie to the rest of the world. Charles provided a list of the plants that were encountered on the remnant strips of Cajun Prairie (Allen et al. 2001). The paper also summarized much of what we had learned about the prairie in general. At that same conference, I also presented 3 papers on varied topics dealing with Cajun Prairie (butterflies and dragonflies, the Cajun Prairie Restoration Project, and the prairie Cajuns) (Vidrine et al. 2001a, b and c). In the next 4 conferences, 2002, 2004, 2006 and 2008, we collaborated on and/or presented more than 10 papers on varied aspects of the Cajun Prairie. Larry Allain and Jim Grace also published a number of papers (Allain and Grace 2001, Grace 1998, Grace et al. 2000a and b) that provided insight into the flora and ecology of the Cajun Prairie. Most recently Larry has prepared a CD on Coastal Prairie Restoration in Louisiana (Allain 2007). Several other papers were done mainly by Charles and me in other venues. These papers will be used in part to create the remaining chapters in this book and to complete this chapter (see Appendix 25).

The Cajun Prairie Remnants

The prairie remnants along the railroad rights-of-way are pivotal in our thinking of the Cajun Prairie. The main kernels of knowledge are centered on the diversity of the remnants: wet/mesic versus dry, seasonal blooming, varied prairies (Faquetaique, Plaquemine, Mermentau, Calcasieu and D'Arbonne), biotic diversity, and degrees and types of insult and damage. Images of these prairies collectively provide views of this variety as we saw them in the 1980s and 1990s on Plates 2-28 and sadly as they appear in 2008 on Plate 29.

The prairie remnants were visually stunning—an ever-changing kaleidoscope of color painted by seasonal wildflowers varying from inches tall in spring to nearly 10 feet in late summer. Stark color changes included the exposed black soil of winter burn, the lush blues and greens of spring, the flush of color of varied wildflowers by the season, the burnt yellow-gold of autumn, and the rare white of snow (indeed rare as compared to the Midwest prairies). Using the Plates and the tables showing phenology (blooming seasons) of wildflowers and the comparison of plants in different prairies, the intent is to provide a visual or series of visuals that permit the reader/viewer to connect the dots and mentally picture vast sweeps of prairies of Louisiana.

Jim Grace, Larry Allain and Charles Allen (2000a) studied vegetation associations in coastal tallgrass prairie in Louisiana. They summarized the associations as follows:

> *"The coastal prairie ecosystem occurs along the northwestern coast of the Gulf of Mexico (Smeins et al. 1992) and can be considered to be the southernmost component of the tallgrass prairie biome in North America."*
>
> *"Today, approximately 99% of the grassland community has been lost to agricultural and urban development. In its present condition, coastal prairie is considered to be one of the most endangered community types in North America."*
>
> *"Perhaps due to its rapid conversion by agriculture and urban development, very few studies have characterized the coastal prairie community."*
>
> *"None of their work included samples from the eastern lobe of the coastal prairie, which occurs east of the Sabine River in Louisiana."*
>
> *"The eastern coastal prairie (also known as the Cajun prairie) has suffered nearly complete elimination. As of 1989, it was estimated that less than 100 of the original 0.9 million hectares of eastern coastal prairie remained (Allen and Vidrine 1989). Considerable interest currently exists in the conservation and restoration of the coastal prairie community (Grace 1998). Despite the fact that management priorities and restoration strategies depend on an accurate assessment of the native characteristics of a community, little of this information is currently available. The most extensive characterization of vegetation in the eastern coastal prairie has been conducted by Allen and Vidrine (1989), who found over 500 species of vascular plants in remnant strips along railroad rights-of-way throughout the interior of the coastal prairie region. In their study it was found that the most common prairie grasses were Panicum virgatum,*

Andropogon gerardii, Schizacharium scoparium, and Sorghastrum nutans. At present, the largest remnants of coastal prairie are located in the southern extreme of the coastal prairie region where they have been protected by development because of their proximity to wetlands. To date, there have been no quantitative assessments of these lowland remnants (proposed common name in the National Vegetation Classification is 'marsh fringing prairie'. In this study, our objective was to characterize the vegetation associations and their relationships to environmental conditions in a lowland coastal prairie."

"The prairie examined in this study is a small (23 ha) remnant located in Vermilion Parish in southwestern Louisiana. This prairie is one of three known examples of native lowland prairie remaining. The prairie exists as a distinct community that is largely surrounded by marsh and is effectively an island in a large landscape of wetlands and agricultural fields."

"A conspicuous topographic feature of the area is the presence of mima mounds,, which are small soil mounds, approximately 5-10 m in diameter and 1 m or less in height, of disputed geologic origin. The density of the mounds at the prairie is quite high, with hundreds of these mounds at the site. The prairie is characterized by sandy-loam soils of Pleistocene age."

"Based on species composition, as well as geographic location, it would appear that the vegetation described in this study represents the hydric extreme of coastal prairie."

"A specific examination of the species associated with mima mounds in this study revealed that they were characterized by a collection of upland and colonizing species, though not a single vegetation association." "Thus, mounds appear to be characterized by a combination of upland and colonizing species, the latter perhaps because of the greater degree of animal disturbance associated with mounds."

"The site examined in this study appears to represent an extreme lowland form of coastal prairie, with predominately wetland species at low elevations and facultative upland species at high elevations. While comparison to other prairie remnants suggests that it is not representative of upland prairie habitats, we feel that it is of considerable conservation value. The degree to which the vegetation of this site is made up of native species (99% of total cover), as well as what is known about the history of the site, suggest that most of the vegetation associations found represent contemporary examples of lowland coastal prairie."

"The results of this study suggest that a considerable amount of the vegetation diversity in prairie is contributed to by topographic features such as mima mounds. Where feasible, restoration efforts should consider

the potential to reestablish topographic variability, either within sites or through the selection of multiple sites."

Later, Charles Allen (2003) discussed wildflowers of the Louisiana prairies, including the Cajun Prairie:

"The prairie includes the large Cajun Prairie near the coast and the small inland prairies many of which are very similar, if not the same, as the Blackland Prairies of Texas, Mississippi, and Alabama. Cajun Prairie once covered 2.5 million acres in southwestern Louisiana but now a few railroad remnants are all that remain of this ecosystem. The three top families in Cajun Prairie are the Grass, Sedge, and Aster. The top grasses include big bluestem, little bluestem, switchgrass, Indian grass, and eastern gama grass that are also common in Midwestern prairies plus slender bluestem which add some uniqueness to the Cajun Prairie. A couple of interesting species include the button snakeroot (Eryngium yuccifolium) with its white flowers and Agave or Manfreda virginica with red stigmas. And in the Bean or Legume family the false indigoes with one white (Baptisia alba) and several yellow (Baptisia bracteata ssp. laevicaulis, Baptisia nuttalliana and the spectacular Baptisia sphaerocarpa). Other beans include Sampson Snakeroot (Psoralea psoraloides) and sensitive brier (Schrankia). The aster or sunflower family include a lot of yellow rayed species like golden asters, hairy sunflower, blackeyed susan, and compass plants, Silphium. But the star of the Cajun Prairie is blazing star or Liatris."

Even more profound, however, are the prairie remnants. Although they are currently either falling into complete demise or suffering from lack of fire and study, they are captured in my photographs from the 1980s and early 1990s (Plates 2-28). Thus these photographs become major glimpses into the past.

While we have no idea what any one prairie looked like 50, 100, 400 or more years ago, we can create a series of images as to what the prairie as a whole might have looked like (see Appendix 5). First, we need several disclaimers. One, the prairie is a natural burn ecosystem, and as such, it naturally burned every year or two providing for a 'phoenix-like' rejuvenation. Thus, fire provides a baseline from which we can begin a discussion. Second, the original prairie was routinely disturbed by large herbivores, e.g. Bison and deer, and more recently by cattle, horses, sheep, pigs and more. These large herbivores have a dramatic impact on the ecosystem. In fact, wet areas like *marais* and *platin* (marsh and pond-like areas) were routinely disturbed by wading/reeling groups of animals. Drier areas were also disturbed by hoofs as they cut trails through the sod—Robin mentioned that the trails were often vegetated by totally different plants than the rest of the prairie. Third, extreme years of drought or rain have dramatic impacts on prairie vegetation. This is obvious in restoring prairies. Wet years made for good years for seedlings and the establishment of plants with wet feet; however, a couple of drought years, and the very same plants simply disappeared. Finally, it is obvious that the prairie ecosystem is dynamic and changes as a result of physical disturbance as well as climatic disturbance. But plants themselves, and of course the pollinators and minute infectious agents, have dramatic co-evolutionary effects, such that a group of plants may establish themselves in an area and make that area more hospitable for other plants. An example that is obvious is a tree that shades a particular area and as such displaces sun-loving plants from that area. It may be easier to explain with pollinators. Some plants require a specific group of pollinators—if the pollinators disappear, the plants as a group become senescent and the plants that depended on the specific pollinator disappear. There is a condition known as soil sickness, which is thought to be the result of phytotoxins released by the roots of a plant and selectively inhibit the growth of similar plants and sometimes many different plants from the root zone (many sunflower relatives do this). Thus, specific communities exist on some prairies and not on others. With all these caveats, it may appear difficult for us to describe the prairie as it was. However, we have data and photographs of remnant prairies from which we can attempt a dramatic portrayal. These remnants were small but they were burned routinely. The remnants included both wet and dry habitats from 3 major named prairies: D'Arbonne, Calcasieu, Mermentau and Plaquemine. Smaller remnants from Vermilion, Mamou, Faquetaique and Sabine were instructional also. Each prairie had some unique vegetational feature: either one or more unique species or a unique community (Appendix 21).

Brown (1997) also pointed to "*natural ponds and marais that form so striking a feature of the prairies of St. Landry and Calcasieu.*" Most of the early authors made a big deal of these wet prairie features and some made a big deal of the pimple or mima mounds (both Darby and Lockett). Thus the prairie contained some very specific 'micro-relief' of several feet. Stories of walking through a *marais* during the wet season mention walking a half mile with water rising to chest-deep in the deepest part. Such areas were simply not traversable in the wet season but by boat. Rolling hills were noted in the eastern part of the Cajun Prairie, but the western part was noticeably flat—great for planting rice. The rolling hills were planted in cotton and corn. All of it was used extensively for pasture and free range for hundreds of years.

We have two distinctive types of prairie: wet/mesic and dry. We have studied five distinctive prairies: Mermentau, Plaquemine, Vermilion, Calcasieu/D'Arbonne (complex forming an extensive prairie west of Bayou Nezpique), and the Faquetaique/Mamou complex (surrounding Bayou des Cannes and Eunice). The latter complex has very small remnants, but the presence of some plants, like Butterfly-weed (*Asclepias tuberosa*), provides a great deal of interest. Not to mention, I grew up and currently live in this region. And the writings in parts of Robin, Darby, and Lockett, specifically related to these latter prairies. From these prairies, we have developed a flora of the Cajun Prairie that should be useful for many kinds of natural history studies and for restoration projects (Grace et al. 2000a, Allen et al. 2001, Allain et al. 2006 and Allain 2007). Further studies of animals, soils, etc. in these remnants are essential. Preservation of these remnant prairies is unequivocally important. How to preserve them, i.e. manage them, is not so clear, but burning these remnant prairies is essential. Remnant prairies are small and imperiled, and restoration projects provide a great potential to preserve the genetic varieties of the varied plants of the Cajun Prairie. Nikolay Valvilov's *Origin and Geography of Cultivated Plants* (1992) and *Five Continents* (1997) heralded the importance of plants grown in their places of origin as major genetic resources that are essential for the future survival of humanity.

The overall findings of our research were: The Cajun Prairie intergrades almost imperceptibly into fresh marsh to the south (Grace et al. 2000a) and into flat pine savannahs to the north (Allen 2003 and 2010) and intergrades with gallery forests and bottomland hardwoods along the streams. Grace (2000b) indicates that moisture gradient and micro-relief have the greatest impact on biodiversity of plants in the Cajun Prairie remnants. The diversity of plants exceeds 600 species and approaches 1000, depending on how one defines the prairie ecosystem. The animals associated with the Cajun Prairie are numerous and only now being detailed. The plant species are now pretty well known, and their ecology is forthcoming.

Gallery Forests, Marshes and Pine Forests

The adjacent habitats to the Cajun Prairie intergraded with the prairies; this is especially obvious in the shared plant and animal species. The gallery forests along streams have largely been timbered, usually more than once. These forests and marshes may be considered as integral parts of the Cajun Prairie ecosystem and indeed functionally divide the prairie into smaller units. The flora of these forests and marshes overlaps with the prairie flora, especially in non-woody plants, but there are a number of non-prairie plants. The trees were of interest to early explorers, who mentioned long lists of tree names (Robin 1807, Darby 1816 and Lockett 1969). The Longleaf pine forests have a story of their own, dramatically told by Earley (2004). Allen and Vidrine (1990b) surveyed the woody vegetation of St. Landry Parish in the early 1800s using land surveyor staking notes from the original surveys in 1803. The woody plants were described in '*Trees, Shrubs, and Woody Vines of Louisiana*' (Allen et al. 2002). The following excerpt presented their perspective on the prairie plants and associated habitats:

"The coastal marsh region lies adjacent to the Gulf of Mexico and

extends inward irregularly. Most of the region is true marsh with wet soil conditions and is dominated by graminoid (grasses, sedges, and rushes), non woody plants. There are a few shrubs, for example Baccharis halimiifolia and Iva frutescens, within the true marsh. A number of other shrubs, trees, and woody vines are found on the elevated areas within the marsh. These include coastal shrub thickets, the chenieres, and the man made spoil banks. The coastal shrub thickets are narrow strips of shrubs vegetation usually parallel to and very close to the Gulf of Mexico. Some shrubs commonly observed include Morella cerifera, Tamarix spp., and Zanthoxylum clava-herculis. Woody vine species include Cissus incisa, Smilax rotundifolia, and Vitis mustangensis. The chenieres are dominated by Celtis laevigata and Quercus virginiana and in recent year is being invaded by Triadica sebifera. Common plants on the man-made spoil banks include many Baccharis halimiifolia and recently many Triadica sebifera."

"The prairie includes the large Cajun Prairie near the coast and the small inland prairies. Cajun Prairie once covered the nearly flat area just inland from the coastal marsh in southwestern Louisiana. A few railroad remnants are all that remain of this ecosystem. The Cajun Prairie, like the coastal marsh, is dominated by non woody graminoids. A few shrubs (Caenothus americanus, Baccharis halimiifolia, Morella cerifera, and Styrax americana) were scattered across the prairie. The prairie was not continuous but dissected by finger like forests along the streams called gallery forests. The gallery forest species are a reflection of the size of the stream, the elevation, and the distance from the Gulf of Mexico. In very large streams at low elevation, one would find Nyssa aquatica, Taxodium distichum, and other wetland species but along smaller streams at higher elevations and especially inland, one would find Liquidambar styraciflua and Pinus taeda, and other more upland species. There are many disturbed areas within the original prairie zone and there are being completely inundated by dense stands of Triadica sebifera. There are also small inland prairies scattered through central and northern Louisiana such as Copenhagen in Caldwell Parish and Keiffer in Winn Parish. These are surrounded by a forest termed the Inland Prairie Forest. Some common trees and shrubs in these forests include Cornus drummondii, several species of Crataegus, Frangula carolinensis, Fraxinus americana, Pinus echinata, Sideroxylon lanuginosum, and Viburnum rufidulum."

This section of their book continued to describe regions in Louisiana including the 'pine region,' 'floodplain region' and 'upland hardwood forest region.' The dry prairie intergrades most obviously with pine forests. The 'coastal marsh region' and the 'prairie region' intergraded extensively as described by

Grace et al. (2000a). I have made no specific effort to list the trees associated with the prairie as these are available in Allen et al. (2002) and in Allain et al. (2006). The restoration of the prairie habitat may require the restoration of gallery forests and other forest features of the prairie ecosystem in order to be sustainable. The restoration of marshes may hinge on the restoration of prairies and *vice versa*, but this is the topic for another effort.

I had the opportunity to view a few wet prairies in the Gueydan area. I mainly remember Fewflower milkweed in areas thick with Eastern gama grass from these visits. Also I remember a trip home from Texas on the Interstate-10 through the Sabine Prairie west of Lake Charles. It was in April in the early 1990s. I stopped to visit a small remnant prairie along the railroad right-of-way just west of Vinton—it was prairie but it was not that diverse. However, on several occasions, I saw Louisiana Iris (probably Giant blue iris (*Iris giganticaerulea*)) blooming south of the Interstate. One location had several acres of blooms. At this juncture in my prairie experience, I was not very interested in wet prairies—that passion would develop later as would my love for Louisiana Iris.

In a natural progression of events, the next move is to attempt to recreate the prairie, even if in a modest way. Charles was thinking about this from the very onset of our research (French 1987) in his quote, "One of our goals is to re-establish a small section of prairie." More than 20 years later, the results of our efforts are evident.

The Restoration Projects

The Cajun Prairie Restoration Project in Eunice, LA

In the long run, the most positive production of our many efforts is our Cajun Prairie Restoration Project in Eunice, Louisiana. I often described it as one of my finest efforts. Walking through it, visiting it each month, and simply thinking about it over the past 20 years have provided me with many profound concepts. Plate 30 depicts the project in different seasons. But the projects also inspired others (Althans 2002 and Pastorek 2003). Marc Pastorek's comments summed it up, *"The Cajun Prairie Restoration Project in Eunice, Louisiana is the best representation of a restored prairie in Louisiana or Mississippi. …Along with wholesale introduction of desirable species into Eunice Prairie, particular attention has been given to the collection of unique genetic variations of individual species from the many of the remaining Cajun Prairie remnant sites. The ten-acre Eunice project has been the impetus for numerous regional prairie restoration projects."*

Since there were no remnant prairies of any size found in Prairie Faquetaique, the prairie recreations with plants from all the other prairies and their genetic varieties served as gardens, e.g., the Cajun Prairie Restoration Project in Eunice and the Cajun Prairie Gardens. While these are naturalized gardens, the possibility exists to restore prairie in the named prairies, e.g., Calcasieu, D'Arbonne, Mermentau, Plaquemine, and Vermilion Prairies, because reasonable amounts of remnant prairie exists. The remnants would then be the sole sources for the new restorations and in essence the creation of prairies true-to-type (and genetics) could occur.

Vidrine et al. (1995 and 2001a) summarized the restoration work in Eunice. Larry Allain (2007) and Allain et al. (2006) detailed the plants of the Cajun Prairie and their habitat preferences. This information was obtained by studying the distributions of the plants in the wild and by following the development of the plants in restoration efforts. The Cajun Prairie Restoration Project in Eunice as an experimental restoration project clearly demonstrated that restoration of Cajun prairie is feasible and has similar results to restoration efforts in the Midwest. The project provides an important repository for locally adapted prairie ecotypes and a model for other restoration efforts being attempted.

In 1988, the city of Eunice leased a 10 acre lot from the Union Pacific Railroad for the purpose of reconstructing prairie in the city limits of Eunice (The Cajun Prairie Restoration Project). Also, the Cajun Prairie Habitat Preservation Society was created to preserve and restore this ecosystem. A number of local agencies cooperated in this effort (Allen and Vidrine 1989 and Vidrine et al. 1995). In September, 1988, the site was mowed and herbicided with glyphosphate. After plants noticeably died and dried, a prescribed burn was conducted to remove biomass.

In the fall of 1988, seeds were collected from local prairie remnants along railroad rights-of-way. Most of the seeds were collected by students from local schools and clubs. The seeds were stored dry at 40^0 F until the day of planting but received no other treatment. Since most of the seeds were collected during the autumn, the majority of the seeds were from summer and autumn bloomers. Sod was rescued from remnants in danger of destruction by hand digging. Plants were also propagated in containers from cuttings and seeds.

On December 9, 1988, the seeds were distributed by hand by the individual who had collected them. This produced a heterogeneous matrix of prairie plants. The area was then harrowed in order to work the seeds into the soil. Plants in containers and sods were then transplanted. The sods added additional species as well as inoculated the soil at the site with the rich soil community of the native prairie.

The process of planting prairies involved both the selection of plants and the proper placement of plants in the site. The plant species selection is referred to as the restoration matrix (concept of Betz 1986, Schramm 1990 and McClain 1997). Appendices 6 and 16 contain the major prairie plants, and these should be in the matrix.

Prescribed winter burns of the site are conducted annually. Minimal weeding is done. Removal of Chinese tallow trees is proving to be the greatest obstacle, requiring a tremendous amount of time. At only 3-5 years of age, Chinese tallow trees suppress fuel and are often no longer controlled by fire. Chinese tallow trees thrive in wet areas where there is often little fuel. If the trees are left to grow, they are very difficult to eradicate. Since birds apparently love their seeds, bird droppings inoculate the area each year providing ample opportunity for this invasive weed to get a foothold.

During those first years of restoration, I recall carrying truckloads of plugs of prairie plants from remnant prairies or from home and installing them in the site. My son Daniel would come with me, and we would pull seedlings of Chinese tallow trees. This was especially easy early in the year or after a good soaking rain even in June. After pulling several hundred seedlings on one's hands and knees, it was time to go home. Bruno Borsari also helped me do this on several occasions. This rather simple task is absolutely necessary until the prairie can build sufficient fuel to burn hot enough to kill seedlings from the previous year.

The project was assessed in 1995 after six years of development (Vidrine et al. 1995). It was apparent that at least one half of the site was predominated by native prairie plants. More than 250 plant species of the nearly 600 known from the nearby remnants (100 miles or less) were established and spreading. The remaining portions of the site, most at lower elevations, had succumbed to exotic vegetation, primarily the Chinese tallow tree. Exotic species, which became established at higher elevations, were eventually excluded from the site. General evaluations in 1999-2000 and in 2008 demonstrated that succession was occurring. A number of aggressive, early succession natives and exotics were either no longer evident or greatly reduced by 2008 (Vidrine et al. 2000 and Allen 2000). It is assumed that the combination of annual prescribed fires, interspecific competition, and other ecological factors have affected the re-establishment of a maturing prairie plant community. Several new species have appeared and flowered in the last 10 years including: Large-leaf milkweed (*Asclepias longifolia*), Pineland milkweed (*Asclepias obovata*), Toothache grass (*Ctenium aromaticum*), Prairienymph (*Habertia lahue*) and Giant orchid (*Pteroglossaspis ecristata*). These were the products of seed either in the sod or sprouting after long periods required to either germinate and/or flower. This dynamic community continues to undergo plant community succession. The newly recreated habitat is especially dynamic, and it indeed rapidly changes in response to the genetic and ecological interactions within and between species, the climate, and the interference of man's activities or lack thereof. The observed increase in insect activity the area is in no small way due to the re-established native plant community (Vidrine and Allen 1993).

It is evident that since the initial planting that the plant community on the higher elevations has shifted from an annual plant-dominated early succession sere to a perennial-plant dominated late succession sere. Borsari and Vidrine (1997) examined the clump-size increase in 6 major species of plants in the site. Ten growing seasons had been completed in 1997 since the original sods were placed at the site. No fertilizer, watering or other kind of intervention other than fire had affected the plants to our knowledge. The results clearly depicted that the sod-forming matrix plants that appear as clumps can be readily examined and measured. The larger clumps have disintegrated in their centers, and numerous other native plants are now growing in and among the separate sprouts. Of the measured species, the fastest growing grass was Eastern gama grass (*Tripsacum dactyloides*), and the fastest growing forb was Clustered mountain-mint (*Pycnanthemum muticum*). Many other species of grasses, e.g., Little bluestem (*Schizachyrium scoparium*) and Slender bluestem (*Schizachyrium tenerum*), formed distinctive small clumps, and they apparently spread by seed such that the clumps reached a specific diameter and maintained that size. The two strategies of competition were obviously important in the overall restoration process as evidenced by events in plant succession at the site.

Borsari and Shirley (1993) examined the soil profiles after only 3 years into the restoration process and clearly demonstrated an accumulation of organic matter in the restoration site. This evidence indicated the impact that the prairie plants have upon soils in maintaining and generating soil fertility nutrients. Vidrine and Borsari (1998 and 1999) developed philosophical models for agriculture and for integrated pest management using Cajun Prairie vis-à-vis the Cajun Prairie Restoration Project for small farms and landscapes. The development of diverse predatory insect and arachnid communities is obvious in the restoration site (Allain, unpublished data, 2000). These topics will be elaborated upon in upcoming chapters.

Successful establishment of a diverse perennial prairie community on the well drained portion of this site clearly demonstrated that restoration of Cajun Prairie is feasible. However, determining site suitability and matching species to site conditions are critical parts of restoration planning (Allain 2007). Most exotic and early succession plant species have been displaced as succession occurs in the restoration (Allen 2000 and Vidrine et al. 2000). Insect diversity is great in the newly created habitat resembling that of remnant prairies (Allen and Vidrine 1990a and Vidrine and Borsari 1999). This indicated that, to some extent, ecosystem function has also been restored.

In 2009-2010, The Cajun Prairie Restoration Project is a well-established prairie restoration, especially on the elevated soils. The lower and wetter areas are experiencing extreme pressure as applied by members of the Cajun Prairie Habitat Preservation Society. Removal of trees, herbiciding, transplanting, paving trails and maintaining the parking lot and a covered metal shelter at the site are continuing projects. These activities are greatly improving the setting for access to the prairie. New signage, plant identification signs and lesson plans for teaching school-age children about the native prairie were developed. The site should be viewed as a garden and as an outdoor classroom for young and old alike. It is certainly a classroom for experienced restoration ecologists and other plant and animal enthusiasts. The project is also a model for other projects involving the restoration of prairies and gardening with native plants in naturalized settings.

The Cajun Prairie Gardens

The Cajun Prairie Restoration Project in Eunice provided results, which served as the basic plan for the construction of the Cajun Prairie Gardens. My wife (Gail), our children (Daniel and Caroline) and I began restoration in 1996 with a small postage stamp prairie garden in the dry part of the yard (Vidrine et al. 2003 and 2008b and Appendix 15). The gardens were developed on a washed-out rice field. The land had been leveled and terraced with levees for rice production providing some topographic micro-relief and drainage. We added 786 cubic yards of clay soil as a foundation for our home and in effect altered the drainage, which created a wet area in the front yard. This became the location of our wetland restoration.

In 1997, we enlarged and divided the garden into 11 rectangular plots that are separated by mowed paths roughly 6 feet wide. The plots have different moisture regimes: dry, seasonally wet/mesic, and wet/hydric. More appropriately for Cajun Prairie, these divisions are: *prairie* (meadow or dry prairie), *marais* (marsh or mesic prairie) and *platin* (pond or hydric prairie), respectively. The eleven plots contain a total of 1.5 acres. Several small, elevated areas were created to mimic 'pimple mounds.' Pimple mounds are specific features where small hills, ca. 3 feet in elevation, appear in ordinarily flat terrain. Plate 31 depicts the project in different seasons.

Native plants were grown from seed or from plugs collected from less than 100 miles in distance from the gardens. We changed the restoration process that had been developed for earlier prairie restorations by neither plowing nor herbiciding the site prior to plugging and interseeding (Packard 1997). However, we did use the annual burn in winter (December or January) as a part of the restoration process.

The primary goal of the gardens was to demonstrate landscapes that are sustainable, i. e. needing neither water nor fertilization. The gardens are laid out and diked with levees. This modification permitted water from rain to pool into *marais* and *platin* restorations. With winter rains in 1997 and 1998, it became apparent that the *marais* and *platin* plots would remain wet most of the winter. These plots were transplanted and seeded with plants that preferred mesic and hydric habitats just as we had done for the prairie plots using the prairie plants that preferred drier habitats. These wetland restorations were highly experimental and provided some of our more important research findings.

In 2002, we estimated the amount of relative area of the available ground in the plot that was covered with native prairie plant basal rosettes and seedlings as percent restoration. For example, if approximately 50% of the ground was covered with native plants, then the percent restoration was reported as 50%; thus, a general estimate of the recovery of the plot could be ascertained. We did this estimate in the spring, preferably before the growing season is well underway and while basal rosettes were easily recognizable and overlap of rosettes did not cause many assessment problems (Vidrine et al. 2003).

The different plots were in varied stages of development and succession. More than 300 native prairie species, 150 varieties of Louisiana Iris, and 100 naturalized plant species (ornamentals, vegetables, fruit trees, hosts for butterflies, and nectar-producing plants) were grown in garden habitats which provide protection for wildlife, mainly birds and beneficial insects (Allen and Vidrine 1990a, Grissell 2001 and Vidrine et al. 2001b). The gardens also integrated elements of sustainable agriculture using the Cajun Prairie model (Vidrine and Borsari 1998). As an outdoor classroom, the gardens provided a live demonstration of natural gardening (Vidrine et al. 2001a, Vidrine et al. 1999 (2001) a, b and c).

Three plots (dry) were treated as prairie restorations. Transplanting and interseeding of these plots mimicked our efforts at the Eunice Prairie during the second and third years of restoration (Vidrine et al. 1995). As of 2002, these plots are 60-70% restored, i. e. covered with native prairie species.

In addition to prairie restoration, we developed wetland restorations (mesic to hydric) (Zohrer, 2001). Five plots are considered as mesic or *marais* restorations and were started in 1998. Two larger plots of the five also had ponds dugout of their centers such that the ponds can hold as much as 3 feet deep of water in winter. As the plots dry, the dugout ponds (*platins*) dry, though they usually retained a little water even in the driest times. These ponds contained water lilies. The ponds have banks that slope gradually along the bottom of the *marais* and extend all the way to the levees or edges of the plots. The largest mesic plot was initially developed as an iris garden with 150 different, wild collected and/or hybridized cultivars of Louisiana iris and blue flag iris. These sites contained many native plant species that we were examining as potential elements of wetland restoration. Incidentally, this was the first attempt at wetland restoration in the Cajun Prairie landscaping experiments. Small, elevated areas have been created in 4 of the 5 plots. These elevations provided topographic micro-relief and thus extending available habitat for dry plants in a mesic setting. These plots were approximately 30-40% restored.

The ninth plot is a raised area that is a 'pimple mound' restoration that was started in 2001. The maximum elevation in this plot is approximately 3 feet. It was only 10% restored.

The tenth plot was a *platin*. It was diked and partially dugout to create a series of deeper basins that can attain a water depth of 3 feet. This pond surrounded the gazebo. It was the first attempt to restore Cajun Prairie in a hydric setting. While it should be permanently flooded, we allowed it to nearly dry out during droughts. The pond was developed in 1998 and was 50% restored. All ponds were stocked with Mosquitofish (*Gambusia affinis*), in order to control mosquito populations. These along with numerous odonate nymphs and other predatory insects control the mosquitoes (Grissell 2001). They seldom stayed wet enough from May to November to retain an aquatic fauna in the last 10 years.

The eleventh plot was the 'butterfly meadow.' It was basically a prairie restoration started in 1997; however, exotic species, including daylilies (*Hemerocallis spp.)*, gladiolas (*Gladiola spp.*) and others are intermixed with Cajun Prairie species. This demonstration was an effort not only to develop a butterfly/insect garden but also to meet conventional landscapers on a middle ground, where natives and exotics were grown together. There was also a trellis, which supported plants that were especially for butterflies and hummingbirds, e.g., Coral honeysuckle (*Lonicera semiperviens*), and passionvines (*Passiflora incarnata, P. caerulea,* and *P. sp.* var. *incense*). As a landscape feature, this plot closely resembled the prairie plots.

The annual winter burns (December, January) have effectively assisted in getting rid of exotics and some rogue natives, including Bahia grass (*Paspalum notatum*), Vasey grass (*Paspalum urvillei*), Curly-dock (*Rumex crispa*), Brasilian vervain (*Verbena brasiliensis*), and the native weedy Bull thistle (*Cirsium horridulum*). The result was a community with more than 300 species of prairie/wetland plants, which were coexisting and changing the above-ground floralscape and the below-ground root system. We have not begun to unravel the intricate community of bacterial, algal, mycorhizzal, earthworm, and other faunistic associates that mobilize soil and water at and below the surface of this clay/sand prairie. The recreated biotic community is home to a diverse insect fauna that resembles the community found on remnant prairies, *marais*, and *platins* (Vidrine et al. 2001b).

Live oak trees and a Hackberry provide food and shelter for birds and many other kinds of wildlife. The gardens are far from completed, and many projects remain not only to be implemented but also to be conceptualized. For example, several postage stamp prairies and iris/daylily gardens are under development on the western edge of the property.

The plant community is very floriferous and needs neither watering nor fertilizer. In fact, efforts to add humus, compost, and topsoil have apparently slowed the restoration process considerably. Small test plots treated as such developed dense, robust, early spring, tall plants that effectively shade the soil inhibiting or slowing the development of seeds and smaller prairie plants. The result is a noticeably retarded restoration process.

The restoration projects at the Cajun Prairie Gardens were clearly successful. The five-year old prairie plots (2002) were as much as 70% restored based upon cover by prairie plants. The four-year old *marais* plots were as much as 40% restored, as is the *platin* plot. The differences in percent restoration can be easily accounted for by time, since the fourth year usually results in a 20% or so increase in cover (Vidrine et al. 2001a). By the sixth years of restoration in each plot, not only the prairie plots but also the *marais* and *platin* plots should be ca. 90% restored. Therefore, the three major subhabitats in the open prairie can be restored in much the same manner, i. e. by transplanting and interseeding with neither tilling nor herbiciding companioned with annual winter burns. A significant limitation to this conclusion was that this land was mowed at least twice annually for nearly a decade; thus, it previously was a lawn. This limitation thus excluded any predictions regarding properties that are infested with Chinese tallow trees or other woody plants.

In 2006, we re-evaluated the Cajun Prairie Gardens landscaping project. In the dry prairie, more than 95% of the surface of the restored area was covered by basal rosettes of prairie plant species (Vidrine et al. 2006). In this study, we statistically evaluated small sample plots within the prairie and lawn. A total of 45 prairie plant species were found in the samples. Ten species were found in both mowed and prairie samples. Four non-prairie species were found in the mowed samples, whereas a single non-prairie species was found in the prairie sample.

The results of other statistical tests compared management techniques, mowing versus burning. The mowing technique showed no significant changes in the diversity of non-prairies and prairies, whereas the burning technique showed significant changes in diversity and an obvious increase in species diversity of prairie plants and an obvious decrease of the non-prairie species.

Restoration of a small-scale lawn to prairie habitat was our goal. The prairie community remains a much more dynamic environment than a typical lawn. For example, the restored prairie plants changed the surface of the soil by providing varying cover, improved draining of water into the soil and reducing water runoff, while preventing swales from holding standing water, which provides habitat for mosquito development. Possibly, the soils in restored prairies improve their chemical and physical attributes as compared to soils in lawns (Borsari and Shirley 1993, Jariel et al. 2006, 2010a, 2010b and Gotte et al. 2010).

The Cajun Prairie Gardens differed from other Cajun Prairie restorations in the following ways: 1. the soil was not immediately tilled prior to the initiation of the projects, 2. the existing plant community was not herbicided prior to the initiation of the projects, 3. the plots follow a specific design with soil dampness varying from dry to mesic to hydric, and 4. the gardens are in a rural subdivision on a commercially-zoned lot. The projects at the gardens are collectively designed to demonstrate Cajun Prairie landscaping as a sustainable method to reclaim lawns as part of a 'native' landscape, a landscape that benefits a wide variety of wildlife and builds new soils. In southwestern Louisiana, these ongoing research projects provide an opportunity for the public to see not only restored landscapes but also the process by which these landscapes are restored.

The wetter habitats provided an additional level of difficulty in restoration in the Cajun Prairie. Restoring *marais* and *platin* habitats in the Cajun Prairie is a difficult task. My efforts were overrun with Alligatorweed (*Alternanthera philoxeroides*) and Uruguayan primrose (*Ludwigia uruguayensis*). Vasey grass is slowly disappearing after 14 years of burning. None of the projects that I have done have completely satisfied my goals, but this is the nature of habitat restoration. Upland prairie habitat seems to be the easiest to restore, and I do feel that the Eunice Prairie and the dry parts of the Cajun Prairie Gardens are good examples of success in restoration at this point in time.

Restoration Overview

Restoration efforts result in biodiversity gardens, in the sense of Aldo Leopold (1949 and 1999) and Nikolay Valvilov (1992 and 1997). Most restorationists are familiar with Leopold's efforts in America, but few are familiar with Valvilov's efforts, which were described in Gary Nabhan (2009). Where Valvilov emphasized cultivated plants (food, fiber and fuel), Buchmann and Nabhan (1996) added emphasis to pollinators and plant symbionts. Biodiversity gardening today takes all of these into account and includes all biota. Nabhan (2009) noted the high degree of overlap of Valvilov's 'centers of diversity of cultivated plants' and the present day Conservation International 'hotspots of wild biodiversity.' Biodiversity gardening overlaps with habitat restoration and conservation; however, biodiversity gardening permits the public to join in the effort to conserve the planet's life. Heirloom gardeners and restoration ecologists share the common goal of preserving and protecting genetic diversity. Our collective efforts to preserve and restore Cajun Prairie in a variety of both public and private projects are part of this global initiative.

The Cajun Prairie contains rare plants and distinctive vegetation types (Allen and Vidrine 1989, Johnson 1996 and Grace et al. 1999), which are in dire need of protection and restoration. Both plants and vegetation types are disappearing at an alarming rate. Restoration projects may be the only recourse for preservation of these unique plants (Allain et al. 1999 and 2000). Appendices 16, 17 and 20 provide a general plan for prairie restoration and some landscaping tips (also see Allain 2010 and Pastorek 2007).

Currently, many viable seed are harvested from the restoration projects. However, at the Cajun Prairie Gardens, spring droughts and aerial agricultural spraying (drift) appeared to severely impact these efforts. During May, pollinator insects literally disappeared. This appeared to have a dramatic impact not only on fruit and vegetable production but also on viable seed production. Our inabilities at the Cajun Prairie Gardens to grow a vegetable garden and an orchard evidenced this phenomenon. Growing prairie plants greatly increased the number of pollinators and predator insects in other areas, e.g., the Cajun Prairie Restoration Project in Eunice (Vidrine and Borsari 1999). Therefore growing prairie plants with vegetables and fruit trees should enhance overall plant production according to the Cajun Prairie agricultural model (Vidrine and Borsari 1998). We are attempting to test this hypothesis at the gardens. As the neighborhood urbanizes, the aerial spraying should dissipate. However, watering in order to alleviate the spring droughts would require elaborate irrigation, an expense and modification of the gardens that we cannot presently justify.

Real weeds, mostly exotics, occur in restorations. They are tough to get rid of, and maybe, in some areas, they are there to stay. Brazilian vervain, Vasey grass, lawn grasses and most ornamentals readily disappeared with burning. Many old field weeds, goldenrods, thoroughworts, ragweeds, Bull thistle, Water hemlock (*Cicuta maculata*) and others disappear slowly. Charles Allen and Sara Thames (2007) considered succession in the restoration of Cajun Prairies, but they emphasized work done with dry prairie restorations. Wet prairie weeds are far more difficult to manage.

The Cajun Prairie Restoration Project and the Cajun Prairie Gardens are models for other restoration efforts in southwestern Louisiana. Similar projects included more than 200 acres by the Lacassine National Wildlife Refuge by Vicki Grafe and Wayne Syron (with Charles Allen). The National Wetland Center hosted Larry Allain's project at the University of Louisiana-Lafayette's Center for Ecology and Environmental Technology in Lafayette, Louisiana. The Dugas Landfill project, with approximately 40 acres, was supervised by Bette Vidrine and Mark LeBlanc for the City of Lafayette and Larry Allain from the USGS National Wetlands Center. Postage stamp garden landscaping models were developed and tested in several locations (Semar and Vidrine 2000, Vidrine et al. 1999 (2001) a, b and c).

Key restoration projects include:

1. Prairie Basse gardens (Bill and Lydia Fontenot)
2. LouAnna (Larry Allain and others)
3. Duralde (Vicki Grafe, Wayne Syron, Charles Allen and others)
4. Gueydan (Larry Allain, USGS and others)
5. Marc Pastorek (Mississippi/eastern Louisiana versions of the Cajun Prairie)
6. Peter and Cassandra Loos (east Texas version of the Cajun Prairie)
7. Acadia Power Partners (Charles Allen, Peter Loos, and others)
8. Hillman's gardens (Tom Hillman)
9. Shallow Lake (Vernon Fuselier's seed only restoration)
10. Harmon Nursery (Martha Harmon)
11. Foreman's prairie (Tyrone Foreman)

Marc Pastorek and Peter Loos played major roles in the development of the Cajun Prairie Restoration Project in Eunice. Both served as Presidents of the Cajun

Prairie Habitat Preservation Society during the pivotal years of redeveloping the mission of the society and securing the property in Eunice. As landscape ecologists/designers, they were instrumental in developing plans for building parking lots and trails. They also spent a great amount of time on other restorations and in search of horticulturally-promising prairie plants with Charles Allen.

The paradigm of restoring prairie habitat was developed by Aldo Leopold (Leopold 1949 and 1999), and we continued his paradigm. It should be expanded to gardens of all types from suburban lawns to farms to landscapes (Jackson 1999, Vidrine and Borsari 1998, 1999, Semar and Vidrine 2000 and Vidrine et al. 2001a). The Cajun Prairie Restoration Project has served for the last decade as a model for at least 10 prairie restoration projects. The projects have ranged from small postage stamp prairie gardens to a hundred hectares. However, there is a great need for more and larger projects. The Cajun Prairie is not only one of the rarest communities on Earth, but it is one that is scantily known. If there ever existed a habitat that might be labeled with a superlative in respect to being threatened and endangered from extinction, the Cajun Prairie would be a prime example.

Habitat restoration, like building a garden, is a multidimensional construct. As in 2 dimensions, one initially constructs a puzzle-like layout of plants to be planted. A random distribution of seed over a diverse piece of property will most likely result in the wasting of a tremendous amount of very valuable and scarce seed. The third dimension is height—seed selection is important as plants vary greatly in height. The fourth dimension is time—plants mature in varying amounts of time, they bloom at different times, and they respond to different kinds of stress at different times in different ways. The first four dimensions are the 'space-time fabric' of the restoration. The fifth dimension is interaction—plants interact with other plants, fungi, bacteria, protists, and animals (this reminds me of the impact of gravity on the 'space-time fabric'). The collection of interactions is extremely important. A simple example would be a butterfly garden—butterflies will be few unless host plants for caterpillars are planted in abundance. Planting just nectar plants yields a garden with many possible visitors but few residents (Vidrine and Hazelton-Robichaux 2003). The sixth dimension is sustainability. Whereas the garden is generally not considered sustainable and as such is replanted nearly every year, restoration takes on this 6th dimension. The 7th dimension is succession. As quickly as a sustainable system is created, it begins a process called succession. The selection of climax sere plants is essential in order to impact a succession toward a particular habitat type. When old fields in southern Louisiana enter succession, the result is usually unlike any nearby natural habitat—the common result is a hodge-podge of rugged natives and even more rugged exotics that collectively provide little habitat for associates like animals, etc.

Understanding the seven dimensions of habitat restoration is a key to success (I often present this to students as the General Relativity theory of gardening where the 'space-time fabric' is affected by the gravity of interacting species—an

Einstein/Newton garden). Each habitat is different. Prairies on different soils or at different latitudes are different. A restoration model for a place is just that—a model for a place that deals with the 7 dimensions in that space and time. The model for a particular place is simply an outline for any other place, and each model must be built upon its own regards. Like multidimensional chess, the game of restoration requires that the restoration ecologist play with several steps ahead in mind—otherwise, it is checkmate before the 5th dimension is in play.

Chapter 7
Abiotic features of the Cajun Prairie: Climate, Geology, Soils, Fire, Hydrology and Agriculture

> *"Fire as a factor in prairie dynamics was not appreciated in* (Clair) *Brown's time; he lived in the era when fire was ecological anathema. Today, fire is considered essential to normal prairie dynamics."* (MacRoberts et al. (1997)).

> "*Today there is an acre patch* (of Cajun Prairie) *here and there that hasn't been plowed because it is in a railroad right-of-way or for some similar reason. The big expanses are gone. Annette Parker, Nature Conservancy botanist, guesses that we could never get them to grow back because the seed stock has long been plowed under or killed by herbicides. This and the virgin baldcypress forest are two habitats I'll never see.*" (C. C. Lockwood (1986)).

> "*Most of us living today hardly know where our foods come from. At best, we are dimly aware of the geographic and cultural origins of the crop genetic resources that form the living foundations of our food supply. We seem to believe that as long as we wish to eat, those resources will be invariably provided to the seed curators, plant breeders, nurserymen, and farmers who make our agricultural supply-and-delivery chain function. But as the seed keepers in Leningrad realized in 1941, we are in a race against time to ensure that the remaining seed varieties on this earth are not extinguished like so many candles in a sudden gust.*" (Gary P. Nabhan (2009)).

The Cajun Prairie is part of the Gulf coastal prairie, which extends from Corpus Christi, Texas, in the southwest to Ville Platte, Louisiana in the northeast (Smeins et al. 1992). The area, like most grasslands, has been converted to agriculture, both ranching and farming. As such the original prairie is no longer intact, but many studies have been done on the varied abiotic factors, including climate, geology, soils, fire and hydrology. These studies were often accompanied by agricultural surveys and followed up with the development of the land into a variety of agricultural efforts. Agriculture, if sustainable, depends on a clear understanding of the ecology of a region. Charles Allen (2006 and Appendix 16) gave 8 reasons that prairies form: (1) low annual rainfall, (2) fire, (3) clay layer below surface of soil, (4) high clay content of soils, (5) wet/drought conditions, (6) grazing animals, (7) wind and (8) mycorrhizae. This chapter will center upon general discussions of the abiotic factors.

Climate

> *"The subtropical climate of the region is dominated by warm, moist, tropical air masses derived from the Gulf of Mexico with prevailing southeasterly winds...Mean annual temperatures vary from 20-23° C (68-73° F) .... The frost-free period ranges from 240 days in portions of Louisiana to more than 320 days on the lower Texas coast" (Smeins et al. 1992).*

The mean annual rainfall is approximately 50 inches at Lake Charles, and the coldest days of the year are usually around January 10-12. As related by numerous authors, this prairie region is basically mild, although the summers are hot and humid, and the winters are cold and damp. The Cajun Prairie seldom experiences snowfall (the white buffalo of John Madson (1993)), and the soil seldom freezes. There are exceptional years, but they are just that. The Gulf of Mexico has a tremendous impact in that gulf showers appear daily in summer, however, hurricanes and tropical storms are a threat each year. The growing season is very long, as long as 300 days in the southern parts of the Cajun Prairie and as short as 240 days in the northern parts—that is about a month for each 25 miles in latitude change. So winter lasts about 2 months in New Iberia and about 4 months in Ville Platte. This change is also evident in water temperatures and in the brooding of mussels with multiple broods during the growing seasons (Parker et al. 1984). Gardeners are experts at using these growing season differences in order to grow a variety of plants—they have very elaborate seasonal charts to effectively display in vivid colors even the smallest differences.

Five recent hurricanes (Lili in 2002, Katrina and Rita in 2005, and Gustav and Ike in 2008) have collectively damaged/destroyed more than 500 square miles of marsh in southern Louisiana. Much of this marsh is considered 'freshwater marsh' and intergrades into the prairie biome. The impacts of future hurricanes are unpredictable as extensive flooding from tidal/hurricane surge is so extensive. Historically the hurricanes were one of the predominant life changing events (Vidrine and Venzon 2006); however, today, the impacts on the marshes and on the livelihood of the people of the prairie are even more evident.

Wind

Temperatures, rainfall and seasonal changes are important aspects of climate. But in a prairie, wind is especially important and usually overlooked. "The wind is maddening" yelled a woman in a movie about life on the prairie. Unfortunately, I don't remember the name of the movie, but I do remember her facial expression as she was yelling. The impact of the wind on people is obvious when you travel to the Midwest. Farmstead homes are nested in an angular pair of groves of trees—one grove on the north side and one on the west side. The prevailing winds are westerlies, and the cold fronts are northerlies. Winter, even in the Cajun Prairie is the wind-time. You can mark your calendar from October to May—the winds will blow several days a week. These winds are sometimes sustained at 20+ mph (miles per hour) and gust to 35+ mph. The Cajun Prairie, however, is decidedly affected by the Gulf of Mexico; thus, south and east winds are common as fronts approach and draw air and moisture from the tropics across Louisiana to the Midwest. The tropical storms provide another dimension, although they occur in summer months. Summer months usually mean 'stillness.' The winds are light and variable and hot. One soon finds himself praying for a thunderstorm to rise from the Gulf and blow overhead—a dazzling air-show during the hoeing and picking of cotton. Wet cotton had to be air-dried before it was ginned, so a storm also meant a break from picking; however, hoeing continued unless it rained enough to flood part of the field, then you were shoveling.

The prairie is a habitat of extremes—too wet or too dry, too cool or too hot, muddy or dusty and windy or windless. The moisture of the Cajun Prairie is something to experience. Growing up in this environment without air-conditioning made a weather expert out of everyone. The temperature commonly hits the dew point—and the humidity rises to 100% and literally it rains and you are in the clouds. Growing up on a farm where cotton was the major crop gave a good introduction. Mornings were shrouded in fog with dew dripping off each plant. Literally, you walked into a cotton field, and you were immediately soaked to the skin. The plants like many plants had leaves that funnel the water to the tips or special hairs that keep the plants dry, but clothes were immediately wet.

The cool winds evaporate rainfall and water so quickly that you can almost literally observe the level of water in a bucket descend. Within a couple of days, a muddy lot will dry to dust or bake hard if the sun accompanies the wind. During the 'dust bowl' days, the Midwestern prairies literally blew away; similar events are repeating themselves in the fields of China now. Plants of the prairie are adapted for wind and rain. Their roots funnel the water into the ground and store large amounts of that water. All of the plants work together in this operation. The entire prairie grows at the same time—nearly a foot each month of the growing season in the Cajun Prairie. The heat of the July and August sun is masked by the prairie plants. They are 5-6' tall and shade the ground. I have walked along the trails in my gardens when the air temperature around my head is nearing 100° F, and my hands can sense the temperature among the plants near their base as at least 10° F cooler—the air is also moist by comparison.

The animals and plants have divided the habitat in a process labeled 'resource partitioning.' Whether the factor is wind, sunlight, moisture or some unknown underground resource, different plants and animals will occupy regions based upon these, often limiting factors. In the end, the organisms modify the habitat in such a way as to permit the most diverse group of organisms to exist in an area. I will deal with this concept in Chapter 9 with animals of the atmosphere.

Geology

Smeins et al. (1992) summarized the geology of the Cajun Prairie and the coastal prairies of Texas.

> *"Most of the region is underlain by Pleistocene deposits, and two major depositional surfaces are recognized: (1) the Beaumont (Texas) or Prairie (Louisiana) formations; and (2) the Montgomery and Bentley formations (Texas)."*
>
> *"The present shoreline became established as recently as 2500 years B. P. Impacts of sea level fluctuations and concomitant alterations in river channels, deposition, erosion, delta formation, and other alluvial land-forming processes could be expected to have no less of an impact on the landscape of the coastal-prairie region than the glaciers had on northern portions of the grassland biome."*
>
> *Quaternary geologic processes of the coastal-prairie region and the resulting deposits and landforms are complex and have resulted in a great variety of environmental suites in an otherwise seemingly homogeneous, subdued landscape. Slight differences in elevation and variation in composition of substrate produce marked differences in soils and associated variation in biologic components of an area."*

Fearn (1995) pieced together a history of the Cajun Prairie region.

> *"Pollen, phytolith, charcoal, and diatom analysis, in conjunction with sediment stratigraphy and radiocarbon dating, of sediment cores from Lake Arthur and Prien Lake, estuaries in southwestern Louisiana, provide 6000 years of data used to reconstruct the late-Holocene vegetation history of Louisiana's natural prairie and to define coastal processes which affected the sites. The Cajun Prairie is a grassland outlier in a climate which normally supports forest. Poorly drained, impermeable soils with poor water yielding capability (water-holding capacity) induce summer drought severe enough to limit the success of trees, resulting in a natural tallgrass prairie.*
>
> *Pollen, phytoliths, and charcoal suggest that the grassland island has neither expanded nor contracted over the last 6000 years and that fire has contributed to its maintenance. Pinus, Quercus, and Taxodium have been components of southwestern Louisiana's vegetation for the entire period of record with a minor increase in pine from 2000 to 1000 B.P.*
>
> *High pollen concentrations followed by an Ambrosia (ragweeds)*

rise and a drop in Taxodium (Bald Cypress) and Pinus (pines) mark the settlement horizon. Low pollen influx after settlement indicates high rates of erosion and clastic input to the lakes. A crash in the diatom flora of Lake Arthur probably relates to maximum pollution of the estuary by agricultural chemicals.

The use of phytoliths in sediment to distinguish the source of Gramineae (grasses) pollen is a promising new technique. Dramatic increases in Gramineae pollen accompanied by high percentages of rondel shaped phytoliths document expanding marsh vegetation around Lake Arthur and Prien Lake as rising sea level initiated drowning of the low gradient rivers by 5000 B.P.

Radiocarbon dates on peat samples indicate relative sea level of -5.5 m by 6000 B.P. followed by slow steady rise at a rate of 9 cm/100 years up to present times. Freshwater diatom assemblages between 4500 and 3000 B.P. coincide with westward progradation of the Mississippi River's Teche Delta Complex, and an abrupt shift to brackish/marine diatom flora at 3000 B.P. signals abandonment of the Teche system."

The prairie developed from a combination of geological and climatological events (Fearn 1995). A thick clay-pan developed in southwestern Louisiana following the recession of the Mississippi Embayment. The region gradually slopes into the Gulf of Mexico. Along the coastal area are salt and fresh marshes that serve as massive breeding grounds for fisheries and birds. The gradual upslope area was grown over by grasses and forbs resembling those of the tallgrass prairies.

Soils

Maps of soils and physical features are commonly available for the state of Louisiana and the Cajun Prairie. William Darby constructed a map of the region based upon his visit in 1805. Samuel H. Lockett, Colonel C. S. A. and Professor of Engineering, Louisiana State Seminary in Pineville and Louisiana State University in Baton Rouge, conducted a topographical survey of Louisiana between 1869 and 1872. Only parts of his survey were published in his lifetime. The entire manuscript was edited and published a century after Lockett began his survey. Lockett's accompanying map went through two editions, 1876 and 1891 (Lockett 1969). His map was used by Brown (1972). Other important maps included Hilgard's 1880 map entitled "Agricultural map of Louisiana" (Hilgard 1884). A 1935 map of the Soil Divisions of Louisiana published by the Louisiana Agricultural Extension Service, LSU, Baton Rouge, LA, is also readily available. Many newer maps are also available in varied atlases (Newton 1972 and Goins and Caldwell 1995).

Map 4 provides a reproduction of one of Lockett's maps, specifically the map of the natural regions of Louisiana from Newton's Atlas (Newton 1972). The extent of the marshes, bluffland and pineland regions surrounding the Cajun Prairie are evident. Map 5 also indicates generalized soil areas of Louisiana.

The Midwestern prairies and the Cajun Prairie do share a similarity in that when settlers first saw the prairie in summer, they thought it was a desert. The dry and baked soils of summer with grasses and plants often forming a tinder box waiting to burn was a scary place. The Cajun Prairie was described as a marsh in winter; the *marais* and *platins* held water up to 3 feet deep (waist-deep on many) for a month at a time. The shrink-swell nature of the soil in part

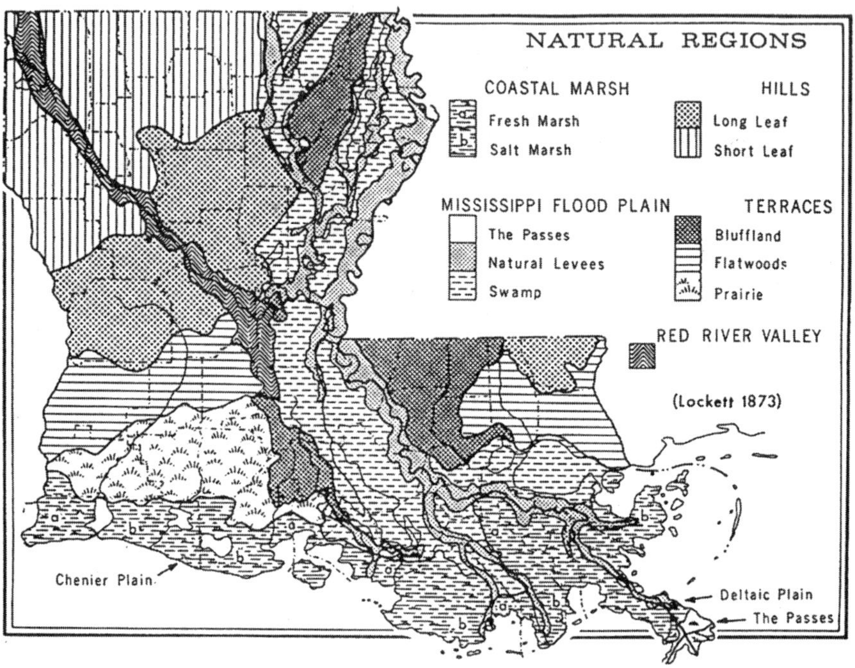

Map 4. Natural regions of Louisiana. Reprinted with permission from Newton, Milton B. 1972. Atlas of Louisiana: A guide for students. School of Geoscience Misc. Publ. 72-1. Louisiana State University, Baton Rouge, Louisiana.

describes the impact of these changes on the soils in the prairies. Although the desert verses marsh scenes are foreboding, the prairies were beautiful and inviting after the winter or spring burns as the grasses and forbs sprang from the scorched earth. Smeins et al. (1992) summarized the soils of the Cajun Prairie and the coastal prairies of Texas:

"Vertisols occur primarily on the coastward Beaumont geologic

formation within Texas, whereas Alfisols are largely located on the Montgomery and Bentley formations on the inland one-third of the region in Texas and within Louisiana. Vertisols cover over 15,000 km² and Alfisols over 10,000 km² (about 40% and 25% of the region, respectively)."

Smeins et al. (1992) continued,

"A characteristic feature of nearly all of these soils is their poor surface and internal drainage, which is due to the low relief and dense clay subsoils. Typically, soils are waterlogged during the winter season when precipitation exceeds evapo-transpiration, although saturation may exist for short periods during any season following heavy rains. During periods of low rainfall in summer the soils tend to be droughty. Since settlement, extensive drainage systems have been developed to facilitate removal of water, but waterlogging and temporary flooding still occur. Drainage systems have, nonetheless, undoubtedly had an influence on soil-forming processes and soil-plant relationships."

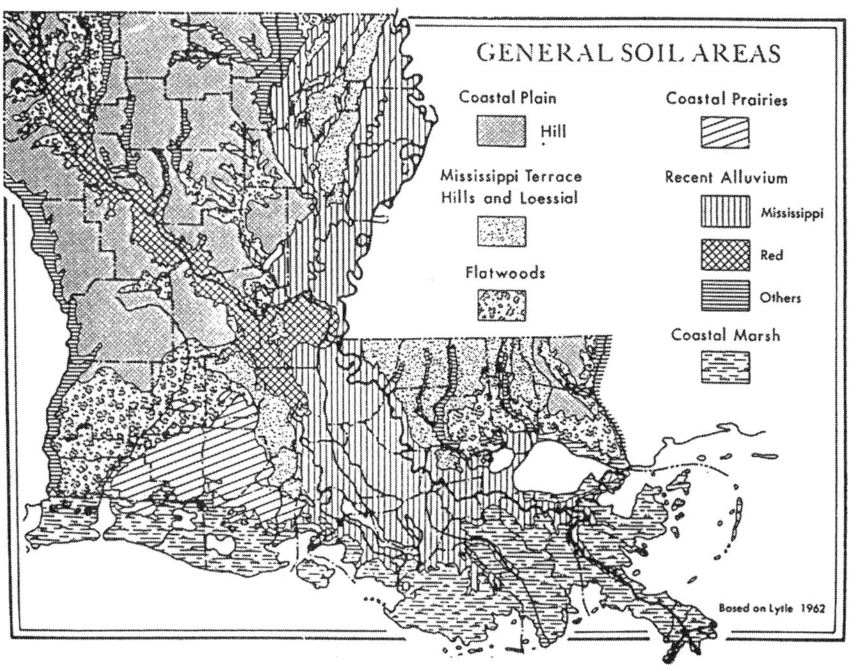

Map 5. General soil areas of Louisiana. Reprinted with permission from Newton, Milton B. 1972. Atlas of Louisiana: A guide for students. School of Geoscience Misc. Publ. 72-1. Louisiana State University, Baton Rouge, Louisiana.

"Alfisols are represented by the Katy, Edna and Crowley series. These soils have sandy, sandy-loam and silt-loam surface layers with clayey, slowly permeable subsoils. Their surface horizons tend to dry much faster than do the clayey Vertisols. Soil reaction is generally between pH 5.0 and 6.0 in the surface layer and from neutral to pH 7.5 in the subsoil. Base saturation varies from 45 to 75% for surface layers, cation exchange capacity from 50 to 350 meq kg^{-1}, and organic matter from 1.0 to 2.5% (Westfall et al. 1971)."

"A typical feature of Alfisols is the presence of circular sandy mounds (referred to as "mima mounds" or "pimple mounds"). There are hundreds of thousands of these mounds, and they cover as much as 25% of a landscape in some areas. They vary from 2 to 20 m in diameter and from a few centimeters to more than 1.5 m in height. ...Mounds tend to follow subdued drainage divides and appear to be products of past coastal environments and associated wave action, dendritic drainage patterns, and wind action. These mounds contribute variation in local relief in an otherwise monotonous landscape, and they have an influence on patterns of plant distribution and abundance."

Orville Touchet, a local soil scientist (January 1999), met with Latimore Smith (Louisiana Natural Heritage Program), Charles Allen and me in the mid-1990s in order to discuss soils of the Louisiana Coastal Prairie. His description of the soils of the prairie based on the notes I took included the following.

'The prairie is 85-200 thousand years old. Acadia silt loam is found in Acadia, Allen, Calcasieu, Evangeline, Vermilion and presumably Jefferson Davis (no recent soil survey in this parish) in gallery forests. Prairie soils are a mix of Crowley silt loam (runoff/convex) and Mowata silt loam (runon/concave) with either mixing with Vidrine silt loam to make a complex where mima or pimple mounds occur. The Vidrine silt loam is characteristic of mima mounds. Thus a Crowley-Vidrine silt loam would be runoff, whereas a Mowata-Vidrine silt loam would be runon. Runon soils are found in marais, whereas runoff soils need levees to hold water. There are 33 natural levees from Pine Prairie (north) to Cow Island (south). Pimple mounds are not evident in the eastern part of the Cajun Prairie as they are covered with glacial till or Loess soils; they become obvious from Acadia Parish to the west and south into the marsh.'

Grace et al. (2000a) provided an aerial photo in which the mounds are prominent and obvious on the surface of the soil. These mound features, so predominant in wetlands south of the prairie proper in Louisiana, provided purchase for prairie plants that prefer drier/drained/runoff soils. As simple as the above description may appear, no less than 14 additional soil types occur, including the following silt loams, Carroll, Jeanerette, Midland, Olivier, Mamou, Richland, Frost, Leton, Morey, Kaplan, Patoutville, Loreauville, Coteau, and Frozard. A shorter list of silty-clay or silty-clay-loam mixes is also involved. However, broad generalizations about soils should be sufficient in this discussion of hydrology, since hydrology will be most important in the end in determining plant distribution (Clark et al. 1962, Murphy et al. 1986 and Touchet et al. 1974).

Vidrine et al. (2001c) specifically described the soils of the Cajun Prairies:

> *"In contrast to the deep layers of loessial soils blown into the Midwestern U.S. systems ca. 12,000 years ago, coastal tallgrass prairie soils are mostly marine-derived. They are tight, calcareous clays, laid down in sedimentary layers as a result of coastal flooding due to the dramatic rises in sea level which characterized the end (interglacial periods) of each of the last several Pleistocene "Ice Age" events which occurred in the northern hemisphere."*
>
> *"The highly plastic (shrink-swell) nature of clay soils (found in Vertisols) is much exaggerated along the Louisiana coastal zone, where alternating wet and dry spells combine to produce prairie soil conditions which alternate between waterlogged (often anoxic) when wet and rock-hard (with cracks) when dry. This situation, combined with frequent lightning-generated fires, served to preclude the establishment of woody vegetation on the prairie proper. On the other hand, herbaceous plant species were able to establish and thrive upon the relatively thin, organically-derived layer of topsoil which gradually formed above the Pleistocene/Post-Pleistocene clays."*
>
> *"For the most part, the Louisiana coastal tallgrass prairie region is a rather featureless plain, ranging from ca. 1.5 m in elevation nearest to the Gulf Coast to ca. 16 m at its northern reaches. However, the region is characterized by two rather subtle physiographic features: 1) occasional shallow water-filled depressions, referred to as marais ("little marsh") or platins (nearly circular ponds) by the Cajuns, and 2) occasional series of low, circular mounds, ranging 2-20 m in diameter and a few centimeters to over 1.5 m in height. In some areas, as much as 25% of the landscape may be covered by these hillocks (Smeins et al. 1992), which are most often referred to as "pimple or mima mounds." There is ongoing conjecture as to the origin of both of these features. Thus far, the most plausible explanation may center around a "differential erosion" concept whereby the lands containing these features were subjected to wind and wave action during historic coastal environments (Smeins et al. 1992)."*

The original soil structures are damaged by leveling and levees but are still readable in horizons by soil scientists. The soils are typically divided into 3 horizons: A, B and C (Savage 2004). Horizon A is the topsoil, B is the subsoil, and C is the parent material. In Alfisols, a thin layer of organic matter called Horizon O covers Horizon A. In prairie, Horizon B is a result of clay accumulation (D. Jariel, personal communication, 2008). Topsoil or Horizon A is commonly called the rhizosphere, 'rhizo' referring to 'roots,' as this is the region where the roots of the plants reside for the most part. Prairie plants form massive underground systems with not only roots but also stems in great diversity (e.g., rhizomes and bulbs are forms of underground stems). Midwestern prairie root systems extend up to 16 feet in depth (Weaver 1954, 1968, Weaver and Fitzpatrick 1934 and Knapp et al. 1998). The plowing of the prairie reduced these roots to organic matter, which completely changed the soil's properties as well as the hydrology of prairie soils. The Gueydan prairie restoration project attempted to rebuild mima mounds to restore this hydrology in part, and as such signaling the general consensus that has developed regarding the role of hydrology and landscape restoration.

Numerous authors dealt with soil. In fact it is a science unto its own known as Agronomy. Soils are often described as alive, in that a single teaspoon of soil may contain hundreds of millions of microbes. Soils are the homes for thousands of kinds of mites, insects, earthworms, and even vertebrates (Savage 2004, Jariel et al. 2006, 2008 and Borsari et al. 1998). Soils are also reservoirs for water, lead and arsenic (Gotte et al. 2010). The rhizosphere, the world of plant roots, is considered one of the least known biomes of the world. Venter's (2007) discoveries of the genetic diversity of the world's oceans potentially pale by comparison to the genetic diversity of the soils of the world. Although we depend so very heavily on the soils for our crops, we know little about their diversity. An early study on Yellow Indian grass grown without mycorrhizae showed that plants grow only 2 feet tall; however, with mycorrhizae, plants grow 6 feet tall. The soil biota is very important and will be treated in Chapters 8 (plant roots) and 9 (soil animals).

Fire

Prairie plants and animals are adapted to fire. Bunch grasses produce most of their flammable biomass near the ground and serve as fuel for prairie fires. Some plants have flammable oils (as do pines and other trees, but the oils apparently are not as flammable as those resins and certainly do not explode into flame as the tree resins do). Fire is essential for survival of prairies in anything close to a natural condition with high diversity. Many of the justifications for fire used in Earley's (2004) discussions of the roles of fire in Longleaf pine forests clearly function for prairies.

The loss of fire (i.e. control burns) in this natural-burn ecosystem may be the most dramatic change (Wright and Bailey 1982). Fire and hydrology apply significant selective pressures in prairie landscapes. Fire is also dramatically needed in our Longleaf pine forests; Finis Harris, the forest silviculturist of the

Kisatchie National Forest in central Louisiana, made this number one on his list of proposed management techniques to maintain the forests (Earley 2004).

The use of fire by farmers to clear brush and reduce weeds along the railroad rights-of-way ironically was the very technique that maintained the prairie remnants along the railroads. These remnants that were created by the railroads and maintained by the farmers provided the material evidence upon which much of the ecology in the remainder of this book is based.

Jim Grace (1998) clearly detailed the importance of fire in the prairie ecosystem, especially in coastal prairies. Coastal prairies have been invaded following the introduction of the Chinese tallow tree into the landscape. This tree is able to grow and spread, via bird droppings, to the point of absolutely collapsing the prairie ecosystem. Fire is deterrent to the spread of this invasive tree and other woody plants. Chinese tallow trees apparently can vegetatively spread in an area. The aggressive, native shrub, Eastern baccharis (*Baccharis halimifolia*), is a woody bully in the prairie. Recent studies showed that fire can assist in the abatement of its spread (Allain and Grace 2001).

Fire is also necessary for vital activities of prairie plants. Fire provides light for young plants and for early spring growth. Fire causes some prairie plant seeds to split and germinate. Fire opens up new habitat and provides for the spread of prairie plants. Fire stresses some plants into bloom, and fire mobilizes potassium, which enhances not only bloom but the movement of nitrogen and phosphorus in the soils. Fire as a management tool, with prairie seeds, as in the Cajun Prairie Gardens, can result in the elimination of a lawn and the production of a prairie (Vidrine et al. 2006 and 2008b).

Fire is also necessary for activities of animals (Jariel et al. 2008). In studies on prairie restorations versus lawns, earthworms showed greater diversity and prevalence in the restored prairies. Soil compaction also played a role, but the impact of fire was evident. Timing of fire in order to protect nesting birds and over-wintering insects is a primary concern. Traditionally, fires were commonly set by Native Americans, *vachers* and farmers in the winter. Because of the dampness of the Cajun Prairie, thorough burns are uncommon.

Annual burning does not negatively affect the plants in the prairie as was once assumed (Jariel et al. 2006, 2010a and b). Burning annually was thought to result in the accumulation of minerals in the soil that might be inimical to prairie plants. In fact annual burns apparently expedite restorations and permit the development of a fairly good restoration in 5 years. The fire assists native prairie plants in out-competing exotic plants—the exact mechanisms are not yet clear, however, the results are obvious (Vidrine et al. 2006 and 2008b).

A friend repeated a story of interest regarding Louisiana iris. To the best of my recollection, he burned his marsh after 20+ years of not doing so. The burn cleared away debris and opened up the marsh. When in April, he came to visit the property, to his amazement; a 'sea of blue' covered the marsh. He had no idea an irises were present prior to the burn. All in flower, the irises provided an amazing view. He was so impressed that he asked his friends to view the spectacle the next weekend. In literal caravan, the group drove over the Intracoastal Canal to see the sight. To his further amazement, a few pockets of blooms remained—a far less impressive site. The biological aspects of this story are important. First, the role of fire was clear. Fire opened the marsh canopy and released the minerals for use by the plants; thus blooming was intense and obvious. Secondly, prior to the burn, the iris plants were so thickly crowded that they bloomed little and were literally invisible among the old growth of marsh stems. The iris plants in the area were probably like most wild Louisiana iris in that the plant scapes (flowering stems) had only 2, seldom 3, bloom positions. Blooms open every couple of days depending on the weather, literally heat and sunshine. And thirdly, all of the plants in this one location, although many acres were involved, were closely related if not the clonal progeny of the same plant. With the few bloom positions and the clonal nature of the plants, they all bloomed at the same time, and the few flowers produced on the scape only opened for a total of 10 days or less. This homogeneity provided his original 'sea of blue' view and the much lesser view only a week later. In just 10 days the vast majority of the irises in this one location bloomed in an exaggerated way as a result of both the burn and the genetics of the plants.

The purpose of this story is twofold. First, the story demonstrated the following concepts: the impact of fire, the genetics of the plants, the natural beauty of prairie/marsh plants, and the timeliness of any view. Second, the story illustrated one of many possibilities that existed in the prairie/marsh region of Louisiana. Charles Allen often spoke of 'Easter eggs.' A rare plant or a rare view of plants in bloom was to him an 'Easter egg!' During the bloom season from March to October in Cajun Prairie, such finds were common as Charles and I routinely visited the remnant prairies. 'Oh my!' is another of his expressions—it also signaled the finding of some rare or beautiful plant or a good photograph of a plant typically showing not only the flower but also the diagnostic parts of the plant and preferably taken by a photographer who lay on the ground at the height of the plant.

The importance of fire in the ecosystem cannot be overestimated. Fires were eliminated as management methods in forests and in prairies, and the results included the loss of forests and prairies and the buildup of flammable materials, such that when a fire occurs, the fire is devastating. Our misunderstanding of fire in nature was accentuated in the movies like *Bambi* (the Disney movie) and in campaigns like *Smokey the Bear* with his '*Prevent forest fires at all costs'* slogan. These not only changed the paradigm of using fire in nature but also endangered habitats and many of the species of plants and animals—some now protected as rare and/or endangered species by state and federal agencies.

Hydrology

The topography of the region is gradually sloping hills and valleys. These hills may be only a few feet in height. Historically, the swales were wet all winter and formed *marais* (marshes) or *platins* (ponds). These wet areas reminded me of potholes of the Midwest prairies and were extremely important to migrating birds and other wildlife. They became the foundation for rice cultivation in the 1880s, and that cultivation led to the elimination of these wet areas by land-leveling. The rice cultivation also led to the creation of rather extensive canal systems in order to control water on the landscape—an essential for weed control in rice cultivation.

Vidrine et al. (2001c) described the hydrological features of the Cajun Prairie as follows:

> *"Another set of associated physiographic features which allowed early coastal tallgrass prairie travelers to more easily navigate and orient themselves within the region is the presence of several bayous (sluggish streams) which dissect it in north-south fashion at various intervals. Riparian hardwood forests developed along the better-drained natural levees of these streams. Ranging in width from 10-1000 m, these strip-like, gallery forests were inhabited and traveled through by Native American and early Francophone settlers alike. Too, these streams with their associated woodlands served to separate the vast grassland into sub-sections replete with various wooded landmarks (see map I) which could be utilized for more efficient orienteering."*
>
> *"A final physiographic consideration, which contributed greatly to the eventual cultural disposition of southwestern Louisiana's prairie region, involves its degree of physical isolation resulting from the ecology/ physiography of the lands which surround the region. To the north, the prairie system is bounded by a substantial pine/hardwood "flatwoods" system possessing sandy soils, which are too nutritionally poor and/or wet to clear for crops. Together with eastern Texas, much of that area is considered the Big Thicket (Ajilvsgi 1979). To the east lies the vast Atchafalaya Basin, one of North America's largest river swamps; and the entire southern edge of the prairies is bounded by an equally impenetrable mosaic of freshwater, brackish, and salt marshes."*

The flow of water through an ecosystem is considered the most important aspect of the ecology of the system. Grace et al. (2000b) clearly showed that micro-relief and hydrology determine which species of plants survive in any local in the Cajun Prairie. But this is also consistent with what is known about other ecosystems. The Cajun Prairie was a wet ecosystem for much of the year. The topography contained natural wet areas and dry mounds and to some extent rolling hills, especially to the east. The area was drained by rivers or bayous in a very distinctive pattern. The prairies ran basically east to west with the drainage. The land has since been canalized and the rivers have been channelized and dredged, so as to do away with any impediments to drainage. However, the land has been leveled and levied to retain water for the production of rice. Erosion is so pervasive that the once sandy-bottomed, clear streams became muddy. The extensive roots of the prairie plants percolated water into shallow aquifers that seeped along hillsides or into streams. These hillside seeps contained unique assemblages of plants, including many carnivorous plants, orchids and lilies. The seeps then dried, and their plant communities disappeared.

Changing stream flow and damaging aquifers are not just the result of farming and modernizing drainage. Urbanization, compacting soils by mowing, loss of prairie plants, and cattle ranching also dramatically change hydrology (Vidrine 2008). New threats to the hydrology and water quality are on the horizon with the potential of 'fracking' for oil and natural gas; contamination of water with petroleum products and other pollutants would not only impact drinking water but also dramatically alter the quality of floodwater for rice and crayfish cultivation. The recent documentary video *Gasland* clearly showed the possible threat to the ecosystem (Fox 2010).

The loss of water flow and natural hydrology is not unique to the Cajun Prairie; it is apparent throughout the Midwest. Manning (1995) made this point clear, "*Removing the cows from a five-mile stretch of a grassland creek in Nevada increased stream flow 50 percent. …Of the 150 fish species native to the west, 122 are either extinct, listed as endangered, or are candidates for listing*." The impacts of cattle on watersheds are well-documented. The amount of feces produced and their tendency to collect around streams for long periods gives ample opportunity for the sedimentation and nitrification of streams. Adding feral hogs to the mix makes for a lethal blow, especially since the hogs feast on mussels, which once filtered every drop of water many times in every stream east of the Rocky Mountains.

As a youngster playing the Bayou des Cannes in summer time, I would occasion a hole with cool water coming out of it. The holes were large enough to stick my arm into up to the shoulder. These springs were continuously flowing even in the dry summer. They were along the edge of the stream. The stream was muddy to clear, the spring was clear. Persistent seeps occurred along the banks that made it difficult to climb out. Alas, all this is dredged and straightened now (since 1971).

At home, a shallow aquifer (Chicot and Evangeline aquifers are deep and very deep, respectively) provided water for our shallow well (ca. 16 feet), where we would draw water by hand. We put in a pump with a tank and had running water piped to the house in 1955. I drank this water throughout my childhood. 'City water' as I call it arrived in the mid-1960s when our shallow well was pronounced dry.

Much of my research over the years has dealt with rivers and lakes and the life that lives within them. The primary streams draining the Cajun Prairie were among those that I spent a lot of time studying. The Mermentau River drains most of the Cajun Prairie; the Calcasieu and Sabine Rivers drain the very western part; and the Teche and Vermilion Rivers drain the very eastern part. My research primarily dealt with mussels and their parasites, but research is seldom linear. And as such, holistic concepts creep in constantly, and hydrology is one of those concepts.

The Mermentau River drainages have received the greatest damage among the rivers of the Cajun Prairie (Vidrine et al. 2006). Most of its tributaries have been dredged repeatedly or otherwise channeled to create canals; massive sedimentation from erosion of farmland in its drainage is evident; the gallery forests along the river and its tributaries have been harvested routinely for the last century, and in many cases, the banks have been destroyed completely by the accumulation of literal hills of dredge-spoil from canalization. The runoff from local fields and homes has brought not only soil as sediment but also pollution with fertilizer and biocides. The displacement of native fish populations by channeling and by the construction of dams along with the lost of the natural stream bottoms providing purchase for mussels and cryptic habitat for fish hatcheries are obvious. A major train derailment followed by several major explosions created a major pollution event in Eunice, Louisiana (June 2000) and resulted in the releasing of (yet unrevealed) numbers and kinds of petroleum pollutants into the 5 mile stretch of Bayou des Cannes that remained unchanneled. In addition, the overall compaction of soil by agricultural equipment and the loss of the native prairie have caused a loss of the water table created by percolation. Numerous springs generated by shallow groundwater discharge fed the bayous, but these springs are no longer feeding the streams, and the movement of water is now limited to surface runoff. Irrigation from canals carrying water pumped from bayous is commonly used in rice/crayfish culture. Many deep-water wells have also been constructed to access the Chicot aquifer, our local underground river, in order to obtain cheap water. With the massive use of winter water plowing and the draining of the fields in July, the silt is sufficient to keep the water murky all year long.

In general, the loss of shallow groundwater discharge, sedimentation, stream alteration and pollution have modified or eliminated the communities of freshwater mussels in the Cajun Prairie ecosystem. With underground irrigation gaining favor among farmers, many canals have now disappeared or have been abandoned. The loss of the canals exacerbates the loss of the mussel diversity and abundance in this ecosystem. The overall impact is a simplifying of the ecosystem and ultimately the disappearance of a once diverse community (Vidrine et al. 2006).

The Cajun Prairie is not the only region with this series of problems. Recently, I presented a paper on the Red River (Vidrine 2008), in which I made some of the same remarks. The Red River historically was surrounded by prairies, which are now lost (Freeman and Custis 1821). The streams flowing into the Red River are nearly dry during the summer season, although they flood in rainy season.

The loss of percolation along the roots of the plants to shallow aquifers results in runoff and flooding followed by drying down of the stream when rain ceases. The other comments in that paper also pertain to the commentary on the Cajun Prairie hydrology.

> *"Restoration of forest and prairie habitats associated with streams like Bayou Pierre would promote the percolation of water and springs in the area. It would also reduce runoff of agrochemicals and urban chemicals and permit some bioremediation. The loss of gallery forests and the compaction of the soil are remediated by restoration efforts. Ultimately, these changes in hydrology and local ecosystems provide a revitalization of the mussel community in the rivers."*
>
> *"Freshwater mussels and mite parasites create a diverse community that not only speaks to the ancient interactions between these species but also to the intricate nature of their interactions. In some locations where cotton or other crops are grown, the mites literally disappear from their hosts as is evidenced in some stations in Bayou Pierre—this is an example of the impact of human activities on these communities and was reported in other streams (George and Vidrine 1993). This disappearance might be a result of the runoff of insecticides from these agricultural areas."*
>
> *"Freshwater mussels are on the decline numerically as well as in diversity as a result of human activity, namely loss of seepage or spring-feeding (shallow groundwater discharge) by agricultural/urban compaction of soil, sedimentation from erosion, channeling of natural streams by dredging, and poisoning of waterways from urban and industrial runoff and dumping. The once-spring-fed streams flow intermittently functioning more as drainage canals as a result of the destruction of prairies by agriculture and urbanization. The return of prairies and forests reinitiates the springfeeding by shallow groundwater discharge to streams and minimizes runoff of agrochemicals and urban chemicals, thus providing an opportunity for revitalizing freshwater mussel communities."*

The mussels as larvae (glochidia) are themselves parasitic and require a specific fish host species—fish populations and diversity have been greatly modified by the alteration of stream hydrology and sedimentation. Further, cow feces with nutrients foul streams and kill fish—a general result from pasturing large herds of cattle. The fish communities are changing, and this has a direct impact on the mussel community structure. The fish community is so directly connected to all the other aquatic life that any change therein cannot be overestimated as to its impact throughout the aquatic biota. Even minor changes may have dramatic impacts on other species.

Connecting dots is central to answering even the simple questions. I was once asked, 'Why do you collect mussels at bridges where roads intersect with streams?' The answer was apparent to me, 'Easy access.' The follow-up question related to biodiversity, 'How can you be confident that you have a good sample of the stream's mussels and mites?'

This question required a bit of thought and research. I certainly did not want to have a huge hole in the logic of my plan for sampling. With a bit of serendipity, I connected two seemingly unrelated pieces of information. First, I made several forays into streams searching for mussels and mussel beds more than a mile upstream and likewise downstream. Occasionally I found small beds of mussels, but usually not. Secondly, several stories that I had heard mentioned early settlers looking for mussel beds because these made good crossings for wagons. I also recalled that when asking people for help to find mussels, they often took me to washed-out and/or dilapidated bridge locations. And suddenly, I knew the answer. Bridges and roads are built on top of pre-existing roads and trails. Early trails led to river crossings, where large mussel beds had once been and in many cases still persist, at least at the time I sampled. Thus by sampling at bridges, I was sampling the largest beds, the best mussel locations, in the streams as found by the early settlers.

I know of no Cajun Prairie stories of mussel beds and bridges, but finding mussel beds in the Cajun Prairie was no different than in other areas of the country. I thus assume the same logic was applied when early settlers were looking for safe crossing areas in streams.

Agriculture

Ranching and farming in all their varieties are keys to the understanding of changes that are occurring to prairies. In the Cajun Prairie, ranchers appear to have made the initial impact with their herds of cattle and horses. Grazers apply pressure and are also necessary to maintain grasslands—cattle provided this service as surrogates for native species. There are differences in grazing styles of cattle and Bison: cattle pull the plants out of the ground and graze the grass to the ground in contrast to the nibbling Bison (Manning 1995). The cattle have a more detrimental impact on the prairie diversity than the bison in the long run and even in the short run. The cattle also stay put and graze locally, whereas Bison migrate (roam) and thus reduce significantly the grazing pressure on any single piece of landscape. Prairies apparently can rebound from the impacts of intense grazing, but research on the ability of the Cajun Prairie to rebound is just not available.

Farming, from row crops to flooded rice paddies to crawfish aquaculture, was a dramatic if not drastic change in the prairie. Entire landscapes were plowed and/or flooded in order to provide till for planting and growing crops. End-rows were places where relict prairie plants could survive, but the development of conventional planting techniques brought the entire field into culture. The implementation of herbiciding, fungiciding and insecticiding in order to control every aspect of the habitat destroyed native plants, mycorhizzal fungi, and pollinator insects—in other words, the entire structure of the prairie ecosystem. Agriculture is now much more than the production of food as was clearly evidenced in the documentary video entitled *Food, Inc.* (Kenner 2008).

Vidrine et al. (2001c) provided a general introduction to agriculture in the Cajun Prairie. An extended quote follows:

> *"Agriculturally, the prairie could be divided into two cultural zones, the "corn and cotton" zone in the east, and a larger "rice and cattle" zone in the west (Comeaux 1983). The reasoning for such a division lies in the soils. As previously mentioned, the bulk of southwestern Louisiana prairie soils are dense calcareous clays derived from repeated flooding/ sediment laying via the adjacent Gulf of Mexico during interglacial periods of the Pleistocene. However, during the same period (ca. 10,000 B.P.), in which massive volumes of wind-blown loess (silt-loam) was deposited over the prairie region of the midwestern U.S., appreciable amounts of this same soil were also deposited over Louisiana's prairie region, ranging from a few centimeters in the western zone to a meter or more in the eastern zone. Thus, eastern zone soils were completely friable and possessed qualities which rendered them suitable for corn, cotton, and sweet potato culture. There, soil quality allowed for settlers to almost immediately begin small-scale cultivation to go along with small-scale livestock (cattle, pigs, poultry) operations. Initially, these small farms supported both cotton and corn, with the former utilized as a cash crop and the latter as food for both the families and their livestock (Comeaux 1983, Conrad 1983, Post 1962, and Ancelet et al. 1991)."*
>
> *"In the western prairie zone, however, cultivation would not come easy. Initially, the Acadians settled along the edges of the gallery forests, as their Native American predecessors did, claiming only small pieces of prairie proper in order to cultivate subsistence fruit and vegetable crops such as okra, melons, beans, and other introduced crops (Comeaux 1983 and Voorhies 1983). Eventually, they initiated the practice of growing "providence rice," so named because such crops were irrigated only by natural rainfall (i.e. "Providence"), within the locally wet depressions (marais and platins) in which they were planted. Rainfall is generous in the region, with average annual rainfall exceeding 1.25 m."*

The interdependence of the above factors (soils, fire and hydrology) and agriculture provide a good segue to a discussion about the future of agriculture in the Cajun Prairie. The notion of sustainability is central to Bruno Borsari's recent papers on agriculture, agricultural education, and the Cajun Prairie. Bruno like many trained agriculturalists first saw the landscape as potential farms and ranches; however, his experiences in African agriculture changed his perspective. His arrival at LSUE not only provided me with a great deal of assistance in prairie restoration but also opened a new dialog about the prairie as a model for agriculture (Vidrine and Borsari 1998). We extended that concept to landscaping, a less common concept.

Borsari (1998 and 2003) and Borsari and Vidrine (1998, 2000 and 2005) and Borsari et al. (2002) extended the discussion into the classroom by assessing students' preparedness to continue farming and ranching in the prairies as far as their understanding of the basic principles of ecology and sustainable agriculture. The students varied greatly in their performances, but in general, students were minimally prepared for agriculture in the next generation. Our general conclusions from the latter paper follow:

> *"The agronomic characteristics of prairie soils contributed to the expansion of agriculture in the Northern American continent and promulgated its initial success. However, the intense cultivation of these fertile soils accelerated the degradation of prairie habitats and the loss of topsoil and biodiversity affected eventually the positive outcomes of conventional farming practices. For these reasons, systemic changes may be needed in agricultural education in order to prepare the new generation of agriculturalists to the challenges of ecological farming and holistic decision-making, when natural resources are manipulated. Therefore, an assessment of students' knowledge in their final semester of study from colleges of agriculture was accomplished. ... Descriptive statistics of the students' test scores substantiated the students' preparedness in valuing the potential of coastal prairie ecosystem to support a sustainable agriculture in this southern region of the United States of America. Although students from the Texan institutions achieved higher test scores than their Louisianian counterparts, this difference was not significant. Despite its limitations, the study illustrates how important it is to impart knowledge in prairie habitat ecology in order to gear the agricultural curriculum towards sustainability and to ameliorate didactic methodologies of instruction in the agricultural sciences."*

Our paper (Vidrine and Borsari 1998) presented the Cajun Prairie as a model for agriculture in the future. We followed up with a paper on the role of beneficial insects in agriculture as part of this model (Vidrine and Borsari 1999). We (1998) stated:

> *"The habitat was destroyed by plowing for modern agricultural development, but recent agricultural models suggest that the restoration*

of native habitats (while they are still marginally extant) and the return to sustainable agriculture practices would, over time; increase crop yields, protect the soil from erosion, and reduce the needs for herbicides and pesticides. The perennial polyculture paradigm of The Land Institute is foremost among these ecosystem/agriculture restoration and preservation models and could serve well as a blueprint for the Cajun Prairie in southwestern Louisiana. Thus, adaptation of the Land Institute paradigm to create the Cajun Prairie paradigm makes coastal prairies a model for farms, ranches, and urban yards. Soil building and soil protection practices are central to the paradigms and to the long-term vision of sustainability for future generations."

"The study of prairie gardens and native prairie in the Midwest and in the coastal prairie clearly depict the beneficial impacts of reduced mowing and/or grazing on the soil. The plant community with its mycorrhizae, fungal root symbionts, reconstructs the soil horizons and creats an underground habitat for numerous kinds of organisms which live in the interstices between soil particles and roots. The underground structure of the prairie is more complex than in other habitat, and a minimum ratio of 2 to 1 is commonly presented for the underground to aboveground plant biomass in the prairie. Thus, the prairie plants are soil builders and control erosion."

"In association with the prairie plants, a diverse biota develops with a myriad of inter-relationships. Faunal, fungal, moneran (bacterial), and protist interactions with plants are well-documented for Midwest prairie ecosystems; however, very little is known of these in Cajun Prairie. With the restoration of plant communities, it is apparent that these associations tend to redevelop. The level of complexity and the full diversity of the historical ecosystems are probably not going to be re-established, but there appears to be sufficient diversity to recreate a sustainable system. In fact, many of these "gardens" that have been created in the Midwest by natural landscaping are advertised as "butterfly gardens" and/or "pollinator gardens" (Daniels 1995, Ajilvsgi 1990, Buchmann and Nabhan 1996, Druse and Roach 1994, Kirk 1995, Packard and Mutel 1997, Roth 1997, Shirley 1994, Thompson, 1992, Wasowski and Wasowski, 1988 and 1992, and Wilson, 1992). The movement to this kind of landscaping is growing in the larger cities in Louisiana (Fontenot, 1992 and Ross, 1994), but it is very slow to start in rural Louisiana. Eunice and LSUE provide a prime location to develop model demonstration gardens for this rural landscape and ecosystem."

"Relating Cajun Prairie to sustainable agriculture is most easily done by reflecting upon the philosophy and the perennial polyculture paradigm developed by Wes Jackson at The Land Institute in Salina, Kansas (Jackson 1980, 1987, and 1994). The perennial polyculture paradigms model

agricultural systems that mimic Midwest prairie ecosystems. These models depict economic systems that not only support a rural livelihood but also build and protect soil. Benefits of perennial polyculture (Soule and Piper, 1992) include:

1. Soil building and erosion control by continual addition of root turnover materials to the soil and by the roots holding moisture and producing humus which holds moisture,

2. Advantages of intercropping and crop rotation since several species are planted together,

3. Preadapted water-use patterns of native plants already synchronized with the local habitat reduced need for irrigation,

4. Direct transfer of nitrogen into soil by the use of legumes reduced costs of inputs,

5. Increased biotic diversity in the field reduced pest and disease incidence and encouraged beneficial predators to reside in the field,

6. Weed management occurred by shading of weed seedlings with the perennial canopy and/or by allelopathy,

7. Decreased fossil fuel consumption in seed bed preparation, cultivation, and fertilizer,

8. Reduced contamination of water supply due to runoff of chemicals used to control pests and to add nitrogen to the crop,

9. And preserved genetic bank of unique adaptations of plants in the ecosystem."

The list of values of prairie, reasons for growing and preserving prairies and reasons for restoring prairies continued to grow as new information was literally unearthed. A list of benefits from Vidrine and Borsari (1998) included:

"1. A large number of native plant species are the same ornamentals currently used in the cut-flower industry all over the country. Our species bloom during seasons different from other members of the same or similar species because of their unique genetic adaptations to this climate and geography. These beautiful plants/flowers provide pleasure and beauty in the home, around the home, and in local businesses.

2. A diverse, native pollinator community, which is itself imperiled by the sheer reduction in the number of kinds and individuals of prairie plant hosts, is sustained and, in turn, the pollinators can sustain this plant community. These same native pollinators replace the naturalized honeybees, which are succumbing to mite infestations and to African dilution of genomes by poorer honey-producers.

3. Native grasses and forbs with not only high forage ratings for grazing ungulates but also high value as perennial hay and pasture crops are among the prairie plants that have been ignored by Louisiana ranchers (Stubbendieck et al. 1997), but these are re-established by restoration and landscaping.

4. The obvious soil building and erosion control capabilities of the plant community are well-documented and previously mentioned.

5. The potential edibility of native plants, like Apios americana, may provide new crops for human consumption (Allen et al. 2005 and Kindscher, 1987).

6. Many native herbs, like Mamou coral bean, which may contain unique genes that lead to new, needed antibiotics and anti-cancer drugs are still surviving in remnants of Cajun Prairie. Other herbs, like Echinacea, provide well-known chemicals which enhance immune functions (Kindscher, 1992 and Foster and Duke, 1990).

7. New landscaping designs that teach our children about our heritage as well as demonstrate the interconnectedness of natural things within local ecosystems are developing with these plants. The reduction in costs of lawn maintenance and the reduction in overall pollution by chemicals in producing natural landscapes is enormous. Meadow and prairie plantings in the Midwest have increased into the thousands clearly indicating the growing interest in this kind of landscaping in an educated, progressive community (Roth, 1997). These also serve as outdoor classrooms for the youth and the mature student of life.

8. Beneficial animals like dragonflies and spiders provide protection from deleterious insects including mosquitoes, love-bugs, and other pests. An intact complex food chain is essential for sustainability in any ecosystem. Insects provide abundant food for birds, including resident and migrating species. The plants produce seeds and other parts that feed a large variety of birds and other animals essential to our well-being and the overall health of the ecosystem (Allen and Vidrine, 1990a, Curry, 1994, Opler and Malikul, 1992, Price et al., 1991, Vidrine and Allen, 1993, Vidrine et al., 1992a, 1992b, and 1992c, Vidrine and Borsari 1999, and Weaver, 1954 and 1968).

9. Seed and propagules of these plants are requested often, and the numbers of requests grow annually, but few local retail sources exist. Jobs can be created and expanded by the development of this habitat. Not only research positions but also small business opportunities abound. Demonstration gardens, for example, can provide cut flowers, seed, propagules, information pamphlets, and other materials for minimal costs. Trained professional and paraprofessional individuals can provide landscaping assistance.

10. The habitat is in the final analysis a colorful and dynamic system with wildflowers whose reputations for attracting tourists are well documented. The restored habitat again serves as an outdoor classroom and as a "people pasture" (Madson, 1982). There is a certain "revery" just being in a wildflower meadow, but the "revery" redoubles with the restoration of such habitat. The gardener senses a deep feeling of "self-worth" that grows with each succeeding year as the gardener observes the plants and their biotic associates redevelop a sustainable habitat."

Marc Pastorek's seed catalog (2007) also provided a list of benefits:

"Growing native prairie wildflowers gives a garden local identity.

Cultivating many tiny prairie plots allows us to promote diversity and to preserve rare ecotypes that straddle the fence of extinction.

Prairie plants have historical significance and are a useful connection to the past.

Prairie grasses provide unique foliage texture. Their linear foliage refracts light and sings as the wind moves through.

Prairies are adorned with colorful flowers through the growing season.

Many prairie plants provide forage for wildlife. Birds feast on the plump fatty seeds of composites like compass plant. Butterflies abound in the prairie, sipping nectar from their sun-loving food sources. Dragonflies hunt using strategic aerial feats to hone in on prey.

Most prairie plants are able to withstand extreme environmental conditions. Troublesome or unsightly areas of the landscape with baking sun and heavy clay soils are often ideal sites on which to establish a prairie. Prairie plantings can be used to reclaim wasteland in many cases.

Prairie plants (especially the grasses) have tenacious root systems that hold soil, filter ground water and prevent erosion.

Prairie plants are tough and resilient once established. Within 3 to 5 years after sowing, most prairie plots are almost self-sustaining. Once established they will produce flowers and valuable seed for many years."

Marc Pastorek (2008) later provided a list of reasons to grow prairies:

"Prairies are ecosystems in a climax stage of development. This means they stay in a permanent state as long as tree and shrub growth is kept in check by mowing or controlled burning. Key to prairie reconstruction is selecting climax species when collecting seed.

Establishing prairie changes the hydrology of the reconstruction site.

Fibrous root systems of grasses and forbs often reach six feet or more in depth. These roots act as channels for water to percolate deeply through the soil, replenishing aquifers and hillside seeps.

Many climax species of herbs and grasses have been shown to consistently remove pioneer species such as Blackberry, Dewberry, Vasey grass, and other noxious perennials. Privet, cedar, and most other woody plants can be kept in check by mowing and burning. Research has shown that these desireable 'pushy perennials' are genetically predisposed to aggressively rehabilitate the ecosystem after disturbance occurs, moving out pioneer species and many exotic invasives.

Indicator plants are showy species that identify areas of relative high richness and diversity or 'seedbanks.' Some common herbaceous indicator plants are Silphium, Eryngium, Pycnanthemum, Liatris, Baptisia, Phlox, and Asclepias. Some indicator grasses are Paspalum, Andropogon, Schizacharium, and Sorgastrum. These plants don't always signal rarities but most always identify a level of species richness.

Using indicator plants locate nearby stands of wildflowers and grasses insures a greater chance of success in seed collection.

Prairies and savannas are some of the most threatened ecosystems in the state. Ethical collection of seed during restoration work can save many genetic ecotypes from the brink of extinction.

Prairie gardening in most Midwestern states is an integral landscape design element. This increases habitat diversity and provides tens of millions of dollars to the economy. Millions of acres of prairie and savanna existed in Mississippi and Louisiana in pre-Columbian times. The stewardship of existing prairie remnants and the restoration/ reconstruction of adjacent and disjunct fields could produce similar benefits here in our home state.

Prairie plants, by virtue of where they exist, are proven to be the most adaptable plants to the most hostile environmental conditions. The most important requirements are lots of sun and woody plant suppression.

Prairie grasses and herbs provide ideal forage for herbivores. They support many insects that feed threatened field songbirds. They provide nectar for pollinators and are host plants for egg-laying butterflies and insects. Grasses provide fuel for much needed fire and add winter interest to the landscape.

Research has shown that mowing a newly established prairie planting as regularly as every three weeks during the first and second growing season will keep rank pioneer species from reseeding and also will prevent these plants from shading out seedlings of desirable species.

Most prairie seeds do not germinate in a uniform fashion. Baptisia seed, for instance, will germinate a percentage the first year, another percentage the second and so on. This insures reproduction over the long term. Many prairie species are viable up to thirty years. This phenomenon insures surprises for the prairie enthusiast for many years.

Patience is the most valuable commodity when gardening 'a la prairie.' However, it is my experience that when planting a tree, one expects to wait three to five or more years to 'harvest' the equity of its beauty. This is also true with herbs and grasses."

I have attached a list of benefits of prairie landscaping that I generally hand out to my students in Appendix 18. The benefits of prairies and prairie gardening/landscaping exceed our imagination at this point, but I am optimistic that we will gain respect for prairies and the gardening and landscaping with prairies as we continue to compile the benefits of existing prairies, restoring prairies, and landscaping with prairies.

Vidrine and Borsari (1998) continued,

"The renewed interest in native grasses and forbs not only resurrects a forgotten paradigm of agroecology long since replaced by the paradigm of modern agribusiness but also provides an opportunity to reform agriculture and place primary emphasis on soil creation and protection. The current trend is the loss of 2 tons of topsoil for each ton of agricultural product. Southwestern Louisiana is no different from the Midwest in this manner of exploitation, and thus, Cajun Prairie is a realistic model for agriculture. There has been no effort to develop this model nor is it yet conceived, in part because we are yet to formally define the ecosystem called Cajun Prairie. The genomes of the prairie plants in southwestern Louisiana are unknown and in dire need of study and protection. Restoration projects become "stop-gap" measures to protect this flora, to provide propagules, and to provide the needed genetic material for study."

With the development of a basic understanding of the abiotic factors affecting the Cajun Prairie, the focus now changes to the biotic factors. The primary organisms include bacteria, protists (protozoa and algae), fungi, plants and animals; however, my focus will be on plants and animals. Ecologically, the first endeavor is to generate an inventory of species, e.g., a flora (a list of plants) and a fauna (a list of animals). These lists quickly develop into lists of not only taxonomic groups but also ecological groups. And naturally, what follows is a discussion of communities, succession, phenology of flowering plants, flight seasons of insects and birds, predation, competition and symbioses (the numerous interactions among members of populations and species).

But first, I want to discuss briefly the 'ecological niche concept.' Each population (within a species) has an ecological niche. This niche is usually defined as the habitat and the ecological roles of the population. The niche is often considered unique for each population, and ecologists spend a lot of time describing them. The habitat portion places the population in a specific location and time in the space-time continuum. With plants, this is very strategic. Larry Allain et al. (2006) listed each plant with descriptors, e.g., wetland classification and wetland coefficient. These are abiotic constraints and part of the niche description of habitat. Gardeners are quite familiar with similar constraints for temperature, sun exposure and wind tolerance. The second component of the niche contains the ecological roles of the population. Identifying these roles is a complex task and includes discovering the numerous biotic interactions of the population with other populations of their own species and those of other species and abiotic interactions with the environment, e.g., plants trap heavy metals in their roots (Gotte et. al. 2010). Thus, the ecological niche of a population specifically unveils the role of specific abiotic factors upon a single population; thus the abiotic factors in a habitat are viewed through the lens of a population. With thousands of species, the task is daunting. But first, the particular population (species) must be identified. This task is well under way, and chapters 8 and 9 center on the identification of biota.

The protists, fungi and bacteria are extremely important groups of organisms. However, they have received insufficient attention, and lists of species for specific regions are usually short. Many are considered ubiquitous, e.g., bacteria, while others are known only in their fruiting stages, e.g., fungi. David and Pat Lewis, active members of the prairie society, are local mycologists. They are always searching for new fungi throughout the region. The sheer diversity of these groups is daunting—literally hundreds of species. As you may have deduced, most of the species are unknown to science or known simply as a group of potential species. Efforts like those of Edward O. Wilson to develop Bioinformatics and identify each species on the planet much as we are doing for each human gene in the Human Genome Project are likewise daunting at first. Molecular techniques for identifying species are multiplying (Venter 2007), and the next 20 years may change the nature of this task completely.

Using common names is often considered undesirable; however, the United States Department of Agriculture (USDA) PLANTSdatabase website is attempting to develop a system of unique common names (Appendix 6). I am trying to use this scheme throughout the text for plants, although some of the old common names of plants are far more colorful and entertaining (and informative) than the new ones. Fairly successful programs for naming birds, reptiles, mollusks, etc. are also in the works. As a scientist, my natural tendency is to use scientific names, but the modern development of molecular taxonomy is causing havoc in the scientific names as well. Many species and higher taxa are being reassigned based upon molecular analyses. The flux is so dramatic that the use of either kind of names is problematic for the general reader; thus, I am attempting to use common names wherever possible, but I retain scientific names in parentheses at every opportunity in order to keep the cross-referencing to a minimum. For consistency, I will do the same for the uncommon animals, but I will simply use common names for the commonly known animals, e.g., American Alligator, Bison, Eastern Meadowlark and Whooping Crane (for animals, see Appendices 10-12).

Chapter 8
Biota: Plants

"April 16, 1990 – approximately 60 Calopogon barbatus (= *Calopogon oklahomensis* Goldman 1995: Bearded grass-pink orchids) *blooming in south Fenton Prairie in burned area."* (M. Vidrine, field notes (1990)).

"*The landscape of any farm is the owner's portrait of himself.*' (Aldo Leopold 1949).

"*The power of the prairie is vested in its roots.*" (Neil Diboll, Prairie Nursery sales catalog 2003).

The focus of the chapter is the development of a flora of the Cajun Prairie. Three early visitors to Cajun Prairie, C. C. Robin in 1802 to 1805 (Robin 1807), Darby (1816) and Samuel Lockett in 1868 to 1872 (Lockett 1969), listed several species observed in this grassland during their travels. They focused heavily on trees and the potential for lumber, I suspect. The objective of our studies was to thoroughly document the vascular flora of the remnant Cajun Prairie strips. This list of plants includes not only natives but also exotics. Clair Brown (1972) provided preliminary lists for a flora, but the current flora is much more detailed. Smeins et al. (1992) mentioned the Cajun Prairie, but they focused on the coastal prairies of Texas. Vidrine et al. (1995) presented a general flora for the Cajun Prairie Restoration Project in Eunice, a garden of prairie plants that had survived 6 years of the restoration process. Larry Allain (Allain et al. 1999 and 2000) provided a colorful flora of the Cajun Prairie/Coastal Prairies. Grace et al. (2000a) provided the first flora for a remnant prairie in the Vermilion Prairie known as the Florence Hunting Club prairie. Allen et al. (2001) provided the first flora for the entire Cajun Prairie. Allain et al. (2006) and Allain (2007) provided an even more detailed flora complete with information on habitat preferences. The latter flora is also important in assisting prairie preserve managers and prairie restoration ecologists in protecting and creating prairie habitat. Plant diversity and plant associations can now come into focus and receive the study necessary to understand how the prairie ecosystem worked and works in Louisiana. Plant names will generally follow the PLANTS database (USDA).

Biodiversity

One of the central themes of ecology is biodiversity. Biodiversity is one of the primary attributes of the prairie. Biodiversity includes the study of 'indicator species,' 'species richness,' 'species evenness,' 'keystone species' and lists of species. The sheer numbers of different kinds of living things that enable the prairie habitat to develop are of great interest. But measuring this diversity is more complex than just listing the species; thus it can become a real exercise in mathematics to attempt to describe the diversity of a place. However, my goal is to be descriptive and not overly mathematical in this discussion.

The first aspect of diversity is the identification of 'indicator species.' A single species contributes to the 'species richness' by its sheer presence, and its contribution is no greater than any other species to an index; however, if that species is encountered in specific communities or if that species is unique to one community or habitat, the significance of that species is obvious, and it is called an 'indicator species.' Indicator species of prairie included several of the grasses, including Big bluestem, Little bluestem, Yellow Indiangrass, Switchgrass and Eastern gama grass. Also many other plants are indicator species as are some animals like Bison. A rare species or a species that is routinely abundant in a community can be an indicator species. Thus routine presence and abundance taken together or separately may indicate that a species is useful to indicate a specific habitat. In our studies, we rely heavily on 'indicator species' in order to evaluate prairie remnants and restorations, but the indices of species richness and species evenness are also useful, however, mathematical ecologists rely on them more than I have.

Biodiversity is usually formally measured using a variety of mathematical indices, which emphasize either species richness or species evenness or aspects of both of these generalized features. Two commonly used indices are Simpson's Index and the Shannon Index. The two measures (richness and evenness) are inversely related, but they can be used to describe communities or to examine the role and relative importance of a species in the community.

The second aspect of biodiversity, 'species richness,' is a measure of simply the number of different kinds of living things (species) in a community in a habitat. The prairie is a community of plants with associated bacteria, protists, fungi, and animals. While no single plant indicates prairie, the quality of the prairie grows as the number of indicator species present grows (this number relates to species richness). Because each plant species has a collection of associates, it logically follows that the addition of a single plant species to the community brings a collection of species among the other groups of organisms, including pollinators, herbivores, parasites and more. As a result, this addition of a single species can dramatically affect species richness.

The third aspect of biodiversity is 'species evenness,' a measure that permits describing a community based upon the relative abundance of specific organisms. A single species contributes potentially much more to an index of species evenness by its increased abundance as compared to other species in the community; thus a species, if abundant, can dramatically increase the species evenness. As species evenness increases, species richness decreases. The Cajun Prairie contained a lot of Little bluestem (*Schizacharium scoparium*) and Yellow Indiangrass (*Sorghastrum nutans*) in higher/drier areas, with Switchgrass (*Panicum virgatum*) and Eastern gama grass (*Tripsacum dactyloides*) in the lower/wetter areas (Smeins et al. 1992 and Allain et al. 2006). Thus these species increase evenness values in their respective areas as they are rather evenly distributed, abundant and they displace a variety of additional species. As such, the Cajun Prairie is ecologically described as a Little bluestem/Yellow Indiangrass prairie in dry areas, and a Switchgrass/Eastern gama grass prairie in the wetter areas. Disturbance, often referred to as catastrophe, usually increased 'species richness.' Natural catastrophes included herds of Bison literally plowing the sod with their hoofs, floods and lightning fires. Generally, tallgrass prairies are predominated by Big bluestem (*Andropogon gerardii*), but the Cajun Prairie remnants did not readily support this idea. However, stories as told by our parents speak of travelers not being able to see one another on horseback except for their hats—such tales indicate that the Cajun Prairie was a tallgrass prairie in some places at least. Could it be that the hundreds of years of cattle grazing caused a dramatic decline in Big bluestem? In fact, we don't know, but this is part of the research. Both measures of diversity permit evaluation and ranking of strips of habitat and facilitate description of these habitats.

A fourth aspect of biodiversity is the identification of 'keystone species.' Keystone species are specifically identified as having ecological roles in a community that are essential in maintaining the community and its habitat. Some authors, e.g. Manning 1995, refer to these species as 'capstone species.' The architectural keystone is a stone laid down by a mason in an arch—the keystone has an inordinate importance and must be placed correctly in order for the remaining stones to be held in place. The loss of the keystone results in the collapse of the arch—in a habitat, the loss of a keystone species dramatically changes the community and potentially the habitat. What are the keystone species of prairie? Big bluestem is an excellent example of a plant species; Little bluestem is also a keystone species. Predators like Red Wolves and rattlesnakes are considered keystones in many habitats. Large grazers like Bison and White-tailed deer are also candidates. There are many unlikely candidates that may not be appreciated as keystones until elaborate research unveils their roles, e.g., earthworms, mussels, prairie chickens and many more. Exotic plants and animals can quickly become keystone species in disturbed habitats and completely change the habitat and its community. Cattle and Chinese tallow tree are two important examples in the Cajun Prairie; others include Fire Ants, dogs, cats and exotic plants. Humans, of course, usually become keystone species wherever they occur—they have the ability however to chose to not be so. The importance of a flora and a fauna for the Cajun Prairie becomes obvious—it is essential in order to clearly identify indicator species and keystone species as well as 'native species' and 'exotic species.' An abiotic example that would resemble a 'keystone' effect would be fire—fire is essential to maintain a habitat/community like prairie. The loss of fire is literally the loss of prairie—the ecosystem dies.

The fifth and final component of biodiversity studies that I want to discuss is the list of species. Plant lists make up the 'flora,' while animal lists are 'faunal' lists. Literally a thousand species of plants form the flora and are found associated with the Cajun Prairie; more than 600 are considered 'prairie plants,' and 253 are 'indicator species' of prairie with Coefficients of Conservatism equal to 5 or greater (Allen et al. 2001, 2006, Allain 2007 and Appendix 6). Thousands of animals constitute the faunal list for the Cajun Prairie. Like the partial flora, a partial faunal list is provided in Appendices 11 and 12, but the discussion of animals will continue in Chapter 9.

Plants of the Cajun Prairie

Plants are used to describe and allocate biomes/habitats. A flora is essential in order to identify and classify not only a habitat but also individual pieces of the habitat. In Chapter 7, we saw habitat classified by geology, soils, agricultural crops, and now plants will be used to classify habitats.

When most people of my parents' generation were asked about the prairie, they referred to the grasses as *paille jaune* (French for 'yellow straw'). The entire prairie in the winter turns to a yellow-golden sea of grass—this is evident in some meadows, where the prairie grasses remain today and in our restoration sites. This is literally the beginning of a flora. The present flora follows Allain et al. (2006) and is presented along with an appendix with prairie plants with Coefficients of Conservatism (C) of 4 or greater (Appendix 6).

Allain et al. (2006) provided the following description:

> *"Little bluestem (Schizachyrium scoparium) and Indiangrass (Sorghastrum nutans) dominate the plant community with switchgrass (Panicum virgatum) and eastern gamagrass (Tripsacum dactyloides) becoming dominant in low areas. Big bluestem (Andropogon gerardii) is present in moist, sandy soils but it is less common than in Midwestern tallgrass prairie (Smeins et al. 1992). Species such as brown-seed paspalum (Paspalum plicatulum), slender bluestem (Schizachyrium tenerum) and ashy sunflower (Helianthus mollis) are restricted to upper coastal prairie and are not found elsewhere in tallgrass prairie. A varied forb component fills out this grass matrix, creating one of North America's most diverse communities."*

The authors developed a Floristic Quality Assessment (FQA) system for the Cajun Prairie. It provides a rather deep view of the flora of the Cajun Prairie based upon our experience on the remnant strips of prairie.

"Native species were assigned a coefficient of conservatism (C) on a scale of 0 – 10 based upon their degree of fidelity to remnant coastal prairies and their tolerance of disturbance. Those species with high community fidelity are limited in the number of communities in which they occur and are considered indicator species. The C value represents the authors' confidence that a plant was collected in a high quality prairie remnant. A species with a C of 10 (conservative species) indicates that we are 100% certain that, if it came from southwest Louisiana, it was growing on a coastal prairie. A species with a C of 0 (early succession or disturbance species) indicates no confidence that a plant came from a remnant prairie."

"In addition to community fidelity, plant species assigned high C values are also considered to be intolerant of disturbance. A C value of 0 – 1 indicates an early succession species adapted to severe disturbance, where as a C value of 8 – 10 indicates a species from a community that has little history of disturbance. A species with a C value of 5 represents a plant from a remnant natural community that may be severely degraded."

The list from Allain et al. (2006) also identified exotic (non-native) species with negative coefficients, with a C of -3 indicating a plant species that *"both invade prairie and displace native species."*

"Aggressive exotics, like Chinese tallow tree (Triadica (formerly Sapium) sebifera) (locally called Chicken tree), can dominate a site and replace a prairie remnant with a forest monoculture in as little as 10 years. Such invasive exotics were assigned a C value of "-3.""

"Conservative species (i. e., "prairie species") with C values of 5 – 10 numbered 254 in this study, which exceeds the number found in other tallgrass prairies (Smeins and others 1992). Much of this diversity can be attributed to the great number of interstitial species. Genera found in true prairie are represented in coastal prairie by more species as exemplified by Paspalum, Panicum, Agalinis, Asclepias, Eragrostis, and a great variety of sedges (Cyperaceae). Because so little coastal prairie remains in Louisiana, it is probable that species richness was even higher. Certainly numerous populations with distinctive genetic composition have been lost."

"High numbers of mid-succession species (30%) with a C of 4 – 6 may be due to the variety and complexity of seral communities found in coastal prairie. The coastal prairie is also an ecotone and shares species with many bordering community types, such as oak-hickory forest, longleaf pine savanna, flooded swamp forests, and salt and fresh

marsh. It is difficult to determine whether some species were a common historical component of coastal prairie or whether they have immigrated from adjacent communities. Complicating these decisions is the degraded state of Louisiana's coastal prairie. Additionally, most remnants occur along railroad tracks which function as corridors for seed dispersal and plant migration. It is often not clear whether rare species occurring in railroad remnants are adventive or remnant individuals..."

"Of the species identified from prairie remnants and restoration sites, 13% were non-natives (exotics). Most nonnative species are uncommon, and few can displace native species."

"Species richness varied from 117 species at Fenton to 77 species at Midland and 94 species at Florence...Both the Fenton and Midland remnants have more species in conservative classes (C = 5 – 10). However, Midland had fewer disturbance species than Fenton, which contained limited areas of disturbance. Floristic quality was highest at Midland where there appeared to be little evidence of disturbance."

"However, Fenton has a higher AFQI (Adjusted Floristic Quality Assessment value – a new technique outlined in this paper, which includes an additional variable that takes into consideration exotics and a quantitative value of their invasive potential), owing to more overall conservative species and greater size. It could be argued that if all other variables are equal (rare plants, threat of disturbance, etc.), the Fenton remnant has greater conservation value."

"Floristic quality analysis is a versatile, relatively easy to use and repeatable system for quantifying habitat quality. It will be of value to restorationists, managers, and others involved in assessing the integrity of prairie remnants and restorations. Further, it will be useful in developing management strategies based on these criteria."

The flora of the Cajun Prairie is distinctive, but not that different from the Midwestern prairies (Allen et al. 2001 and Appendices 23 and 24) and coastal prairies in Texas (Tveten and Tveten 1993). Traditionally, a flora was prepared for each parish in Louisiana, e.g., Allen Parish (Allen 1984). A flora for the state of Louisiana was prepared by Thomas and Allen (1993, 1996 and 1998), as well as books on specific groups, e.g., grasses (Allen 1992 and Allen et al. 2004), trees (Allen et al. 2002), and edible plants (Allen et al. 2005 and Fontenot 1992). Regional plant guides exist that list plants in the prairie region; two recent books are good examples (Nelson 2006 and Neyland 2009). Plant names, both scientific and common names, can be researched at the USDA PLANTS database online. The database is attempting to simplify searches online for plants by formalizing common names and scientific names—where common names are regional and scientific names are in constant flux as new discoveries are made.

These are indispensable resources for appreciating and understanding the flora of Louisiana.

Lauren Post (1962) used the phrase "Cajun prairies" on several occasions, but he does not refer to the entire region as the 'Cajun Prairie.' He referred to each individual prairie, e.g., Prairie Faquetaique or Prairie Mamou, as a separate location. This has apparently cultural and possibly ecological significance. Allen and Vidrine (1990b) examined the land surveyors' notes in the courthouse in Opelousas for St. Landry Parish. The surveyors noted trees, which could be used to create a flora for the region; however, the majority of the entries were labeled 'stake in ground' signaling that no trees were around. In other words, prairie could easily be outlined. We observed that not a single tree was evident in the current city limits of towns like Eunice and Crowley based on the surveyors' notes (Allen and Vidrine 1990b).

Allen and Vidrine (1989) used the phrase "Cajun Prairie" as a descriptor of the entire region, i.e. all the individual prairies in southwestern Louisiana combined. While this is useful, there may be drawbacks. When we restored prairie in Eunice, we collected seeds from all the prairies and not from just Prairie Faquetaique in which Eunice is located. As a result, the restored prairies contain a great mix of the genetics of the plants of varied prairies.

As I am preparing this book and examining the thousands of photographs/slides that I have, it becomes apparent that the remnants (from the varied prairies) are each different in appearance based on the plant inventories and bloom seasons. While I think this is apparent, a thorough evaluation of the data that we have would be necessary to substantiate this. Nonetheless, the restored prairies serve as reservoirs for the variety of plants that once occurred in the Cajun Prairie. The first Plate shows some of the colorful blooming plants, while Plates 30 and 31 show our restoration efforts in Eunice and at my home. Plates 2-28 depict views of the remnant prairies in the Cajun prairie in the 1980s and early 1990s. The remnant prairies were re-evaluated in 2008 (see Appendix 22); Plate 29 depicts views of the remnant prairies in 2008. The blooming seasons of many of these plants are graphed in Appendix 9, while Appendix 10 provides a list of factors affecting blooming times.

One of the major goals of this project is to provide the reader with a view of the past. This view would be a naturalistic one (Appendix 5). The prairie now essentially gone was a thing of beauty and ecological significance. As a foundation for the future study of this environment, it is essential that a baseline of the natural area be created. Unlike many of the earlier chroniclers of the natural prairie whose emphasis was on how to make the prairies better and more profitable, my purpose is appreciate them for what they were and what they provided to the then and future inhabitants of the region. I also extend my thoughts to the prairie ecosystem, such that the so-called prairie is not just the rolling hills with grasses and forbs but also the marshes, lakes, and rivers and further even the gallery forests that line the streams and lakes and marshes of the region.

While I do not wish to diminish the economical aspect of this ecosystem, as this would be folly, I want to extend the overall perspective of the economics of the ecosystem. There are many values that have gone unnoticed and even demonized in regards to this prairie. Cajun Prairie intergrades almost imperceptibly into fresh marsh to the south (Grace et al. 2000a and b) and into flat pine savannahs to the north (Allen 2003).

Smeins et al. (1992) described the coastal prairies in Louisiana and Texas. They spend little discussion on Louisiana; however, they do discuss general ecology, geology and soils. This discussion is excerpted as follows:

> *"Abundant native ungulate populations existed at the time of settlement by Europeans, and by the early 1800's populations of feral cattle and horses were already large. The first scientific observers (at the end of the 19th century) saw the results of over 100 years of overgrazing. Consequently, they misinterpreted the potential vegetation of the area. Later investigations of remnant grasslands and well-managed ranches suggested that the coastal prairie be identified as the "Schizachyrium scoparium—Sorghastrum nutans (Little bluestem—Yellow Indiangrass) association.""*
>
> *"Estimates suggest that less than 1% of the grassland remains in a relatively pristine condition. Although historical changes greatly altered the character of the region, the contemporary acceleration of change (including land drainage, damming of rivers, overgrazing, cultivation for rice, soybeans, cotton, sorghum and sown pasture, energy development, and urbanization) may soon destroy its last vestiges."*
>
> *"The modern flora of the climax grassland communities of the upper coastal prairie consists of about 200 species in 40 families......" "Thus, the modern flora of the upper coastal prairie clearly reflects a close relationship with the flora of eastern and central North America, but little affinity for the western flora."*
>
> *"The grasslands of the coastal-prairie region are similar in their main features to those of the true prairie (tall-grass prairie). Schizachyrium scoparium is the primary dominant in all grasslands of Texas and also throughout the true prairie. Sorghastrum nutans is more important in Texas grasslands, and particularly in the coastal prairie. Therefore, an appropriate designation for the coastal prairie would seem to be the Schizachyrium scoparium—Sorghastrum nutans association."*
>
> *"Andropogon gerardii appears to be a less important component in grassland of the coastal-prairie region, as compared to other Texas prairies and the true prairie."*
>
> *"An important distinguishing floristic characteristic of the*

grasslands of the coastal-prairie region appears to be its variety of interstitial graminoids. The diversity of Paspalum species... and species of Dichanthelium, as well as many genera and species of sedges contribute to its uniqueness. Another unique aspect is the great diversity of seral species and seral pathways. Thus, whereas climax grasslands of the coastal and true prairies are quite similar, seral communities appear to be different and more complex in the coastal prairie, as compared to the true prairie."

As a result, we can reconstruct an image of the prairie based upon what little eyewitness information that we have and based upon the remnant prairies that we have documented. Our list of plants and animals is restricted by our documentation. We are also able to extrapolate from the surrounding habitats to some extent, but we are hampered by simple rumor. For example, *marais* and *platins* are considered as natural wet areas in the prairie grassland; however, a number of authors contend that these were created and maintained by cattle milling around in an area. A few of these wet areas remain along railroad tracts with the prairie remnants, and as such, we can speculate that the flora in these remaining wet areas is similar to what was in the *marais* and *platins*, whether they are natural or not. Dry prairies between Eunice and Opelousas were described as covered with grasses the height of a man on horseback in summer and as so wet that wagons were buried to the axle in winter. It remains obvious that the prairie was a mosaic of wet and dry prairies, and images of these prairies are mired in the context of events that may or may not be conducive to reasonable recollection.

The Cajun Prairie, like the Midwestern prairies, can be described generally by looking at a few plants that singularly and collectively indicate the prairie proper. Harris (2003) had a nice introduction to the Midwestern prairies, and she included descriptions and great write-ups on a number of 'very-prairie' plants (indicator if not keystone species), including the following: Big bluestem, Little bluestem, Switchgrass, Yellow Indiangrass, Purple cone flower, blazing stars, gayfeathers, Black-eyed susans, Compass plants, Button snakeroot, Butterfly-weed, bergamonts, wild-indigoes, and goldenrods. Our wildflower article (Allen and Vidrine 1989) provided a similar but brief introduction to the same plants as the floral assemblage of the Cajun Prairie dry prairies. While the plants listed here are indicative of what most prairie enthusiasts think of as prairie plants, the prairie ecosystem contains wet areas (often called potholes or seeps in the Midwest) with very important components to be considered as part of the prairie flora.

Wet prairies in the Cajun Prairie contained a different assemblage of plants including: Fragrant spider-lily, Eastern blue-star, Fringed sneezeweed, Fewflower milkweed, Eastern gama grass, Texas brown-eyed Susans, Crimson-eyed rose mallow, Zigzag iris, White-top sedge, Whorled milkweed, Indian plantain and Maryland meadowbeauty. Seeps contained orchids, sundews, sunbonnets, candyroots, drumheads and colic-roots (Allen et al. 1989). Larry Allain (Allain et al. 2006) listed these plants using Wetland Classification as OBL (obligate) and FACW (facultative) in wetlands.

The study of local plant populations in the Cajun Prairie provided some very interesting observations including:

1. Butterfly-weed locally tolerates clay and some wetness—not so anywhere else in the state.
2. Different color stems or leaves in Big bluestem, Little bluestem, Eastern gama grass and Yellow Indiangrass are present.
3. Flower color variations in Fewflower milkweed (scarlet, red, orange, yellow and bicolor), Downy phlox (white, blue, pink and red), Eastern blue-star (white, blue and purple) and Crimson-eyed rose mallow (white, pink and rose pink).
4. Plant species with white flowered variants: Eastern blue-star, Dense blazing-star, Bergamont, Downy phlox, Finger false dragonhead and Blue sage (Louisiana populations are usually white unlike those of Texas).
5. A variety of species identified by time of blooming or habitat preference.
6. Plants with a variety of pollinators.
7. Plants that serve as hosts for butterflies (Allen and Vidrine 1990a, Fontenot 1992, Vidrine and Hazelton-Robichaux 2003 and Ross 1994) and other insects (Buchmann and Nabhan 1996).
8. A phenology for blooming plants in the Cajun Prairie.
9. Successional notes on varied populations in a variety of habitats.
10. Propagation notes on varied populations for container and garden growth.

Phenology

A flowering phenology shows which plants are blooming during specific days, weeks, months and seasons for a given year. Our first paper (Allen and Vidrine 1989) showed a phenology for Cajun Prairie plants in 2-month intervals. A more thorough phenology is required for plants in the prairie, but many factors alter the onset of flowering, e.g., burning time, temperature and water (see Appendix 10). We prepared a more detailed phenology for blooming in remnant prairies in 1987 (Allen and Vidrine 2003) and for restored prairie in 1995 (Vidrine et al. 1995). I have kept data each year at the Cajun Prairie Gardens, so a more detailed phenology is available (Appendix 9). In the 2003 paper, a comparison between the phenology of Cajun Prairie species and Midwestern species, where the same species occurred in both locations, was made. It is obvious that the

Midwestern plants begin blooming later, in some cases 3 months later, and terminate blooming earlier, in some cases 3 months earlier, than the Cajun Prairie plants. The Cajun Prairie bloom season begins in late February and continues until the end of November. The Midwest blooming period usually begins in late April and apparently ends in October among the species compared. With almost 10 months as a blooming season in the Cajun Prairie as compared to the 7 month blooming season in the Midwest, significant differences were obvious.

Watching prairie plants bloom provides an intricate clock, month by month, week by week, time is evident, while seasons are obvious even to the novice. However, there is always a caveat. If you burn the prairie at any time in the year other than the dead of winter, the clock can be reset causing some early spring plants to rebloom. The stress this puts on plants in the ecosystem is not well understood. In fact, such burns can effectively serve as herbicidal control methods for controlling woody vegetation like brambles of *Rubus* spp. Routine burning, year after year, is purported to result in the increase in the sheer volume of grass species/stems in the prairie (raising species evenness), whereas reducing the number of burns enhances the number of kinds of forbs/wildflowers in the landscape (raising species richness). A balance is necessary as weedy species can successfully colonize a restoration or a remnant if fire is minimized as a management tool. A significant amount of research needs to be done in Cajun Prairie dealing with varied burn regimens in order to evaluate the impacts of such variety upon the habitat.

Succession in the Prairie Ecosystem

Plant communities (and other communities, e.g., protists, fungi, animals, and even bacteria) undergo a process called ecological succession. The process, unlike phenology, may take years to complete. In a specific region (habitat), plant associations essentially 'grow-up' into a specific assemblage after years of development. The mature assemblage is called the 'climax sere or climax community.' In the prairie, the grasses and forbs form the climax community, but this community is maintained in nature by fire (a natural fire ecosystem). Other climax communities are also maintained by varied abiotic factors, e.g., rainfall or the lack of it (desert), elevation, day length and seasonal extremes in temperature. Some climax communities occur as a result of a biotic limiting factor, e.g., presence or absence of animals (pollinators), presence or absence of predators or parasites (pine beetles eliminating pine forests), and presence or absence of other plants or fungi or bacteria. A famous experiment where Yellow Indiangrass is grown with verses without fungi associated with its roots (mycorrhizae) clearly shows 6 feet tall plants verses 2 foot tall plants, respectively. Charles Allen discussed this for prairies (see Appendix 16).

Charles Allen and Sara Thames (2007) described the succession of plant species in four Cajun Prairie restoration projects.

> *"The early successional dominants in the Cajun Prairie habitat are mainly annuals with variation in species composition tied to moisture, location, and history of agricultural use. Most annuals disappear in less than two years but goatweed (Croton capitatus) often persists for three or four years. The second stage of succession includes three groups of perennial species: native non-prairie, introduced, and Cajun Prairie. With time, the Cajun Prairie species increase in abundance with a corresponding decrease in abundance of the other two groups."*

They provided lists of species commonly found in the first two stages of succession in restoration projects. Vidrine et al (2003 and 2008b) described succession in the Cajun Prairie Gardens, but with a different approach. The first paper was simply descriptive, but the latter paper compared restored prairie to adjacent mowed lawn. The ability of Cajun Prairie plants to grow and colonize a plot under a restoration regime of fire is impressive.

Edible Plants

Plants as food items received a large amount of attention during pre-settlement times by the Native Americans and by the early settlers (Allen et al. 2005). Anything edible was monitored and noted for future use. The search for edible plants is Charles Allen's special passion—he often eats ('grazes') while he walks through the forest, and he routinely give talks on wild plants as edibles and teas. Wild plants as edibles remain a popular topic (Kindscher 1987, Allen et al. 2005, Fontenot 1992 and Harris 2003). Native Americans developed an agricultural system for growing native plants: The 'Three Sisters agriculture system' (Harris (2003). Many prairie plants are edible and useful for other purposes.

Bartram (1792) traveled through Louisiana and wrote of strawberries, "*On the more fertile borders of the plains, adjoining the surrounding forests, are Sideroxylon, Pyrus coronaria, and Strawberry vines (Fragaria) but no fruit on them; the inhabitants assured me they bore fruit in their season, very large, of a fine red colour, and delicious and fragrant.*" Wild strawberry is now rare in Louisiana and is on the Louisiana Natural Heritage Program rare plant list (MacRoberts et al. 1997). It apparently was common in pre-settlement times. Peter Custis (of the famous Freeman and Custis Expedition of 1806 up the Red River—much like Lewis and Clark went up the Missouri River at the request of Thomas Jefferson following the Louisiana Purchase in search for the Northwest Passage to the Pacific Ocean) found it in the prairies along the Red River in 1806 (Flores 1984). Le Page Du Pratz (Du Pratz 1758) was so impressed by its abundance in the state that he wrote at length on the matter no less than five times in his history, telling us in each case that the prairies were literally red with fruit. Clair Brown (in 1942) made a special search for wild strawberries (MacRoberts et al. 1997). However, there is the possibility that some bias was the basis for these strawberry accounts. Many early commentators emphasized the more positive aspects of the landscape. Strawberries were also important in the Midwestern prairies (Kindscher 1987 and 1994). The Indian strawberry (*Duchesnea indica*) is introduced in the Cajun Prairie region and found in yards of homes and in open woodlands (Allen et al. 2005). The Virginia strawberry (*Fragaria virginiana*) is native and may have occurred in the Cajun Prairie (Thomas and Allen 1998).

American plum (*Prunus americanus*) was an abundant tree in the Midwestern prairie—reports of wagon trains stopping next to a natural orchard and loading the wagons full of plums were made. They reported eating plums for days after the fateful encounter (Kindscher 1987 and 1994). The American plum occurs in Louisiana, and we had three plum trees along the fencerow at the back of the farm of my youth; however they may have been Chickasaw plums (*Prunus angustifolia*) (Allen et al. 2005). I recall feasting and harvesting them when school let out in May at school yearend. I have tried to grow plums, including the American plum, but aerial spray drift herbicides kills them. Remnant prairies are sprayed from all directions: railroad companies spray along their tracts, the Department of Transportation and Development (DOTD) sprays along ditches and roads, farmers spray fields and local mosquito control districts spray insecticides along roads and ditches, and these chemicals drift over entire landscapes. It is impossible to ascertain the insult upon nature of the wholesale herbicidal drift as airplanes apply herbicides and insecticides. These biocides are so effective that few pollinators occur during the months of April to June in some regions, e. g., the Cajun Prairie Gardens. Both host plants and the insect pollinators are eliminated. Many are surprised when they find out that male mosquitoes are important pollinators—notwithstanding any discussions of biomagnifications into birds and other plant and insect consumers.

Other Plant Topics and Uses

Holmes (1990) provided the Francophone names for a number of plants, mainly trees (Appendix 7). Many of these plants can be used in *pot pourri*, e.g., the mountain-mints, bergamonts, Spotted beebalm, Sweet goldenrod, goldentops (*Euthamia* spp.) and flowers/leaves/fruits of many other plants. Bill Hudson, my contact for the Welsh Prairie remnant, knew a florist who routinely harvested blazing stars from the railroad rights-of-way for making flower arrangements back in the 1980s.

Myriam Stanford reminded me that Chinaberry fruits were used in making soap. These ornamental trees, which grew quickly and provided shade near many Cajun homes, were also famous for producing berries that were used as 'shot' in young boys' crafted 'sling shots.' I recall doing a bit of this myself, as we had a large tree next to our garage and one near the barn. I used to read on top of the tin roof of the barn under the shade of the overhanging limbs.

Sue Forest, former head of the LSUE Library, often made flower arrangements with prairie plants. She also had a knack for making beautiful blankets with flower motifs. She donated a prairie wildflower blanket to the Cajun Prairie Habitat Preservation Society and surprised Gail and me with a 'Louisiana Iris' blanket after visiting the Cajun Prairie Gardens during the April splash of iris in the marsh.

What about *Castilleja ludoviciana*—Louisiana paintbrush? This was considered a rare, Cajun Prairie plant, possibly unique. It remains rare in Louisiana, but it has been placed into synonomy with a widespread species across the eastern United States (*Castilleja coccinea* (L.) Spreng.—the Scarlet Indian paintbrush) (Charles Allen, pers. Comm. 2008).

Lawn grasses, including Bermudagrass, St. Augustine grass and Bahia grass, are exotics. Their pollens are powerful allergens. As the number of cases of severe allergies grew, it became apparent that many of these cases are linked with exotic grasses. Appendix 19 lists some of the more obvious exotic plants in the prairie region of Louisiana.

As an overview of plants used by Cajuns and locals, Vidrine et al. (2001c) provided a discussion:

> *"Over recent years, much interest has been manifested regarding potential traditional uses of native southwestern Louisiana prairie plant species by Cajun settlers. Alas, once again, documentation regarding this particular folkloric tradition is mostly lacking, limited mostly to unpublished, recorded conversations between field researchers and their lifelong Cajun Prairie informants. From these as-of-yet informal documents, it is quite apparent that Cajuns utilized a substantial number of native prairie plant species for both food and medicine. Regarding the traditional Cajun healing arts in particular, it is very important to begin any discussion thereof by first distinguishing between personages and processes involved. While native medicinal herb use seemed fairly ubiquitous throughout the prairie region, it appears that in most cases, no particular personages arose within each community whom might be identified as specialized practitioners; and in the relatively few cases which have proven to be exceptions to this statement, the practitioner was most likely a person of full or at least mixed Indian blood. The explanation is simple: Local Native Americans had been regularly utilizing native plants for food and medicine for untold generations prior to the appearance of European settlers in the region. Once the Cajun settlers had come into contact with these indigenous peoples, who themselves were obviously willing to assist in treating the various maladies with which the Cajuns*

experienced, the Cajuns probably saw no use in "reinventing the wheel." Over time, however, it is apparent that a substantial number of Cajun households collectively and routinely utilized herbal cures shown to them by the Indians."

"On the other hand, there arose very special Cajun personages known as traiteurs ("treaters or healers") who cured via a particular form of "faith healing." For the most part, these healing sessions were accomplished very quickly, with no extraneous gestures or spoken words or prayers. The traiteur simply touches the affected part of his/her patient, and evokes a silent prayer. No goods or money are offered or accepted, and the words of the silent prayer are never be revealed by the traiteur. In all cases, the vocation of traiteur is passed along from an older to a younger family member; never to anyone from outside of the family. The "receiving" family member is sworn to utmost secrecy, and can never reveal the exact nature of the process - not even to fellow family members. Traiteurs are obligated to treat anyone and everyone who requests their services, and can never accept good or money in return. A substantial number of authentic traiteurs continue to practice throughout the region today."

"Thus, the practice of the Cajun traiteur stands in obvious contrast to that of medicinal herb practitioners, gris gris practitioners, or even Protestant faith healers. By necessity, traiteurs were not (and are still not) forthcoming with even the most anecdotal background information regarding their vocation, it is thought that the practice is probably Roman Catholic at its root. Traditionally, the vast majority of Cajuns were devout Roman Catholics, and even today the majority of them continue to follow this tradition."

"Regarding traditional medicinal and culinary native plant usage by Cajuns, several references might be consulted for more detailed information: Holmes (1990), Fontenot (1992), and Vidrine et al. (1995). Regarding the translation of Cajun French names of various native prairie plants into proper botanical nomenclature, Holmes (1990) and Reese (1992) are good references. In addition, several excellent listings have been "passed along" throughout the region over the past few decades. Unfortunately, most of these do not possess even the most rudimentary data (author, date, names and addresses of informants) for purposes of formal documentation. An exception exists in U.S. Fish and Wildlife Service field biologist John J. Lynch's "Glossary of Local, English, and Scientific Plant Names" which he included in a 1942 report entitled <u>*Louisiana's State Wildlife Refuges*</u> *(Lynch 1942)."*

"As further testimony to the "staying power" inherent within the Cajun prairie culture, it should be noted that perhaps as many as two

dozen herbal remedies continue to find somewhat ubiquitous usage today. Of these, some of the better known include teas of eastern coralbean (Erythrina herbacea; locally called mamou) for upper respiratory infections, red bay (Persea barbonia/palustris; called laurier) for sinuses, gastric distress, and to mask the unpleasant tastes of other teas, and groundsel bush (Baccharis halimifolia/angustifolia; called manglier) for fevers. Toothache tree (Zanthoxylum clava-herculis; also commonly known as prickly ash) and toothache grass (Ctenium aromaticum) are now used more for entertainment of audiences visiting the prairie, but it is easy to demonstrate the immediate oral numbing sensation produced by the plants." Recently I learned that American elderberry (Sambucus nigra ssp. canadensis) is used in numerous herbal remedies (George Broussard, personal communication 2009).

"Other plants were used by the Cajuns for varied purposes. False indigoes or locally named pock-pock plants (Baptisia spp.) were not only used as dyes for cloth but also as dye for eggs in order to continue the tradition of pocking colored eggs at Easter. Many of the plants were edible including berries, cherries, nuts, "Indian potatoes," etc. The prairie grasses and forbs both as green pasture and as hay provided the forage for their cattle and other grazing animals. Other plants gain use in religious activities as palms for Palm Sunday. Spanish moss (Tillandsia usneoides) was especially important for filling mattresses. Finally the plants signaled the frontiersman of potential danger; this is best exemplified by the plant called lizard's tail (Saururus cernuus), a common and obvious plant in marshy areas where snakes and alligators might lay in wait. The common name of this plant "hérbe à malo" literally means "plant growing where harm lies."

Plant Essential Oils as Antimicrobials

Jameel Al-dujaili and his students have collected and evaluated the essential oils of varied Cajun Prairie plants in an effort to assess the antimicrobial actions of these oils (Al-dujaili and Vidrine 2008). Vidrine et al. (2008a) reported tests on the antimicrobial properties of essential oils of prairie plants:

"Prairie plants have a history of use in a large variety of ways. One of the most intriguing ways is the use of plants as medicine. Escherichia coli O157:H7, a common pathogen found in meat products, and Listeria monocytogenes, a common pathogen found in vegetable salads, were used as target species for the study of antimicrobial activity of several prairie plants (native Cajun Prairie plants) and selected exotic plants (Old World plants). Antimicrobial properties of varied plants (often called herbs) and

Plate 1. Top: Orchids: Bearded grass-pink, Snowy orchid, Giant orchid and Spring ladies'-tresses. Middle: Fewflower milkweed, Butterfly-weed, Woods poppymallow and Dense blazing-star. Bottom: Finger false dragonhead, Purple cone flower and Coralbean.

Plate 2. Top: Frey prairie remnant. Plaquemine Prairie. April 1988. Lance-leaf tickseed (yellow),Virginia spiderwort (blue) and Crowpoison (white). Bottom: Frey prairie remnant. Plaquemine Prairie. August 1987. Ashy sunflower (yellow), Button snakeroot (white in foreground) and Indian plantain (white in background).

Plate 3. Top: Estherwood prairie remnant. Mermentau Prairie. Wet prairie. April 1990. Fragrant spider-lily. Bottom: Estherwood prairie remnant. Mermentau Prairie. Dry prairie. August 1, 1987. Compass plant (yellow) and Dense blazing-star (red).

Plate 4. Top: Estherwood prairie remnant. Mermentau Prairie. Dry prairie. August 2, 1996. Compass plant (yellow) and Dense blazing-star (red). Bottom: Estherwood prairie remnant. Mermentau Prairie. Dry prairie. August 2, 1996. Compass plant (yellow) and Dense blazing-star (red).

Plate 5. Top: Estherwood prairie remnant. Mermentau Prairie. August 1, 1997. Compass plant (yellow) and Dense blazing-star (red). Bottom: Estherwood prairie remnant . Mermentau Prairie. September 7, 1987. Compass plant (yellow), Scaley blazing-star (red), Shiny goldenrod (yellow flat top) and Justiceweed (white).

Plate 6. Top: East Midland prairie remnant. Mermentau Prairie. August 1987. Slender rosinweed (yellow), Dense blazing-star (red) and Lindheimer's beeblossom (white). Bottom: West Midland prairie remnant . Mermentau Prairie. Dry prairie. March 1988. Narrow-leaf springbeauty and Crowpoison (white), Eastern yellow stargrass (yellow) and Eastern blue-eyed grass (blue).

Plate 7. Top: West Midland prairie remnant. Mermentau Prairie. Dry prairie. April 1988. Yellow wild-indigo (yellow). Bottom: West Midland prairie remnant. Mermentau Prairie. Dry prairie. August 1987. Dense blazing-star (red), Flowering spurge (white) and Slender rosinweed (yellow).

Plate 8. Top: West Midland prairie remnant. Mermentau Prairie. Dry prairie. September 1986. Dense blazing-star (seedheads in foreground) and Switchgrass (background). Bottom: East Mermentau prairie remnant. Mermentau Prairie. Dry prairie. August 1986. Clump of Switchgrass behind Charles Allen, Compass plant (yellow), Dense blazing-star (red), Lindheimer's beeblossom (white in foreground) and Indian plantain (white in background).

Plate 9. Top: East Mermentau prairie remnant. Mermentau Prairie. Dry prairie. August 1989. Button snakeroot (white), Indian plantain (white), Compass plant (yellow), Dense blazing-star (red) and Shiny golden-rod (yellow). Bottom: East Mermentau prairie remnant. Mermentau Prairie. Dry prairie. August 1990. Compass plant (yellow) and Dense blazing-star (red).

Plate 10. Top: East Mermentau prairie remnant. Mermentau Prairie. Dry prairie. August 1990. Compass plant (yellow) and Dense blazing-star (red). Bottom: East Mermentau prairie remnant. Mermentau Prairie. Dry prairie. February 1990. Little bluestem and Switchgrass.

Plate 11. Top: West Mermentau prairie remnant. Mermentau Prairie. Dry prairie. August 1987. Button snakeroot (white), Dense blazing-star (red) and Black-eyed Susan (yellow). Bottom: East Mermentau prairie remnant. Mermentau Prairie. Dry prairie. August 1989. Button snakeroot (white), Dense blazing-star (red), Ashy sunflower (yellow) and Florida paspalum (foreground).

Plate 12. Top: West Mermentau prairie remnant. Mermentau Prairie. Prairie tour, August 1988. Gail and Daniel Vidrine, Joann Quillman. Button snakeroot (white), Dense blazing-star (red) and Indian plantain (white). Bottom: Welsh prairie remnant. Calcasieu Prairie. Wet prairie in distance. May 1991. Texas brown-eyed Susan (yellow in background) and Slender rosinweed (yellow in foreground).

Plate 13. Top: Welsh prairie remnant. Calcasieu Prairie. Wet prairie. May 1991. Texas brown-eyed Susan (yellow in background) and Slender rosinweed (yellow in foreground). Bottom: Welsh prairie remnant. Calcasieu Prairie. Wet prairie. July 1991. Ashy sunflower (yellow) and Indian plantain (white).

Plate 14. Top: Welsh prairie remnant. Calcasieu Prairie. September 1993. Bruno Borsari. Kansas gayfeather (red) and Ashy sunflower (yellow). Bottom: Welsh prairie remnant. Calcasieu Prairie. Dry prairie. May 1991. Purple cone flower (red), Woods poppymallow (scarlet), Multibloom hoarypea (white) and Black-eyed Susan (yellow).

Plate 15. Top. North Iowa prairie remnant. Calcasieu Prairie. February 1988. After annual burn. Bottom: North Iowa prairie remnant. Calcasieu Prairie. July 1989. Finger false dragonhead (pink).

Plate 16. Top: North Iowa prairie remnant. Calcasieu Prairie. September 1987. Big bluestem (tall grass in backround), Ashy sunflower (yellow) and Little bluestem (short grass in foreground). Bottom: North Iowa prairie remnant. Calcasieu Prairie. September 1988. Big bluestem and Switchgrass (tall grasses in backround), Wrinked leaf goldenrod (yellow), Sweet goldenrod (yellow) and Lanceleaf blanketflower (center, yellow).

Plate 17. Top: South Woodlawn prairie remnant. Calcasieu Prairie. June 1988. Finger false dragonhead (pink) and Rough coneflower (yellow). Bottom: South Woodlawn prairie remnant. Calcasieu Prairie. August 1989. Indian plantain (white) and Slender gayfeather (red).

Plate 18. Top: South south Fenton prairie remnant. Calcasieu Prairie. February 1988. After annual burn. Bottom: South south Fenton prairie remnant. Calcasieu Prairie. March 1988. Nodding wild-indigo (yellow), Downy phlox (pink), Carolina anemone and Crowpoison (white).

Plate 19. Top: South south Fenton prairie remnant. Calcasieu Prairie. June 1987. Rough coneflower (yellow) and Big bluestem (circular clones in background). Bottom: South south Fenton prairie remnant. Calcasieu Prairie. July 1987. Rough coneflower (yellow) and Big bluestem (circular clones in background).

Plate 20. Top: South south Fenton prairie remnant, western side of railroad tracts. Calcasieu Prairie. May 1988. Slender rosinweed (yellow) and Lindheimer's beeblossum (white). Bottom: South south Fenton prairie remnant, eastern side of railroad tracts. Calcasieu Prairie. May 1988. Slender rosinweed (yellow) and Lindheimer's beeblossum (white).

Plate 21. Top: South Fenton prairie remnant. Calcasieu Prairie. Natural seep area. April 1988. Hairy-flowered spiderwort (blue in foreground), Fragrant spider-lily (white) and Eastern bluestar (blue in background). Bottom: South Fenton prairie remnant. Calcasieu Prairie. Natural seep area. Early September 1987. Kansas gayfeather (red), Indian plantain (white), Slender rosinweed (yellow) and Big bluestem (tall grass in background).

Plate 22. Top: South Fenton prairie remnant. Calcasieu Prairie. Natural seep area. Late September 1987. Sweet goldenrod (yellow), Kansas gayfeather (red, mostly seedheads), Indian plantain (seedheads), Florida paspalum (tall grass scattered in scene) and Big bluestem (tall grass in background). Bottom: South Fenton prairie remnant. Calcasieu Prairie. Natural seep area. October 1987. Bushy goldentop (yellow in foreground), Kansas gayfeather (red, mostly seedheads), Indian plantain (seedheads), Florida paspalum (tall grass scattered in scene), Switchgrass and Big bluestem (tall grasses in background).

Plate 23. Top: South Fenton prairie remnant. Calcasieu Prairie. Natural seep area. November 1987. Little bluestem (short grass in foreground), Switchgrass and Big bluestem (tall grasses in background). Bottom: South Fenton prairie remnant. Calcasieu Prairie. Natural seep area. July 1987. Charles Allen in circle of Big bluestem.

Plate 24. Top: South Fenton prairie remnant. Calcasieu Prairie. Natural seep area. September 1987. Charles Allen in circle of Big bluestem. Bottom: South Fenton prairie remnant. Calcasieu Prairie. May 1990. Dry prairie. Purple cone flower (red), Woods poppymallow (scarlet), Black-eyed Susans (yellow), Sensitive brier (pink), Narrowleaf mountain-mint (White cluster at center), Downy phlox (pink) and Flowering spurge (dots of white).

Plate 25. Top: South Fenton prairie remnant. Calcasieu Prairie. May 1990. Dry prairie. Purple cone flower (red), Black-eyed Susans (yellow), Sensitive brier (pink) and Flowering spurge (dots of white). Bottom: South Fenton prairie remnant. Calcasieu Prairie. May 1988. Dry prairie. Woods poppymallow (scarlet) and Flowering spurge (dots of white).

Plate 26. Top: South Fenton prairie remnant. Calcasieu Prairie. September 1, 1993. Dry prairie. Kansas gayfeather (red), Ashy sunflower (yellow), Shiny golden-rod (yellow), Lindheimer's beeblossum (white), Button snakeroot (white) and Florida paspalum (grass on right edge of scene). Bottom: South Kinder prairie remnant. D'Arbonne Prairie. April 1988. Circular clones of Big bluestem.

Plate 27. Top: South Kinder prairie remnant. D'Arbonne Prairie. May 1988. Wet prairie. Eastern blue-star (blue). Bottom: South Kinder prairie remnant. D'Arbonne Prairie. June 1988. Dry prairie. Multibloom hoarypea (red in foreground), Kansas gayfeather (new growth), Slender rosinweed (yellow in background) and Flowering spurge (dots of white).

Plate 28. Top: South Kinder prairie remnant. D'Arbonne Prairie. April 1988. Wet prairie. Largeleaf wild-indigo (white). Bottom: South Kinder prairie remnant. D'Arbonne Prairie. February 1988. Dry prairie. Little bluestem (died back in winter).

Plate 29. Prairie remnants without fire maintenance: Top row: Estherwood prairie remnant. Mermentau Prairie. Dry prairie. August 17, 2008. Tom Hillman. Middle row: Mermentau prairie remnant. Mermentau Prairie. Dry prairie. March 20, 2008 and August 17, 2008. Bottom row left: Mermentau prairie remnant. Mermentau Prairie. Dry prairie. March 20, 2008. Bottom row right: Welsh Prairie remnant. Calcasieu Prairie. Wet Prairie. May 18, 2008.

Plate 30. The Cajun Prairie Restoration Project in Eunice Louisiana (Cajun Prairie Habitat Preservation Society). Top left: annual burn. Top right: April 2008. Yellow wild-Indigo (yellow). Middle left: May 2009. Bruno Borsari. Slender rosinweed and Black-eyed Susans (yellow). Middle right: June 2008. Rough coneflower (yellow) and Finger false dragonhead (pink). Bottom left: September 2008. Kansas gayfeather (red). Bottom right: December 2008. Little bluestem.

Plate 31. Cajun Prairie Gardens near Eunice, Louisiana (Gail and Malcolm Vidrine). Top left: March 2007. Greening up after annual burn. Top right: April 2005. Louisiana irises in wet area. Middle left: April 2007. Lanceleaf tickseed (yellow in foreground), Yellow wild-indigo (yellow in center) and Fragrant spider-lily (white in background). Middle right: June 2007. Crimsoneyed rose-mallow (white and rose-pink) in wet area. Bottom: May 2007. Butterfly-weed (orange), Lindheimer's bergamont and Bergamont (pink), Slender rosinweed (yellow) and Largeleaf wild-indigo (white).

Plate 32. Top: Bayou des Cannes (undredged), south of Fournerat Road. August 1989. Pat Mire, handfishing. Bald cypress. Middle: Left to right: Rough green snake, Eastern box turtle and Monarch butterfly on Fewflower milkweed. Bottom: Left to right: Marc Pastorek, Charles Allen, Peter Loos, Bill Fontenot, Malcolm Vidrine and Larry Allain (photo by Tom Hillman; used with permission).

their components are found mostly in the essential oil fractions. In studies with E. coli, we clearly demonstrated that native prairie plant essential oils were more effective that Old World plant extracts at inhibiting the growth and development of bacterial colonies not only in agar media but also in ground meat. In further experiments, L. monocytogenes colony formation in agar media and in salad was also inhibited with essential oil extracts from several prairie plants. The prairie plant essential oils strongly inhibited colony development in two major bacterial pathogens contaminating our food supply."

"Native prairie plants studied and used for essential oil extraction:
Monarda fistulosa (bergamont/wild bergamont)
Monarda lindheimeri (Lindheimer's bee balm)
Monarda punctata (spotted bee balm/horsemint)
Pycnanthemum tenuifolium (narrowleaf mountain mint)
Pycnanthemum albescens (whiteleaf mountain mint)
Pycnanthemum muticum (clustered mountain mint)
Solidago odora (sweet goldenrod)
Nothoscordum bivalve (crowpoison/false garlic)."

"The main constituents of essential oils are responsible for the fragrant and biological properties of herbs. The inhibitory effect of essential oil against food-borne bacteria has been reported in microbiological media and in real food (Al-Dujaili and Vidrine 2008)."

"Dry plant powders and essential oils have been used against E. coli O157:H7 in earlier studies (Al-Dujaili and Vidrine 2008). In the MSA, native mints killed the bacteria completely, and crowpoison killed 95% of bacteria population. However, other herbs, including sweet goldenrod and Old World species, only moderately reduced the bacterial population in SMA. In ground beef, four selected herb essential oil extracts were tested. Whiteleaf mountain mint killed up to 80%, crowpoison killed up to 60%, and dill and sweet basil killed up to 40% (Al-Dujaili and Vidrine 2008). The results of the two sets of experiments clearly suggest that native plants from the Cajun Prairie are at the very least equal to if not better than their Old World species at inhibiting growth of the bacterium E. coli O157:H7."

"The results of additional experiments indicate that the essential oils also inhibit the growth of L. monocytogenes. In agar diffusion and salad, crowpoison and clustered mountain mint exhibited the greatest inhibition. We achieved very high levels of inhibition with the oils in specialized media in controlled experiments, but somewhat less inhibition in the more complex mini-ecosystems of ground meat and salad. These results confirm

those of other authors (Al-Dujaili and Vidrine 2008), but do not reduce the potential of these oils in food preservation/protection."

"These collective results indicate a potential for these native plants/herbs to be used in the food and medical industry. Further investigation needs to be done to identify the chemical, physical, and biological characteristics of these herbs and their oils. The future roles of these plants in public health are intriguing."

Plant Associations and Assemblages

With plants, it is all about associations. Plants associate with bacteria on their roots for nitrogen fixation, with fungi on their roots for phosphate mobilization, with animals for pollination and transport (seed dispersal), with protists as lichens and more. Plants also associate with other plants, e.g., Spanish and Ball mosses (*Tillandsia* spp.) and orchids as epiphytes aboveground; however, they also associate below ground as plants entangle roots and trade various molecules. Some molecules are nutrients, while others are allelopathic, i.e. the molecules inhibit growth and development of adjacent plants not only of their own species but also of other species. Obvious dead zones occur around these allelopathic plants. Animals are all about associations, too. We don't like to admit it!

We studied both natural areas (remnants and forests) and man-made areas (rice fields and pastures). Rice fields provide a tremendous habitat for a great variety of life forms. One hundred and twenty-three non-rice, weeds were reported from rice fields (Allen et al. 1989). In 1994, Allen et al. (1994) did an analysis of the woody vegetation of a beech forest area in the Louisiana Arboretum and provided a flora for that habitat. In 1984, Allen (1984) published a flora for Allen Parish, which contained the northwestern edge of the Cajun Prairie and flat pine woodland literally dotted with savannas.

What are the impacts upon the flora of the cattle, horses, sheep, pigs and chickens? We are yet to study these impacts. We can however surmise from Midwestern studies that the impacts are extreme if not devastating to the diversity. What are the impacts of mowing, fungiciding, herbiciding and insecticiding? Again, the impacts are not yet studied but evidence in general shows that the impacts of these biocides on the community in many habitats clearly reduced biodiversity. What are the impacts of the introduction of exotic weeds, mowing, urbanization, agriculture and the oil industry? The introduction of Chinese tallow trees and the impacts were discussed by Jim Grace (1998), but we have just begun to consider the impacts of all the other insults upon the prairie ecosystem. However, biodiversity of the Cajun Prairie can be discussed based upon the remnants that were studied.

Each prairie remnant was floristically different, and thus we can surmise that each of the named prairies may have had different assemblages of plants and animals. Very little work has been done detailing these differences. In fact, several named prairies lack remnants of any significant size. Differences were rather obvious among large blooming plants. The table (Appendix 21) can be summarized by comparing two major stretches with remnant prairies that were studied in the 1980s. The two primary prairies are: 1) Calcasieu/D'Arbonne Prairie remnants (Kinder, Fenton, Woodlawn, and Iowa remnants along U. S. 165 between Kinder and Iowa and the Welsh remnant west of Welsh, and 2) Mermentau Prairie remnants (Estherwood, Midland, and Mermentau remnants along U. S. 90 between Estherwood and Mermentau and Morse prairie remnants. However, two additional prairie groups (Plaquemine Prairie remnants and Faquetaigue/Mamou Prairie remnants) are also listed on the table in order to add to the comparison. Rather than enumerating their similarities among the major blooming plants, the differences are listed and notably the absences.

The Calcasieu and D'Arbonne Prairie remnants lacked American aloe agave, Butterfly-weed, Clustered mountain-mint, Dense blazing-star, Green-flowered milkweed, Nuttall's prairie parsley and Western horsenettle. The Mermentau Prairie remnants lacked Bearded grass-pink, Blue-flower eryngo, Butterfly-weed, Canadian louse-wort, Carolina anemone, Drumheads, Evening rain lily, Finger false dragonhead, Fringed sneezeweed, Giant orchid, Gulf cordgrass, Hairy spiderwort, Kansas gayfeather, Maryland golden-aster, New Jersey tea, Nuttall's wild-indigo, Pinkscale gayfeather, Purple cone flower, Rayless goldenrod, Rough cone flower, Snowy orchid, Single-stem scurfpea, Tall tickseed, Talus slope penstemon, Texas ironweed, Toothache grass, Virginia spiderwort, Whiteleaf mountain-mint, White colic-root and Whorled milkweed. More differences are available if smaller species, grasses, sedges, etc. were also examined. However, the point can stand upon this data.

If the remnants are representative of the original prairies of their regions, the Calcasieu/D'Arbonne Prairies were the most diverse, but they lacked some impressive wildflowers. For example, two blazing stars, Dense blazing-star and Kansas gayfeather, are almost identical. The former was abundant on the remnants of the Mermentau Prairie—it bloomed with abandon throughout the month of August. Kansas gayfeather was common to abundant in the Calcasieu/D'Arbonne Prairie remnants and in the pine flatwoods—it bloomed as August ends and bursted into magnificent bloom in September into October. Charles Allen often commented that blazing stars and gayfeathers (*Liatris* spp.) were the most common wildflowers of the Cajun Prairie, thus they provide an excellent comparison. Traveling through the Mermentau Prairie in early August, the rider would have seen large swaths of blazing stars; however, the same rider would have seen none in the Calcasieu Prairie. The same trip in September would have resulted in the exactly opposite views. Smaller landscapes, like gardens in the prairies, are likewise very different depending on the variety of species of plants. However, other regions currently without remnants may have been very different. For example, the Welsh Prairie remnant in the Calcasieu Prairie contains a good number of Compass plants, Aldo Leopold's favorite prairie flower, unlike the other remnants in the Calcasieu Prairie. Generalizations always have exceptions, and the exceptions are sometimes maddening.

Restoring prairies may require thinking about individual named prairies in more detail, and in contrast to our current efforts, restorations within named prairies may be planned and developed with only local seeds and propagules. At present, restoration is carried out using seeds and propagules collectively from all named prairies. This is a result of the limited expanses of remnants and the limited resources currently allocated to restoration activities. Larry Allain and others are attempting to grow plants specific to prairies in a large scale manner for the future restoration of prairies along specific genetic lines. A new series of projects is absolutely necessary. These projects would entail the restoration of prairies in each of the named Cajun prairies with plants native to those prairies. This would give us an even clearer perspective as to what the original prairies looked like.

The prairie ecosystem contained many marshes and ponds as well as rivers. The vegetation of these watery (hydrosphere) habitats is not central to this effort, but there are some species of algae that are commonly found attached to mussel shells and other substrates (Curry et al. 1981). These algae include: *Basicladia* sp., *Oedogonium* sp., *Cladophora* sp. and *Spirogyra* sp. The algae and protists of the region have been studied in some detail for the Cajun Prairie ecosystem, but they are omitted from this discussion.

Among the aquatic plants, there are a number of interesting plants like American lotus, cattails, water lilies, water-shield, duckweeds and Louisiana Irises. The latter group of five species is one of the most botanically and horticulturally famous plant groups that originated along the southern edge of the Cajun Prairie (Caillet and Mertzweiller 1988 and Caillet et al. 2000). In a marshy, moderately forested area south of Abbeville (Vermilion Prairie in Vermilion Parish), the 'Abbeville iris' (*Iris nelsoni*) or 'Giant fulva' was discovered by W. B. MacMillan (Dormon 1958). This species is a natural hybrid between the Red iris (*Iris fulva*) and the Giant blue iris (*Iris giganticaerulea*). The former is common in the Atachafalaya Basin, while the latter is abundant in the marshes along the Louisiana coast. A third species, Zig-zag iris (*Iris brevicaulis*), is still occasional in roadside ditches and remnant prairie wet areas and may also have contributed some genetic material to the Abbeville iris. In fact, a fourth species (*Iris hexagona*) is also known from the marshes. All of the 5 species are currently called "Louisiana irises" and collectively formed the largest natural iris marshes in the world extending from the New Orleans area to the Sabine River along the Louisiana coast. The marshes like the prairies have been drained and otherwise heavily damaged or changed by development of the oil industry, draining for cropland and pasture, urbanization, hurricanes, and erosion (Note approximately 25 square miles of Louisiana coastal marshes are lost each year—the double hit by Katrina and Rita in 2005 eliminated 250 square miles in one year). These same 5 species currently form the foundation for a very impressive cultivation enterprise not only in the southern United States but also in California and in Australia. More than 1000 varieties have been identified from wild collected material and from crosses made by the dozens of hybridizers currently working with these irises. Flower colors extend from white to black (the darkest purple), and the rainbow of colors from red, orange, yellow, green, blue, indigo and violet. The blues and reds come in nearly every shade imaginable. Cultivars also emphasize blooming times, scape shape and flower placement, flower shape, hardiness, etc.

Caroline Dormon, Mary S. DeBailion, W. B. MacMillan and Ira Nelson were among the early collectors (1920s-50s) (Caillet and Mertzweiller 1988 and Caillet et al. 2000). Their stories provided insight into early botanizing in the marshes of Louisiana. Unfortunately, far less attention was paid to these plants in the prairies, although various orchids and other horticulturally-interesting plants existed there. Louisiana Iris were used to make 'tincture of iris,' a medicinal concoction. Many other plants were used as medicinals (Fontenot 1992, Kindscher 1992 and Foster and Duke 1990).

Much of the literature on plants as well as most of our common knowledge regard the aboveground plants. In many cases, we know next to nothing about the roots of the plants. My wife, Gail, has a common complaint about my gardening. She contends that I am overly prone to digging plants up and to taking them out of their containers to examine the roots. Roots have always fascinated me, and the prairie is famous for having the most fascinating underground root systems. Weaver's (1954 and 1968) research literally unearthed the realm of the roots in the prairies. His shadow-box drawings are phenomenal.

My favorite stories about roots relate the differences between the aboveground verses underground parts of several plants. Coralbean is an extraordinary example. The aboveground plant is annual in the Cajun Prairie, but the underground part is a huge perennial ball of a size up to a foot wide on a modest plant. I have seen large plants that I could not dig with a shovel for the sheer size of the underground stem. Compass plant sends a stem up each year that blooms the classic-looking sunflower. The stems are usually under an inch in diameter and extend to nearly 6 feet in height. Digging a Compass plant is an ordeal. The taproot is 3-4 inches in diameter at the soil surface, and it slowly tapers as it grows deeper—reminiscent of a 3 foot carrot. Roots and stems of Eastern gama grass are so thick that a shovel can barely cut the mass apart—not a job that I would want to do routinely.

The plants produce one of the wonders of the world—a massive, underground root system. This root system is depicted for prairie plants from Iowa in the Figure 1—a drawing that I usually simply call *Roots*. *Roots* depicts a comparison of lawn (Kentucky blue grass at the far left) to a variety of prairie plant species. The lawn has literally 'no roots' and fails to provide many of the benefits of the roots of prairie plants. The roots of prairie plants contain as much as 70% of the biomass of the plants. When the winter fires clear the aboveground brush of a year's growth, the roots are all that remains. The roots are a misnomer as many of the underground parts are actually stems, but that is another botany lesson that we will discuss shortly. The great variety of root systems depicted reveals that the diversity of plant growth strategies seen aboveground is mimicked underground by a variety of rooting behaviors. The underground biomass is the source of topsoil, when it degrades. The root systems are also called the rhizosphere—a location where numerous kinds of organisms varying from bacteria to protists to fungi to animals make a home (see discussion of biota of rhizosphere in Chapter 9). The same roots are responsible for trapping many kinds of pollutive chemicals and remediating them. Water percolates as it is filtered through this maze of roots and enters the aquifers; the water is also stored in the roots and released slowly during droughts. Many of these roots are edible or medicinal. Woods poppymallow is reported to taste like carrots—I would not willingly eat one since the plant is so beautiful in bloom. Roots of Coralbean are famous for their medicinal use—I not only remember my mother collecting seeds and roots for making 'Mamou cough syrup,' but also I remember being served fairly large spoons filled with this concoction when I was ill.

The roots are extraordinary for vegetative propagation and prairie restoration. Many of the structures that we think of as roots are really modified stems with a propensity for propagation. A single underground stem can be cut into a dozen pieces and a dozen clones will grow—each identical to the other. A single shovelful of prairie sod may contain the stems and roots of a dozen different species of plants and the essential mycorrhizae and bacteria to assure their survival. Plugging prairies with shovelfuls of sod exponentially increases not only the potential success of the restoration effort but also the species richness of the restoration site. I am convinced that many species simply do not germinate from seed in sufficient quantity to create a diverse plant community without an inordinate amount of effort by the ecologist.

We have a prairie flora, although it may be a conservative one. The intense grazing, plowing and the removal of fire from the landscape certainly eliminated numerous populations and probably a number of species. It is probable that the intense grazing by cattle greatly reduced the density of big bluestem, once considered the 'redwoods' of the countryside. Herbicides (and other biocides) further reduced the diversity. The loss of herbicide-sensitive species as well as the species dependent on pollinators eliminated by insecticides or mycorrhizae eliminated by fungicides may have been exacerbated by the increases in plant species that are encouraged by the elimination of competitive species. Even the loss of predatory insects by use of insecticides may result in the increase in herbivorous insects gaining opportunity to reduce or eliminate prey species. These integrated biological relationships are readily manipulated in the garden but no less so in the natural habitat. Even the application of fertilizer, often thought to be beneficial to all plants, specifically retards root growth and development in many native prairie species. Further the fertilizer application changes the soil pH and ion capacity not only affecting the flora but also affecting soil organisms, e.g., earthworms. The smallest change in pH, an exponential index, makes for dramatic change in the microhabitat.

The flora that persists on the remnant prairies is made up of the 'survivors' of

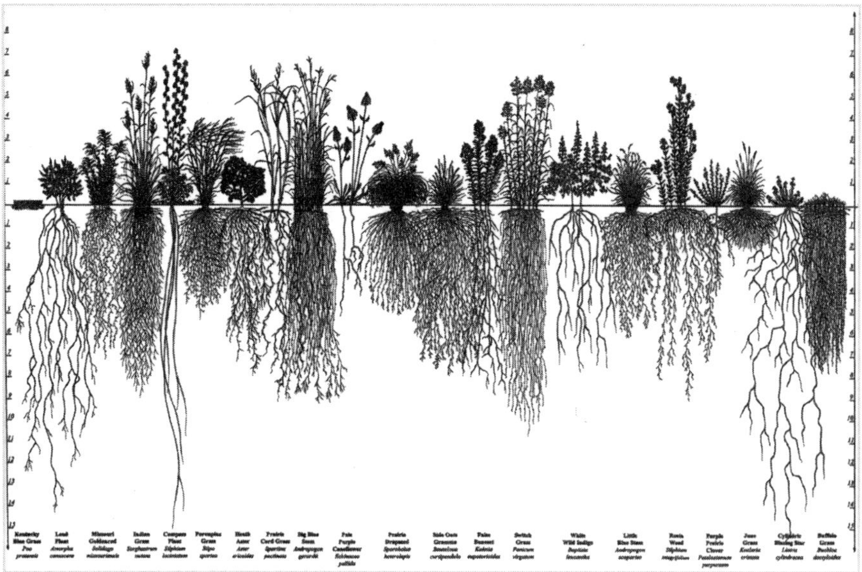

Figure 1. Root systems of Prairie Plants. 1995. Artwork by Conservation Research Institute, Heidi Natura. Reprinted with permission.

a multitude of catastrophic insults applied by man. Not only his plow, his cattle, his introduced plants, his long list of chemicals, his concrete structures, his mowing, but also his contribution to the long term changes linked to climate change and changes in hydrology have contributed to the loss of a flora. The increasing salinity resulting from massive withdrawal of water from underground rivers but also from the sea itself as the hurricanes bring sea water deep into the Cajun Prairie both up low-lying streams and over the land itself in inland surges is affecting major plant communities. Recent reports of the loss of huge iris marshes following the Katrina-Rita one-two punch exemplify the impacts of salinity inland.

Changes in the flora are not limited to the plants alone. All of the other kinds of living things depend upon plants. The loss of a plant species may mean the loss of a butterfly species, as many butterfly larvae are host-specific and can feed on only one kind of plant. So just as the loss of a pollinator can signal the loss of a plant species, the loss of plant species can signal the loss of an animal species. Orchids often are dependent on a specific kind of fungus to serve as mycorrhizae for feeding and likewise dependent on a specific kind of bee for pollination. The grasses of the prairie are among the most suitable for large grazing animals. Bison initially, and later cattle depended upon the prairie species. Deer usually eat wildflowers, insects eat nearly everything, birds eat seeds, etc. Everything in the natural prairie was recycled. Although it was not a closed energy system, as millions of migratory birds and butterflies (and Bison) each year moved energy laterally, the bulk of the energy moved vertically first aboveground as captured sunlight and then below ground to roots in a yearly cycle of plant growth and senescence followed by fire. The soil was alive, and organisms of a myriad of kinds were moving energy and building topsoil. Many of these organisms were animals. Thus our discussion of prairies continues into the next chapter on animals of the Cajun Prairie.

Chapter 9
Biota: Animals

"*The human assault on biodiversity has been recognized, though not its scale, extent, and imminent threat.*" (Sachs (2008)).

"*To keep every cog and wheel is the first precaution of intelligent tinkering.*" (Aldo Leopold (1949)).

"*Dung beetles roll up a little ball of cattle dung, put an egg in it, and bury it.*" (Paul Raeburn (1995)).

My first thoughts as a budding zoologist were on our farm in the Mamou Prairie. The pasture was grazed by our cattle, and I would sit there in the summer and watch the Dung Beetles (*Phanaeus vindex* MacLachlan) roll cattle dung into nickel-sized balls. A single patty could be effectively rolled-up in a few hours by these industrious, if not comical, beetles. The ideas of the beetles depositing eggs into the ball before they buried it and beetle larvae eating the ball for sustenance and development never crossed my 8 year old mind, but I couldn't stop observing these singularly minded insects. I have not seen one in nearly 40 years—I am not sure whether this is because I am not around cows much or they have disappeared from the Cajun Prairie. My interests fanned out from these insects to invertebrates of many kinds, but most of my friends were focused on animals that they could catch, shoot, rear or eat.

The people of Louisiana pride themselves for living in a state described as the 'Sportsman's Paradise.' The waterfowl, deer, fishes, frogs, turtles, alligators, nutrias, shrimps, crabs, crayfish, squirrels and other animals that are hunted, trapped and netted are seemingly limitless. Producing a faunal list for Louisiana is quite different from producing a flora. Zoologists usually are very specialized and lack interest or knowledge of the names of the vast array of animals. Among the zoologists, you can probably rattle off their specialties: mammalogist, entomologist, ornithologist, ichthyologist, herpetologist, lepidopterist, odonatologist, etc. and there are many more including myriopodologist, malacologist, acarologist, myrmecologist, protozoologist, etc. In any case, you get my drift. Animals are treated differently when it comes to developing lists, and no single source provides a near complete faunal list, in contrast to examples of a flora (Allain et al. 2006, Thomas and Allen 1993, 1996 and 1998). Thus, forming a faunal list of species for the Cajun Prairie has proved to be more difficult than a flora.

Additional difficulties to building a faunal list arose because of changes in the natural system resulting from:

1. the legion of introduced animals,
2. the obvious loss of keystone species,
3. massive habitat alteration,
4. the small size of remnant prairies, which is too small for large animals,
5. the loss of host plants for animals like butterflies and
6. the loss of animal and plant hosts for symbionts and parasites.

Animals are also difficult to enumerate because of the influence of introduced animals, e.g., cattle, sheep, horses, pigs, chickens, dogs, cats, fire ants, nutrias, etc. Each of these introductions wreaked havoc of a different kind to both the plant and animal worlds. Even today the introductions continue with things like Asiatic Clams (*Corbicula*), Zebra Mussels, Formosan Termites, Africanized Killer Bees, Mediterranean Geckos, Banana Spiders (a South American orbweaver (*Nephilla clavipes*)), Asian Tiger Mosquitoes (*Aedes albopictus*) and more. We conducted varied experiments with rotifers and even questioned whether rotifers are native or exotic, because of their prevalence in rice fields (Vidrine and Hazelton-Robichaux 2008). Nine-banded Armadillos and Cattle Egrets extended their ranges to the Cajun Prairie (Strecker 1926, Lowery 1974a and b). The number of exotic introductions is almost impossible to keep track of or manage.

Habitat alteration has caused dramatic impacts upon plants and even more dramatic effects on animals that rely on plants and undisturbed habitat. Dredging, sedimentation, drainage projects, pollutants, changes in hydrology in general, and the pouring of concrete have collectively altered every stream in the Cajun Prairie. Aquatic fauna is under such extreme stress not only in the streams but also in the entire marsh. Natural events like hurricanes have taken on an even greater importance ecologically as natural levees and barrier islands are disappearing.

Keystone species like Bison, Red Wolves, prairie chickens, Passenger Pigeons, native parakeets and prairie voles have disappeared from the terrestrial habitat (Allain et al. 1999 and 2000). The aquatic habitat has fared even less well with the loss of massive mussel beds and a variety of fish, turtles and numerous invertebrates. Several animals and a photograph of local prairie enthusiasts are found on the final Plate in this book. The final Plate in this book also depicts an undredged stretch of Bayou des Cannes, where Pat Mire is handfishing. Scenes like this one are rare as most streams have been dredged, cleared and/or straightened.

The Animals of the Cajun Prairie

Thousands of different species of animals occur in the Cajun Prairie ecosystem. Colorful guides to the animals and plants of southern Louisiana do exist (Allain et al. 1999 and 2000). Nancy Camel (2006), Charlie Hohorst and Marcelle Bienvenu (2009) and Brian Miller and William R. (Bill) Fontenot (2001) provided beautiful introductions to the birds of Louisiana. C. C. Lockwood (1981, 1984, 1986, 1995, 2005 and 2007) provided colorful essays with photographs of the Atchafalaya and other parts of western Louisiana. Gay Gomez (2008) introduced plants and animals of the marshes. These books collectively introduced the sheer beauty and some of the unique biota of the region. Numerous other books likewise introduced the prairie habitat in North America and/or other regions of Louisiana and the Gulf coast region.

Fortunately, major books and studies have been published on animals of Louisiana with lists reflecting the region including the Cajun Prairie. Among these are works on birds (Lowery 1974a, Miller and Fontenot 2001, Huner 1998 and Ingold 2008), mammals (Lowery 1974b and McPherson 2008), reptiles and amphibians (Dundee and Rossman 1989, Johnson 1999, Shively and Vidrine 1984, Vidrine and Hatler 1995 and Hardy 2008), fishes (Douglas 1974, Hoese and Moore 1977 and Douglas and Hoover 2008), butterflies (Lambremont 1954 and Ross and Lambremont 1963), dragonflies (Bick 1953, Strickland and Strickland 2008 and Mauffray 1997), crayfish (Walls and Black 2008 and Walls 2009), mussels (Vidrine 1993, 1975, 1989, 1990 and 1993 and Roback et al. 1980), snails (Vidrine et al. 1987a), rotifers (Vidrine et al. 1985 and Vidrine and Hazelton-Robichaux 2008), mosquitoes (Chapman and Johnson 1986, Darsie and Ward 2005, McLaughlin 1884, McLaughlin and Vidrine 1984a, b, c and d, 1985, 1986, 1987, and McLaughlin et al. 1987a, b and 1988), water mites (Vidrine 1996a, b and Vidrine et al. 1987b), clam shrimps (Vidrine et al. 1987c) and many more. Even tardigrades are receiving attention from McNeese State University scientists, Juliana Hinton and Harry Meyer. Other works cover larger regions and enumerate many other animal groups, e.g., many insect orders. However, many groups have never been studied in any detail. Collections of these animals are located in museums and await the development of faunal lists. Some of these will be discussed in this chapter, but I will de-emphasize those groups for which books are already written as space prohibits repeating their efforts even in a scaled-down approach. Rather my emphasis will be on animals in general and on the communities that they constitute. The animals of the Cajun Prairie are found in three distinct realms of the Biosphere: the Rhizosphere, the Atmosphere and the Hydrosphere. In other words, animals are found in the soil, air or water. At first it may seem trivial to separate the animals as such, but I did not want to ignore nor in any way minimize the role of the organisms in the soil. In giving soil organisms equal status with the organisms in the air and water—those that most people are aware of—the roles of the soil and its biota are emphasized.

The Rhizosphere

The soil is generally thought to be something inert that roots grow into and somehow magically produce plants. The roots collectively create a small universe called the rhizosphere. In prairie habitats, the rhizosphere becomes highly significant, as underground plant parts, roots, rhizomes, corms, bulbs, and other stems, clearly make as much as 70% of the biomass of the plants, and 100% after a fire. The prairie is often described as an underground system. The plants have already been mentioned in Chapter 8, even though little attention was placed on the roots and stems, neither above-ground nor below-ground. By the same token, little attention will be placed on other 'Earthlings' living underground except to mention them as a flora and fauna.

The soil is alive (Nardi 2007, Brady and Weil 2007, Nancarrow and Taylor 1998, Stewart 2004, Logan 1995, Tompkins and Bird 1998 and Baskin 1997). Several papers in Knapp et al. (1998) explored the biota of prairie soils and streams in grasslands. James B. Nardi's (2007) book, *Life in the Soil*, provided a thorough introduction to the diversity of life in the soil for naturalists and gardeners. It is easy for me to directly connect the taxa in this book to those in the Cajun Prairie soils.

William Bryant Logan (1995), in his book *Dirt*, described the prairie soil as both a most desirable soil for agriculture and a 'root' (yes, pun intended) cause for the possible collapse of society—if not for the biota of the soil then maybe for the people who depend on it for agriculture and maybe for both the soil biota and man!

> *"But the most desirable virgin soils are not from the forest. In a forest, at least two thirds of the organic matter resides above the level of the soil. In a prairie, the proportion is reversed. The great prairie soils of the American Midwest and of the Russian steppes are rich with the remains of millennia of dense, sinuous roots that have lived and died in a soft mineral accretion formed from windblown silts high in calcium and other bases. Not only has the root mass given its organic residue for fertilizer; its sugars have also attracted a vast microflora, and its polysaccharides have glued smaller grains of soil into larger aggregates that permit easy passage of food-bearing water and air."*
>
> *"In the 1930s, Hans Jenny studied virgin prairie soils side by side with cultivated soils in Missouri. He found that after only sixty years of cultivation, with zero erosion, the farm soils had lost one third of their organic matter. As a result, there was no longer the same level of glue to bind the soil into aggregates, and instead of rising like yeasted dough, it was collapsing into heavy slabs."*

The nitrifying bacteria that convert ammonium to nitrate, the nitrogen-fixing bacteria that convert atmospheric nitrogen gas to usable nitrates or ammonium, and the fungi that mobilize phosphates are examples of the major connections of the microflora and biota of the soil with the plants. The rhizosphere is rich with examples of plants and organisms interacting beyond our scope of imagination (Baskin 1997). Yvonne Baskin went on to spell-out the nature of the vitality of the soil.

> *"If all the microflora were to disappear, life on earth would quickly come to a halt, wilting and perishing into a never-to-rot compost pile spread across the land. Yet even these vital creatures do not work in isolation. The soil community includes ranks of invertebrate animals, some of which graze on bacteria and fungi, assuring the release and recycling of energy and materials tied up in even the smallest life forms. Many of these soil animals also speed the work of decomposition by shredding, consuming, digesting, and excreting organic debris, turning it into more accessible crumbs."*
>
> *"The tiniest of the soil animals are the microfauna, such as nematodes (roundworms) and protozoa that live mostly in water films around soil pores. Although some are plant parasites that cause serious crop damage, other graze on bacteria and fungi. They are joined by intermediate-sized soil animals, the mesofauna, which include highly specialized invertebrates, such as mites and springtails, that occupy air-filled soil pores and prey mostly on fungi. A square meter of soil may support populations of 10 million nematodes, 1 billion protozoa, and, depending on its organic matter content, 200,000 to 400,000 springtails and mites."*
>
> *"Largest and most noticeable of the soil invertebrates are the macrofauna: earthworms, ants, termites, millipedes, woodlice, beetles, insect larvae, and others. All are large enough to alter the physical structure of the soil and fragment the litter as they tunnel and feed, aerating the soil and forming channels for infiltration of water. Each earthworm can ingest and excrete up to thirty-six times its own weight of soil each day. Earthworm populations may consume from ten to five hundred tons of soil per hectare per year, excreting it as dark and fertile castings. In the tropics, ants are the chief earth movers, and they run a close second to earthworms in cold temperate forests as well."*

In the prairies of the Americas, the major native earthworms belong to the genus *Bimastos*, and not *Lumbricus*—the introduced worms (Nardi 2007). However, the introduced worms that came in with the Europeans quickly overtake an area and actually consume the leaf litter and soil as much as 5-10 times faster than the native worms. This may be good for the gardener, but in the natural habitat, leaf litter and soil turnover rates that are drastically changed may cause dramatic ecological disturbance. Nardi (2007) cautioned his readers that little is known about the inter-relationships between leaf litter and ground cover as they relate to native plants. Further the roles of earthworms escape most people, who actually know little of soil organisms and soil itself. But this is changing as people recognize that soil erosion and soil fertility are concepts that become important to societies on the verge of collapse.

An entire industry is based upon earthworms, vermiculture, and dozens of books are available. This is an understatement since agriculture itself is totally dependent in the long run on earthworms. Amy Stewart (2004) made a strong case for the earthworm as a 'keystone species.' She also introduced earthworms along with Charles Darwin,

> *"It seems strange, then, that most scientists before Charles Darwin didn't consider worms worthy of study. Very little was known about them in the nineteenth century, when Darwin emerged as a sort of champion of worms, devoting his last book to a painstakingly detailed research on their physiology and behavior." Darwin's (1881) book, The formation of vegetable mould, through the action of worms, with observations on their habits, was the result of more than 40 years of observation.*

My ecology professor at LSU, Walter J. Harman, used to laugh and repeat, *"Most people are shocked when I tell them what I do for a living, I tell them that I sit back and think about earthworms, literally for most of my waking hours."*

Bruno Borsari et al. (1998) recently studied the behavior of earthworms in restored Cajun Prairie as compared to agricultural fields. Domingo Jariel et al. (2008) studied the earthworms and microarthropods of soils in restored Cajun Prairie as compared to adjacent lawn. Jariel et al. 2008) concluded:

> *"Burning enhances the regrowth of grasses and wild flowers in Cajun prairie, which attracts soil-dwelling invertebrates, such as annelids and microarthropods. Soil organisms are necessary for plant litter decomposition and soil aeration. A portion of St. Augustine grass (Stenotaphrum secundatum) lawn at Louisiana State University at Eunice (LSUE) was converted into a Cajun prairie restoration project in 1990. Since 1991, the lawn was mowed every week whereas the prairie was burned every January. We hypothesized that these soil-dwelling organisms would be more abundant in burned prairie than in mowed lawn. The burned prairie and the mowed lawn had indistinguishable soils with respect to temperature, moisture, bulk density, particle density and micro-arthropod abundance, but the prairie had greater abundance of annelids and a greater porosity. The similarity of these soils suggests restoration has little effect on soil physical properties in 15 years."*

With conventional agriculture, no attention is given to earthworms and soil biota. In fact, the soil is considered as dirt, a lifeless thing that can be manipulated with a variety of chemicals into a medium for culture (Logan 1995).

In the Cajun Prairie, little needs to be said about ants, except to note that the fauna has changed dramatically with the introduction of Fire Ants—insects that made Edward O. Wilson famous. These ants are common; they even get into your home (Taber 2000). Their mounds are everywhere; however, a massive effort was made to eradicate this pest in the 1960s. It was a disaster, ecologically! The weapons of choice were initially heptachlor and then mirex, both chlorinated hydrocarbons (Taber 2000).

Mirex, the major ecological catastrophe maker, was first synthesized in 1946 but was not used in pesticide formulation until 1955. Mirex is a bait, which is a stomach insecticide. The insecticidal use was focused on southeastern United States in order to control the imported Fire Ants (*Solenopsis* spp.). To combat the problem, approximately 250,000 kg (1 kilogram equals 2.2 pounds) of mirex was applied to fields during 1962-75. Most of the mirex was in the form of 4X mirex bait, which consists of 0.3% mirex in 14.7% soybean oil mixed with 85% corncob grits. Application of the 4X bait was designed to give a coverage of 4.2 g mirex/ha and was delivered by aircraft, helicopter or tractor. Use of mirex as pesticide began in 1962 in the United States, and all uses of mirex as a pesticide was banned in 1978 (see 'mirex' in Wikipedia 2008 and Taber 2000).

I recall seeing the huge airplanes flying over in 1962 and then again in 1967. Their objective was to cover every square inch with mirex and take out the Fire Ants. Unfortunately, evolution happens, and the .01% survivors were resistant and re-colonized. Re-spraying did little good, however it cause massive fish die-offs and probably had many other deleterious impacts. Again, there is no way to ascertain the losses in biodiversity as a result of this insult to nature. The great aerial application planned for 1972 was halted out of fear (this time based on evidence in the laboratory at LSU) that the required elevated dosage of mirex would completely wipe out the aquatic biota. As an interesting aside from Robert Murray, biologist for Fort Polk during the 1950s-1980s, regarding Fire Ants, he recalled the appearance of the Fire Ants and the concomitant disappearance of ticks. Prior to the Fire Ants' invasions, any trip into the forests led to a tick infestation during certain times of the year. After their invasion, tick infestations became rare. My many trips to work into the wilds of Fort Polk occasioned very few tick encounters. The prairies were notorious for ticks historically, but they are few now. However, Fire Ants are everywhere. Taber (2000) reported that not all ticks are eaten by Fire Ants.

The more interesting part of Stephen W. Taber's (2000) book is the role of Fire Ants (some 20 kinds) in the environment. They interact with every organism that they come into contact with, potentially making them 'keystone species' in the Cajun Prairie of today. Not only do they eat Lone Star Ticks (*Ambyomma americanum*), a pest in the woods and lawn but also a vector of STARI, tularemia and Rocky Mountain spotted fever, but also they eat termites and many pests that would ordinarily attack crops. Aside from the pestiferous nature to man and his animals, the ants kill many beneficial organisms including butterflies, parasitic wasps, walking sticks, and earthworms. They however alter the soil chemistry and serve as food for many animals, including my chickens. These latter two are often considered benefits of Fire Ants. In the long term, they altered the dynamics of the Cajun Prairie ecosystem. Taber related one story I had not read before regarding dragonflies eating the abdomens of queens (queens are females as in bees that lack a stinger—the modified ovipositor—and rather have a large abdomen full of developing eggs in their ovaries) during their nuptial flight. The male abdomens are apparently also eaten as the females fly with a swarm of males. The number of potential Fire Ants destroyed by such a dragonfly feast is astronomical.

Protists that destroy Fire Ant mounds may be a harbinger of future changes in the prairie. Once a colony is infected, the microsporidian protozoan (*Thelohania solenopsae*) debilitates the queen, the workers and even the larvae. The disease shortens the ants' life spans and raises the mortality of sexual females. This work was done at Texas A and M University (Taber 2000). In Houston, Texas, a newly introduced ant (*Paratrenicha* sp. nr. *pubens*) (2002) provided another intriguing story. These miniature beasts eat Fire Ants, but they render havoc in human dwellings and electrical equipment (Ball 2008). They also feed on beneficial plants, ladybugs, hatchling Attwater's Prairie Chickens and humans. They are spreading across the coastal prairies of Texas and should arrive in the Cajun Prairie on short notice.

The Nine-banded Armadillo (*Dasypus novemcinctus*) was first reported north of the Rio Grande in Texas by John K. Strecker, but it apparently had arrived in Texas and Louisiana in the 1910s (Strecker 1926 and Lowery 1974b). John Strecker was a famous naturalist, who also studied fresh-water mussels among other things. Armadillos are nocturnal and are voracious omnivores eating earthworms, insects, fruits and roots. They are notorious for digging large burrows, which are dangerous for cattle and horses. My experience with armadillos dealt mainly with sticking needles into them, drawing blood, autopsying and otherwise general all-around husbandry. I spent 3 years, 1973-76, as the colony supervisor and biologist at the Gulf South Research Institute armadillo facility. The main thrust of the research was to study the development of human leprosy, Hansen's Disease, in an alternative host. *Mycobacterium leprae*, the bacterial causative agent of leprosy, is notorious for its difficulty to culture and propagate outside the human body. The armadillo with its low body temperature made an excellent vessel for the growth and development of the organism. In fact so excellent that armadillos had natural infections (Walsh et al. 1975). The armadillo, another exotic animal, is an example of a range extension where an animal moves into new habitat by seemingly natural modes.

The Cattle Egret (*Bubulcus ibis*) is another example of a recent, natural arrival. It arrived from the Old World in 1942 in Florida; by 1955, it was reported in the Cajun Prairie in Jefferson Davis Parish (Lowery 1974a). Its predilection for hanging around cattle accounts for its name—the presence of cattle in the New World made for an instant hit (niche) for this bird. Natural introductions like Nine-banded Armadillos and Cattle Egrets hint at the nature of communities to evolve. Any exotic species introduction, whether man-made or natural, changes the relationships of all other organisms to one another within an ecosystem. The changes may be minor for some species, but the changes are major for others.

Crayfish, locally called crawfish or 'mudbugs,' create a major feature in the prairie habitat—mounds with extensive underground caverns (Walls 2009). Not only are these crustaceans major terrestrial organisms, but they are among the most cosmopolitan inhabitants of streams and ponds. Their diversity is extensive (Huner 2000, Walls and Black 2008 and Walls 2009). Walls (2009) said, "The old prairie is home to *Fallicambarus macneesei* (Black 1967) and at one time may have had other distinctive species." Aside from the extensive ecology and biology of the crayfish, these organisms are a menu item. The Lenten season is essentially synonymous with 'crawfish season.' Extensive studies have been done to test the edible meat from *Procambarus clarkii* and *P. acutus acutus* for residues of heavy metals and pesticides, including crayfish from the Atchafalaya Basin and from ponds near Crowley, Louisiana. Lead, mercury and cadmium, if present, were present in concentrations below the detection limit (Finerty et al. 1990). Pesticide organochlorine (DDT) residue concentrations were very low (Madden et al. 1989 and 1991). For many years, the aquaculture aspects of the crayfish fishery have been monitored and developed by faculty and extension faculty of the major Louisiana universities, including Southern University, Louisiana State University and the University of Louisiana-Lafayette.

Crayfish can be found walking in the prairie, in shallow water anywhere, underground, in ponds and streams. Their mounds dot the landscape in moist prairies and lawns. I recall harvesting them from rice fields and bait-collecting them in the ditches on the side of the roads as a youngster. Boiled or in an etoufee, this dish may be second only to Cajun gumbo in Cajun cuisine.

In nature, it is not entirely possible to separate the rhizosphere from the atmosphere and the hydrosphere, nor is it possible to separate the biota entirely. Both plants and animals, as well as fungi, protists, and bacteria, move from one sphere to another routinely; some organisms make this a strategy. Our most common earthworms hunt the surface of the ground for leaves and other organic matter during the night filling their burrow with the leaves. Many insects live underground as larvae and emerge to fly as adults; in Louisiana, Love-bugs (*Plecia nearctica*), a fly with an obvious destiny to mate, emerge with a vengeance in mid-May and mid-September—their numbers often are astronomical. A similar but more elaborate life cycle is noted in the cicadas—large and noisy insects that leave exoskeletons hanging by the hundreds in nearby trees while they sing away their lives. Mosquito and midge larvae live in the hydrosphere and emerge into the atmosphere also in astronomical numbers. Thus all three spheres are intimately interconnected and indeed inseparable, except in textbooks.

The Atmosphere

Vidrine et al. (2001c) summarized the large terrestrial biota as follows:

> *"Historical documentation of southwestern Louisiana prairie fauna prior to the turn of the 20th century is similarly lacking. Fortunately, Louisiana State University zoologist George Lowery, Jr. amassed as much pre-twentieth century bird and mammal data as he could, including it in volumes which he later authored. Of the historical bird life on the prairie, most noteworthy from Lowery (1974a) were several references characterizing the Whooping Crane (Grus americana) and Attwater's Prairie Chicken (Tympanuchus cupido attwateri) – both presently hovering on the brink of extinction – as "common" and "abundant," respectively, in the region. Similarly, Lowery (1974b) recounts the details surrounding the discovery of both the red wolf (Canis rufus) and the Louisiana prairie vole (Microtus ochrogaster ludovicianus) – both thought to be presently extinct – within the prairie region of southwestern Louisiana. Much like that of the region's plant life, it is generally believed that historical bird, amphibian, reptile and mammal life in this region was extremely diverse (Lowery 1974a and b, Huner 1998, Johnson 1999 and Dundee and Rossman 1989). Both the diversity and abundance of contemporary mammal life is substantially high. During the winter months in particular, notably large numbers of raptors, waterfowl, shorebirds, sparrows, blackbirds, and other birds still migrate into the prairie region, now almost totally supplanted with rice, sugar cane, cattle, and crayfish farms (Vidrine et al. 1995, Huner 2000)."*

When I think of insects, my first thoughts are of mosquitoes, butterflies and dragonflies. These flying insects were my central focus of study for a decade, the mid-1980s to the mid-1990s. In several papers, my colleagues and I (Vidrine et al. 1992a, b, c, 2001b and Vidrine and Borsari 1998) summarized these flying insects and their activities in the Cajun Prairie ecosystem. Vidrine et al. (2001b) recorded:

> *"During the last half of the twentieth century, the insects of southwestern Louisiana received some long overdue attention. Prior to the mid-1900s, there were few studies of insects of southwestern Louisiana.*

The region was generally overlooked by entomologists as a result of lack of accessibility due to isolation by the Big Thicket to the west and the Atchafalaya River Basin to the east. The extent of damage resulting from agriculture was such that mid-century surveys of Louisiana (Bick 1957 and Lambremont 1954) considered the ecosystem so impoverished as to have little native fauna remaining. In contrast, we conducted surveys of this ecosystem mainly between 1986 and 1991. Nearly 100 species each of Odonata (dragonflies and damselflies) and Lepidoptera (butterflies and skippers) have been identified in the region. Long summers and brief winters extend the local flight seasons of the insects. The prairie plants provide a nearly year-round supply of nectar and pollen. Loss of suitable habitat and extensive use of biocides apparently decimated most insect populations; however, we found relict populations of many species on remnant prairies, in undredged streams, and in relatively old-growth gallery forests."

" Coastal prairies and surrounding habitats have been surveyed for Lepidoptera and Odonata. Lambremont (1954) and Ross and Lambremont (1963) did intensive surveys of butterflies and skippers of Louisiana; however, their works excluded most of southwestern Louisiana. Understanding the butterflies of southwestern Louisiana involves understanding the Cajun Prairie ecosystem, which is part of the coastal prairies of Texas and Louisiana (Allain et al. 1999). Butterflies and skippers of Texas (Neck 1996, Tveten and Tveten 1996) have recently been reviewed. Opler and Malikul (1992, 1998), Glassberg (1999) and Opler et al. (1995) provided recent species lists of butterflies and skippers for Louisiana—none of these provided many records for the Cajun Prairie. A popular article focused on butterflies of the Cajun Prairie region (Allen and Vidrine 1990a). Vidrine et al. (1992c) provided flight records of butterflies from numerous locations in southwestern Louisiana in or adjacent to the Cajun Prairie creating a list of butterflies in southwestern Louisiana."

"Dragonflies and damselflies of the southern United States are rather well known (Westfall and May 1996, Needham and Heyword 1929, Needham and Westfall 1955, Needham et al. 2000, and Dunkle 1989, 1990, and 2000). Those species that occur in Louisiana were described by Bick (1957) and Mauffray (1997). Mauffray (1997) reported 91 dragonflies and 33 damselflies from the state and included numerous references to the Cajun Prairie. A popular article focused on dragonflies of the Cajun Prairie region (Vidrine and Allen 1993)."

"Between 1986 and 1991, we observed 73,742 adult odonates and 8,517 adult butterflies (skippers excluded) (Vidrine et al. 1992b, c). Based upon our records and those listed by Mauffray (1997), 92 species

of dragonflies and damselflies occur in the Cajun Prairie ecosystem. The state list has 91 dragonflies and 33 damselflies for a total of 124 species of odonates in Louisiana. Based upon our records and those listed in Opler et al. (1995), 54 species of butterflies occur in the Cajun Prairie ecosystem. Their state list has 79 butterflies and 58 skippers for a total of 137 species. The names have been updated to match the currently accepted nomenclature."

Vidrine et al. (2001b) also provided tables of flight seasons of dragonflies and butterflies based upon sight records, and these are modified by using common names in an appendix (Appendix 13). The authors continued,

"Records for species determined to be overly difficult to separate as sight records are combined into species groups. Relative numbers of individuals and seasonal adult activity are readily perceived from the tables. The tables also list habitat preferences based upon the locations of most of the flight records for each species. Most of these adult insects either prefer the prairie region or commonly visit the region; although larval development may occur in other habitats."

"The search for remnant prairies led to a similar search for populations of native insects, mainly butterflies and dragonflies. The results of these searches include 92 species of Odonata (dragonflies and damselflies) and 54 Lepidoptera (butterflies, excluding skippers and moths) in the prairie ecosystem and associated habitats. In April 2008, a Mourning Cloak was sighted in the Cajun Prairie Gardens; thus, now 55 butterflies are known from the region. The region contains a diverse assemblage of these insects; however, many species are locally abundant and/or extremely rare based upon our sampling."

Vidrine et al. (2001b) continued:

"The long summers and brief winters extend the flight seasons and activities of the insects. Prairie remnants and prairie restorations provide a nearly year-round source of nectar and pollen resources. Loss of suitable habitat and extensive use of biocides apparently have apparently decimated most insect populations; however, relict populations of many species have been found on remnant prairies, in undredged streams, and in relatively old-growth gallery forests."

"Relict populations persist for a number of species of butterflies and dragonflies in southwestern Louisiana. The continuing threats to these populations by urbanization, intense sedimentation from not only urban runoff but also from the continuous activities of farmers employing

extensive plowing, and intense and continuous applications of massive amounts of biocides to massive monocultures of rice, soybean, and turf grass are evident. Some relict populations of butterflies have survived on small plots of native wildflowers with minimal sources for nectar and safe harbor from insecticides. The rice fields and drainage canals have provided habitat for many common species of dragonflies and damselflies, but the dredging of streams and the leveling of marshes and other wet habitats have robbed rarer species of suitable habitat. Many of these insects survive on the periphery of the prairie."

More recent assessments of the dragonflies and damselflies are available (Abbot 2005 and Strickland and Strickland 2008). The study of dragonflies readily converges with studying butterflies, as both of the aerial acrobats fill the Cajun Prairie skies during the growing season.

Butterfly gardening, pollinator gardening, and predator gardening are rapidly growing as pastimes in southwestern Louisiana (Ross 1994, Allen and Vidrine 1990a, Vidrine and Borsari 1998 and Vidrine and Hazelton-Robichaux 2003). Companion these activities with habitat restoration and the result can be extraordinary. Insects are able to reinvade and colonize a garden, even a water garden, and the end-result may be a very diverse habitat with many native species.

Extinct animals in the Cajun Prairie include Louisiana Prairie Vole, Bison, Red Wolves, Whooping Cranes and Attwater's Prairie Chickens (Smeins et al. 1992). Rare animals include Ornate Box Turtles and American Alligators. Common animals include White-tail Deer, Coyote, Turkey, various birds, crayfish, skunks, opossums, etc.

Clifford Fontenot Jr. (2004) provided some interesting Cajun-French common names for Louisiana amphibians and reptiles that I grew up using. Many areas named for animals and plants (Native American names and French/ Spanish names) (see Appendix 7). Bill Fontenot's weekly piece in The Lafayette Advertiser on wildlife in the Cajun Prairie provides an excellent natural history format for keeping up with local natural events.

Mosquitoes, economically and medically important insects, have received much attention. Louisiana mosquitoes include 60 species, both native and introduced from all over the world (Chapman and Johnson 1986 and Darsie and Ward 2005). The most recent introduction is Asian Tiger Mosquito (*Aedes albopictus*), which is thought to have arrived in the early 1980s in a series of shipments of automobile tires from Japan—back in the day when we gave away our trees in exchange for old tires and considered that to be 'fair trade.' The mosquitoes arrived in California and started being collected along each of the major interstates across the country. My first encounter was in 1985 on my porch, when I slapped one up under my pant leg and popped it into my hand for identification with my hand-lens. I remember distinctly yelling, 'What the heck is this?' only to find out with a quick call to my old mosquito control district that they had just collected the same. Like many mosquitoes, this one has a nasty reputation for transmitting some disease or another, likely dengue fever and West Nile encephalitis.

I had the opportunity to work with mosquitoes in two venues. I served as Assistant Director and Research Director for Jefferson Davis Parish Mosquito Abatement District (1980-1984), and I worked with Roy McLaughlin, my mentor in agricultural/insect field research, who was a Research Entomologist with the Gulf Coast Mosquito Laboratory (a branch of the Agricultural Research Service in the U. S. Dept. of Agriculture (ARS: USDA)). With these opportunities, I studied the diversity of mosquitoes in the Cajun Prairie, their biology and their population control. I had the opportunity to literally walk along the levees sampling the water of more than 1000 rice fields. I found a significant variety of biota and continue to publish on them. In the end, I severed my service with the district to teach at LSUE. My view on mosquito control as it is currently done was greatly affected by the work of my mentor, Roy McLaughlin.

Roy McLaughlin (1984) summarized his findings from a comparative study of the mosquito population biology in two parishes with widespread rice fields in a report. Jefferson Davis Parish had a modern, in fact progressive, mosquito abatement district, whereas Acadia Parish did not have a mosquito district in 1984. By comparing large numbers of light trap collections and field samples as well as standard landing counts of mosquitoes on pant legs, he stated: "*The peak number of adults caught in a trap varied but when all data from traps were pooled over parish, the r-value was still extremely high in all instances. This suggests that the rate of increase was similar also for each parish and only the maximum peak value may have been affected by control practices.*" He noted that the parishes have similar mosquito populations during the early part of the growing season (April-July), and this may be due to minimal spraying in Jefferson Davis Parish as compared to no spraying in Acadia Parish—these being indistinguishable treatments. However, in the later part of the growing season (July-October), spraying was exponentially increased in Jefferson Davis Parish. In spite of all that spraying, "*We consider the data to be sufficient to warrant the hypothesis that growth rate is intrinsic and occurs regardless of control practices.*" In the end of the study, it was obvious that control strategies that would reduce early season populations may significantly reduce the number of female mosquitoes that would lay eggs, but there was no way to determine when and where spraying might effect a significant reduction. Models, as the one developed to support the use of spraying to reduce the numbers of mosquitoes, are dependent upon tightly controlled testing in settings that meet specific assumptions. In the laboratory verses nature, these models either rely on too few variables (sin of omission), too many variables (sin of commission), incorrect assumptions, unknown variables, averaged variables and the measurement of linear outcomes (Pilkey and Pilkey-Jarvis 2007). In nature and field tests, the models make predictions that are seldom realized, except in their most generalized outcomes. Based upon elaborate tests in the field, mosquito control in rural areas generally failed to control intrinsic population growth; however, local, short-term effects of control were commonly observed. In the light of an epidemic outbreak of a mosquito-borne disease, the economics of the system might bear justification. Dozens of examples of insect eradication clearly show a reduction in disease incidence in specific locations. However, literally very little follow-up research upon wildlife is done in an effort to show the impact of the application of metric tons of active ingredients of these insecticides.

The insects (other than mosquitoes, butterflies, dragonflies, bees and wasps) that readily come to mind as major prairie insects are the grasshoppers, katydids, walking sticks, preying mantids and crickets. These insects literally manage the prairie as major herbivores and predators. Matt Dakin, my entomology professor, spent some time out in the prairie remnants collecting grasshoppers and their kin. Interestingly, my fondest memories of Dr. Dakin are related to an aquatic adventure in which the limnology class discovered a rare behningiid mayfly larva in Bundick's Creek near Kinder. *Dolania americana* represented a rare find in the eastern United States—a find in Louisiana was unique (Dakin and Felder 1981). The terrestrial insects will have to wait for someone to chronicle their variety in the Cajun Prairie ecosystem.

I promised back in Chapter 6 to discuss 'resource partitioning' as an ecological concept. The atmosphere is an excellent place to present a concept that occurs in all the spheres because you can see it without much exercise. Dragonflies are my best atmospheric example of 'resource partitioning,' although most vertebrate biologists would immediately begin a discussion of birds as they are also excellent examples. The air is divided vertically and horizontally by the dragonflies. Vertically, the ground and surface of shorter plants of the prairie are patrolled by damselflies and small dragonflies. They eat insects that occupy this space. Medium-sized dragonflies patrol areas from roughly 3' to 10' in height. These are the dragonflies that you see on your clothesline or fences or sitting on the tallest tips of grasses. The area of the atmosphere greater than 10' is patrolled by several large dragonflies, including darners, saddlebags and gliders. They swarm, sometimes by the hundreds, and pluck individuals from other swarms of insects, including mosquitoes, midges and love-bugs. They seem to follow the sun in that they fly near the surface of the plants in early morning and late evening, but as noon approaches, they are seen flying ever higher to 50' or more in the air. Horizontally, the atmosphere is divided by dragonflies in respect to objects in space. A stream or a pond will be patrolled by specific kinds of dragonflies, while a trail or path will be patrolled by other kinds. Some species have long flyways and fly up and down along the flyway for hours. Others have specific perches, where they fly a few times then simply wait. Other dragonflies patrol a specific radius around a perching site. The pennants and skimmers perch on the tips of grasses or branches and follow potential prey with their multifaceted eyes then dart and capture their prey. Literally every stem can have a dragonfly in some areas. A single tree of any size provides for another dimension in this natural theatre of diverse strategies for obtaining prey; mating also changes everything. The diversity of strategies for dividing the airspace is seemingly endless. The nymphs of these very same dragonflies partitioned the space in the hydrosphere as they were developing. Evolution seems to increase the efficiency of each plan for each species over time; thus allowing ever more opportunities.

Before I leave the atmosphere and 'resource partitioning,' I want to remind you that in addition to spatial considerations, temporal considerations also apply, and a 'space' can be occupied by two or more species during a growing season. I also want to remind you that the same dividing processes occur underground as roots, earthworms and many other organisms divide the soil both vertically and horizontally. The '*roots*' drawing (Figure 1) depicts the spatial relationship of roots of varied plant species in the prairie. As I mentioned above with nymphs of dragonflies, the water column of the hydrosphere is also partitioned by the varied kinds of life. My own research centers on mussels and the communities that they create in streams and ponds as they sort themselves along the bottoms. Small differences in current speed, sand or silt content and clarity of water greatly affect which species are present. Even the mites that occur in the mussels are host-specific and if they occur in the same host, they partition the tissues of the mussel for egg-laying and feeding. Most parasites likewise attack specific tissues and locations on their hosts. Fishes are extraordinary at 'resource partitioning' the hydrosphere.

The Hydrosphere

Rivers, lakes, canals, rice fields, crayfish ponds, swamps, borrow pits and canals are evident throughout the Cajun Prairie. A visit to GoogleEarth (http://earth.google.com/) will provide a quick glimpse of the diversity of watery habitats in the region. Many are man-made, but the original waterways remain, however greatly modified. The rivers have histories that are an essential part of the ecology of the region (Davis 1968).

The Cajun Prairie is drained mainly by the Mermentau River and its tributaries. However, the western edge is drained by the Calcasieu River system, while the eastern edge is drained by the Bayou Teche/Atchafalaya River system. The southeastern edge is drained by the Vermilion River system, a small river now draining the urban area surrounding the city of Lafayette. These rivers were once sandy, but now they are all mud bottom, except for the main channel of the Calcasieu River. The benthic communities of organisms in these streams are central to our understanding of the biotic diversity and ecological health of the aquatic ecosystem.

Freshwater community structures of animals are poorly known in southwestern Louisiana. General diversity studies exist on major groups like fish, mussels, crayfish, mosquitoes, etc. Several papers and/or books described benthic community complexity (Curry et al. 1981, Smith 2001, Thorp and Covich 2001, Vidrine 1996a and b, Huner 1998, Klemm 1985, George et al. 1993, Howells et al. 1996 and Walls 2009). We described the general benthic community structure of streams (bayous or rivers) in the region and emphasized the freshwater mussel community of this habitat in southwestern Louisiana (Vidrine et al. 2006). Similar studies have been done in the Red River in Louisiana and the upper Mississippi River, two rivers once lined with prairies (Vidrine 2008 and Vidrine et al. 2010).

I studied a great variety of animals in the prairies, including insects (mosquitoes, butterflies, and dragonflies), mollusks (mussels and snails), rotifers, clam shrimps, mites, armadillos, reptiles, and fish. I worked with the great variety of domestic animals, including cattle, horses, pigs, chickens, sheep, dogs, cats, etc. Although I have not done work with birds, except chickens, I took George Lowery's ornithology class in 1968. During my career, I have had the opportunity to study a number of aquatic systems in the Cajun Prairie, including streams, lakes, marshes, rice fields, etc. As a result, I have a number of papers and reports that provide lists of animal species. The following paragraphs from Vidrine et al. (2006) provided some insight into the diversity of the aquatic systems in the Cajun Prairie.

> *"Freshwater mussels (Bivalvia: Unionidae) once cobbled the stream bottoms in the Cajun Prairie ecosystem in southwestern Louisiana. The development of canal systems for above-ground irrigation of rice fields during the last century provided additional habitat—often the canals were likewise cobbled with mussels that formed diverse communities. This paper introduces the thirty-one known species, their general ecology and their distributions. These mussels develop a 'reef'-like benthic community that is home to a variety of plants and animals as parasites and haptobenthos. Freshwater mussels are on the decline numerically as well as in diversity as a result of human activity, namely loss of seepage or spring-feeding (shallow groundwater discharge) by agricultural/ urban compaction of soil, sedimentation from erosion, channeling of natural streams by dredging, and poisoning of waterways from urban and industrial runoff and dumping. The once-spring-fed streams flow intermittently functioning more as drainage canals as a result of the destruction of prairies by agriculture and urbanization. The return of prairies and forests reinitiates the springfeeding by shallow groundwater discharge to streams and minimizes runoff of agrochemicals and urban*

chemicals, thus providing an opportunity for revitalizing freshwater mussel communities."

"Freshwater mussels are among the most endangered animals in fresh waters (Strayer et al. 2004). Freshwater mussels (Bivalvia: Unionidae) once cobbled the stream bottoms in the Cajun Prairie ecosystem in southwestern Louisiana (Vidrine 1993). The development of canal systems for aboveground irrigation of rice fields during the last century provided additional habitat—often the canals were likewise cobbled with mussels that formed diverse communities (Vidrine and Vidrine 1987, Vidrine and Quillman-Vidrine 1994, Vidrine and Borsari 1994). This paper introduces the thirty-one known species, their general ecology and their distributions."

"These 31 mussel species comprise a complex community that hosts a diverse haptobenthic fauna and flora as well as a diverse symbiotic/parasitic community (Curry et al. 1981, Vidrine and Vidrine 1987). Many of the mussel species were found in all the major streams in the Cajun Prairie, but some are limited in distribution."

"By examining the mussels of the Mermentau River system, which is the main stream draining the central portion of the prairie, we can discover the primary fauna. Noticeably absent are several mussels that would have been expected to be common in this river system, since they occur in adjacent rivers and are relatively widespread nationwide: A. plicata, M. nervosa, O. reflexa, O. jacksoniana, T. donaciformis, and T. truncata. The absence of these mussels may reflect alteration in the stream, which have resulted in the loss of riffle habitat and sand bars. Further the loss of fish host habitat may contribute to these absences."

"The freshwater mussels in these benthic communities represent the large reef-building (megafauna) of the benthos—many other kinds of organisms live in or on these mussels. The community is a diverse and complex one. A more detailed study of the community begins with the parasitic mites, which have received detailed attention (Vidrine 1996a). Thirty species of water mites have been found in mussels in the Cajun Prairie region. Also two species of aspidogastrid trematodes were common in mussels. These parasites create a diverse community that not only speaks to the ancient interactions between these species but also indicates the nature of these interactions. In many locations where cotton is grown, the mites literally disappear such that mussel communities persist without mite populations (George and Vidrine 1993)—this is an example of the impact of human activities on these communities."

"The mussels are often parasitized by one or two species of aspidogastrid trematodes (Hendrix et al. 1985) and one or more species of 30 known species of unionicolid and/or najadicolid water mites in

the Cajun Prairie region. These associations are elaborate and often host specific. Trematodes and mites are commonly found parasitic in freshwater mussels (and snails) of the Cajun Prairie ecosystem."

"Many organisms including fish, muskrats, otters, turtles and crayfish are routine predators or consumers of freshwater mussels. Historically, humans ate lots of mussels and fed them to their pigs. Mussels have also been extensively harvested for mother-of-pearl or pearls or seeds for the pearl industry. Exotic species, including the Asiatic clams (Corbicula), have also affected native mussel community structure. The recent invasion of Zebra mussels is yet to be assessed in southern Louisiana."

Benthic stream bottoms are occupied by a great number of organisms in addition to the mussels. I did a report itemizing many of these organisms for Jan Boydstun (Water Quality Management Division, Louisiana Department of Environmental Quality) (Vidrine 1992). The specimens were from the Houston River near Lake Charles, Bayou Mallet near Eunice, and Bayou Queue du Tortue. The number and diversity of aquatic oligochaetes (earthworms) was astounding.

Mike Loden, one of Walter Harman's students, helped me identify the numerous aquatic earthworms—these are among the most critical of organisms used as indicators of habitat quality in polluted streams. Mike described a tubificid oligochaete (aquatic earthworm) from Louisiana and named it after Dr. Harman; it has since been found in Europe and placed in a different genus *Varichaetadrilus harmani* (Loden 1979) (Timm 2006). This worm was the most abundant worm in the Houston River samples. Much of Dr. Harman's work dealt with earthworms and aquatic oligochaetes, except ironically for his report of the exotic clam *Corbicula* in the Calcasieu River (Dundee and Harman 1963). Appendix 11 includes many of the other organisms found in the 1992 analyses, but those few samples are too few to permit many generalizations regarding the benthic diversity in the aquatic systems of the Cajun Prairie.

The rice fields provided excellent habitat to discover a wide range of aquatic organisms. From 1980 to 1985, I spent long days each summer sampling rice fields for mosquitoes. I carried an eyedropper (standard pipette) and a bottle in my pocket. With each dip of water, my eyes quickly identified and counted the mosquitoes, and then the other organisms were examined. Any interesting critter was placed in the jar; there were many, including rotifers, which I developed a passion for. Other likely candidates included mites, snails, algae, protists, crayfish, other crustaceans, dragonfly and damselfly nymphs, other insects, clam shrimps, etc.

Free-swimming colonial rotifers were common to abundant in rice fields in the Cajun Prairie ecosystem in southwestern Louisiana during surveys conducted in the mid-1980s. Seven species were discovered during those surveys. The species are: *Conochilopsis causeyae* (Vidrine, McLaughlin and Willis 1985), *Lacinularia flosculosa* (Mueller 1773), *Lacinularia ismailoviensis* (Poggenpol 1872), *Lacinularia elliptica* Shepard 1897, *Sinantherina socialis* (Linnaeus 1758), *Sinantherina semibullata* (Thorpe 1889) and *Sinantherina spinosa* (Thorpe 1893). In the summer of 2004, four of the species were recovered in a survey of first crop rice fields. The reduction in the diversity and number of rotifers in 2004 may have been a result of intense rains and low temperatures. The unusual nature of the high diversity of these rotifers in the ecosystem and the possible use of these rotifers in physiological research remains a great interest for me (Vidrine et al. 1985, M. Vidrine and Hazelton-Robichaux 2008 and C. Vidrine and Hazelton-Robichaux 2010).

Native Americans apparently ate large amounts of Atlantic rangia (*Rangia cuneata*), a common clam that tolerates freshwater but requires salinity for larval development. Large mounds of relict shell are found along the coast and in the marsh (Hoese 1973). In the same habitat, crab, oyster and shrimp were harvested. In more inland areas, handfishing for large catfish, turtles and Bullfrogs (*Lithobates catesbianus*) was common (Mire 1992). The Cajuns adopted some of these habits, but apparently not all of them. Stories of giant catfish, giant Paddlefish, and giant Alligator Snapping Turtles are true—I witnessed them.

The rice fields served as crayfish ponds and a steady source of Bullfrogs in my youth. When the water was removed from the rice, harvesting of these critters was in full swing. Unfortunately, many of today's rice fields lack the crayfish and bullfrogs of my youth. Changes in the manner of farming and in the use of biocides have had a dramatic effect on all wildlife associated with rice fields. Crayfish farming has developed as an entirely separate agricultural practice in aquaculture (Huner 2000). This same habitat proved to be visited by legions of vertebrate organisms, which Jay inventoried (Huner 1998) (see Appendix 12). Few studies of the vertebrate populations over time in the remnant and restored prairie are known to the author, but Williams and Trahan (2009) conducted a study on the LSUE prairie restorations and observed changes in the vertebrate populations over a single growing season.

Urbanization had a tremendous impact upon wildlife. The sheer loss of prairie and woodland habitat by its replacement with streets, parking lots and buildings is dramatic. The massive cultivation of cats and dogs is a major side effect of human activity. These predators literally kill off every last bird and small animal in an area. Even hummingbirds readily fall prey to cats when the birdfeeders are too close to the ground or in a good location for the cat to get at the birds. On the other hand, the love of Purple Martins and the building of houses for them have led to a severe decimation of dragonflies and butterflies. Rumors of martins eating lots of mosquitoes still abound in spite of failure by intense research to prove this. Stomach contents of martins literally contain only dragonflies and butterflies. Like so many other mosquito myths, the actions taken to eliminate mosquitoes all too often exacerbate the problems caused by mosquitoes. I have the better part of 100 such stories as a result of my 4-year tenure as the assistant director of mosquito control in Jefferson Davis Parish in the early 1980s. There are an equal number of stories regarding snakes, and I have seen many of these likewise discarded by simple research. When I think of mosquitoes and snakes, it reminds me of prairie restoration—these projects are daunted by rumors of hoards of mosquitoes and snakes, but all of the evidence is to the contrary.

In zoology, two final aspects are essential to unraveling the faunal diversity in the prairie ecosystem. The first aspect is symbiosis, which includes commensalism, mutualism and parasitism. The notion is that any single species has one or more, usually 10 or more, symbiotic associations. My experience with mussels makes this extremely obvious to me. Even the most inconspicuous of animals has many parasite species. Thus any faunal list will be only a hint at the complexity of the diversity. The second aspect is predation, including herbivory and carnivory. Animals eat plants and animals in what is generally constructed into a food chain for a diagrammatic presentation. Each animal has preferences, but the system can be seasonal, such that one food works for one season and another for a different season. If the animal undergoes dramatic changes in lifestyle, e.g., tadpoles change into frogs, mosquito larvae change into mosquitoes, caterpillars change into butterflies, then the food chain is even more complex. And if the animal lives in one or more habitats (e.g., hydrosphere, atmosphere and/or rhizosphere), then the food chain takes on added dimensions. Further, if a species during its development changes from a parasitic to a free-living form or vice-versa, the presence of hosts is required. In the end, the faunal list is not just a list of species; it serves as a primer for a list of interactions (ecology). Appendices 11 and 19 provide an introductory list to the animal species of the Cajun Prairie ecosystem and exotic species, respectively. Jay Huner's inventory of vertebrates for a 600 acre farm is provided in Appendix 12.

There is a need for a book on the animals of the Cajun Prairie. The modest diversity of the animals that we know now is large enough to fill a substantial volume. Any attempt to illustrate even a fraction of the diversity quickly fills a volume, as was evidenced by Miller and Fontenot (2001) and Gomez (2008).

As we unravel these ecological systems, we have an opportunity not only to understand their complexity but also to attempt to recreate/restore the complexity. The nature of conventional agriculture and contemporary human culture is to simplify our surroundings, including our food, our interactions, our community and our security. However, ecological principles clearly dictate that simplicity is lethal; it is usually the last step in succession leading to collapse of an ecosystem (Diamond 2005). Jared Diamond discussed the conundrum of the 'tragedy of the commons.' When resources are limited and resource utilization is essential to survival, the tendency of a society is to use the last of the resource with abandon. Each user attempts to maximize his share, and in the end, the resource is completely depleted. Whether the resource is trees (as on Easter Island), fisheries (as in our oceans), or prairie (as in the Cajun Prairie), the end has been the same. In the Cajun Prairie, we plowed literally every piece of land, and we attempted to eliminate every plant species for which we did not have an immediate use. We dredged every stream, we drained the marshes, and we completely changed the hydrology of the prairie. We literally hunted every prairie chicken to extinction. In the end, our natural system is in danger of collapse. The solution is management based upon sound ecological principles—preserving and restoring resources would permit the redevelopment of a sustainable system.

The most important animals on the prairie are humans and their domestic animals. The impacts of our actions are the most important, and we should be aware of this.

Chapter 10
The Future of the Cajun Prairie

"The forest is a peculiar organism of unlimited kindness and benevolence that makes no demands for its sustenance, extending generously the products of its life's activities; it affords protection to all beings, offering shade even to the axe-man who destroys it." (Buddha's often quoted clarion environmental insight) (If you substitute the words 'prairie' for 'forest,' "soil' or 'grass' for 'shade,' and 'plowman' or 'cattlemen' for 'axe-man,' this quote becomes pertinent to the Cajun Prairie).

"The Coastal Prairie is a unique and vital part of the biosphere that has almost vanished within the last 100 years. Much has been lost both in terms of land coverage and native species, and what remains is in need of protection and rehabilitation. Because so little remains, the future of Coastal Prairie depends on restoration. Americans can help in this effort to protect and restore Coastal Prairie by supporting or participating in restoration efforts. Even a small backyard prairie garden (12' x 12') provides a piece of native ecosystem. Thousands of such gardens dot the Midwestern countryside, providing a refuge for native plants, insects, and birds, and an alternative, sustainable landscape." (Allain et al. (1999 and 2000) (Second place (Blue Pencil award) for Brochure entitled PARADISE LOST. Award granted March 8, 2001 by National Association of Government Communicators at the annual conference in Denver, Colorado)).

The first habitat restoration (a Midwestern prairie recreation) was carried out under the watchful eye of Aldo Leopold in the 1930s at the University of Wisconsin Arboretum. Cottam (1987) described the Wisconsin restoration as follows: *"By the 1950s, a quarter of a century after restoration began on the Curtis Prairie, the 24 ha community included more species than any other area of comparable size in Wisconsin, and its rich mixture of species, including more than 300 prairie natives and several dozen exotics, was providing unmatched opportunities for research on the dynamics of prairies."* Our Cajun Prairie Restoration Project and the Cajun Prairie Gardens in Eunice, LA, are just a bit younger but parallel Wisconsin's efforts in statue and importance for our Cajun Prairie ecosystem.

In the early chapters of this book, the Cajun Prairie has been introduced in a variety of ways. First, we geographically and culturally located the prairie. Using my genealogy, I tried to link the people to the prairie, clearly showing that a variety of people from Native Americans to the French, Spanish, Germans, English, but especially the Cajuns came to this land and thrived on it. They gave the prairie its unique culture with music, food and social events of international importance.

The latter chapters of this book discussed the soils, hydrology, plants and animals of the Cajun Prairie. These chapters will be pivotal for this final chapter, which discusses the current status of the prairie remnants and the future of the Cajun Prairie. An essay on the Cajun Prairie as it may have looked several hundred years ago is provided in Appendix 5.

The Status of the Cajun Prairie as an Ecosystem

The remnant prairies

Upon re-examining the remnant strips of railroad prairie, it appears that a new regimen of maintenance of railroad rights-of-way has developed in the Cajun Prairie region (Vidrine and Hillman 2008). Three methods (mowing, herbiciding and removal of fire) are involved and are extremely expensive considering the cost of fuel and the potential problems with trees falling on the tracts. As a second theme in this book, I want to propose a return to burning as the primary method to maintain these rights-of-ways. It is cheap, natural and perfect for the native plants and animals. It provides the extraordinary result—native prairie—a magnificent habitat for wildlife. The negative publicity for fire in nature has been so effective that people no longer consider fire as a natural phenomenon nor as a method to maintain habitat.

It is easy to compare the conditions of the prairie remnants. Plates 2-28 depict the remnants in the 1980s and 1990s, whereas Plates 29 depicts the remnants as they appeared in 2008. Damage and death to the remnants were obvious in 2008 (Appendix 22). Some of the varied insults included:

1. use of large amounts of herbicides and other biocides
2. topsoil removal
3. scrapping of trees and soil into piles of debris
4. packed trails of four-wheeler traffic
5. complete destruction in order to build roads, airstrips, etc.
6. lack of burning and resultant overgrowth of a variety of woody plants
7. mowing/routine mowing
8. crosstie replacement
9. heavy equipment causing compaction and deep ruts, and
10. changing of hydrology by draining of wet areas and/or creation of new wet areas

With the loss of the remnants, we can say that the prairie is LOST (Allain et al. 1999 and 2000). Here is a tentative ranking of prairie remnants as made on March 20, 2008 as to possibility of preservation by fire. The higher ranks indicate the remnants that are more likely to recover quickly with fire. The ranks are:

1. south Fenton
2. south Kinder
3. north Iowa
4. Welsh
5. south south Fenton
6. Frey
7. east Mermentau
8. east Midland
9. Estherwood
10. south woodlawn
11. west Midland
12. west Mermentau
13. Elton
14. north Woodlawn

Fire is essential. Restoration of habitat is essential. The Cajun Prairie and its adjacent ecosystem components exemplified by streams and gallery forests have been reduced to less than 1.0% of their original area. The few refuges that persist are in eminent danger of eradication by a variety of threats. As an example, restoration of habitat is the only recourse available to flying insects of the Cajun Prairie. Recent efforts at gardening for butterflies and beneficial insects also provided an opportunity for these essential organisms (Fontenot 1992 and Vidrine and Hazelton-Robichaux 2003); however, the habitat must be recreated in large tracts in order to permit sufficient habitat for the success of viable populations. As important as these issues might be, the recent outcries to protect the pollinators like native bees (Buchmann and Nabhan 1996) and to protect and to propagate the variety of natural predators were but a continuation of the present effort to protect diversity and develop sustainable lifestyles and habitats (Vidrine and Borsari 1999) and biodiversity gardens (Nabhan 2009). Similar passionate pleas for the return of fire as a major management tool in order to preserve and create Longleaf pine forests in Louisiana and the southeastern United States literally bookended (began and ended) the magnificent book on these forests by Lawrence Earley (2004). I too want to be clear in my recommendation for fire as the central and first tool to use in the preservation and restoration of our Cajun Prairie.

The Restorations

The Cajun Prairie Restoration Project in Eunice and the Cajun Prairie Gardens are models for the next incarnation of the Cajun Prairie, a series of biodiversity gardens designed to protect and maintain native species of plants, animals, fungi, protists and bacteria. They currently work like gardens, but they can be further enhanced to improve hydrology and serve as a model for agriculture and for landscaping. The system is sustainable, as evidenced by the gardens that have been created, and as such, the new habitat can be sustainable. Sustainability is the first criterion to measure in this new age—as the cost to annually recreate agricultural systems, even landscapes, is quickly going to be prohibitive. This prairie habitat greatly reduces allergens and produces many potential plant remedies for modern illnesses. The prairie species contain genes that are extremely diverse and have never been studied. The restorations and hopefully the preserved prairie remnants provide the hope that the next generation can not only sustain the habitat but also study its details further unraveling its idiosyncrasies.

The overall findings of our work are: The Cajun Prairie intergraded almost imperceptibly into fresh marsh to the south (Grace et al. 2000a) and into flat pine savannahs to the north (Allen 2003), and it intergraded with gallery forests and bottomland hardwoods along the streams. Grace et al. (2000b) indicated that moisture gradient and topographical micro-relief had the greatest impact on biodiversity of plants in the Cajun Prairie remnants. The diversity of plants exceeded 600 species and approached 1000 species, depending on how one defined the prairie ecosystem. The animals associated with the Cajun Prairie are numerous, and they require much more study. The fungi and other groups of living things are poorly known. We, the restoration ecologists and gardeners, have an ecological vision of what the Cajun Prairie was and what it can be.

With both a working fauna and flora for the Cajun Prairie, the biologists, restoration ecologists, and the naturalists can begin the tedious but exceptionally rewarding tasks of preservation and restoration of this ecosystem. The absence of a working list of fungi, bacteria, and protists, specifically for the Cajun Prairie is a fundamental weakness, but lists of the biota are available for general regions like the state of Louisiana and the southern United States. Our sheer lack of appreciation for the latter groups is not limited to Cajun Prairie, but rather to those groups in general. Their small size, inconspicuous lives and known negative attributes have gained them little in the realm of importance; however, they are ignored at our peril. The number of restoration initiatives in spite of setbacks is growing. Restorations following the model created by Charles Allen (Appendix 16) apparently regenerate prairie conditions for plants and animals in five years—the model can be extended to almost any prairie habitat in Louisiana, Mississippi and Texas, with modifications (see works of Larry Allain, Bill Fontenot, Peter Loos and Marc Pastorek).

The Cajun Prairie as Part of the American Prairie Ecosystem

The current central role of prairie ecosystems is to grow food and fuel. The Cajun Prairie is not really different from the Midwestern and coastal prairies in the overall context in that it is a grassland. The decision to use grasslands as commodities was made as soon as humans entered the prairie, even the Native Americans ate plants and Bison and burned Bison 'chips' in their campfires. More than once, hysteria has overtaken the prairie and its inhabitants. The latest hysteria is the 'ethanol' boom and the soon to follow shortages of grain. Prairies are grasslands, and corn is a grass. Switchgrass and sheer prairie plant biomass are currently getting attention for ethanol production. Cellulosic ethanol made from prairie grasses is a possible future fuel source. The prairies provide massive amounts of biomass each year, and the system thrives on fire. Harvesting this biomass and burning the remaining mass would be a potential agroindustrial model for reviving our prairies before they are totally lost.

Biomass harvesting followed by the generation of ethanol and biogasoline are currently being studied (Jordan et al. 2007). Cellulosic ethanol and biofuel, including the new biogasoline, are potentially renewable fuels. The biomass from restored prairies can now be harvested and converted directly into fuels. The problems with corn-based ethanol are only now obvious, including loss of foodstocks, damaging habitat to grow more corn, increased costs of meat and grain and the small gain in overall fuel resources (as it takes almost a gallon of petroleum fuel to make the same amount of ethanol). The use of biomass (cellulose in any form) is far more sensible in the long run.

Jordan et al. (2007) presented some of the general considerations for the energy model:

> *"A 'bio-economy' based on agricultural biomass is emerging in the United States that offers an avenue toward energy independence and a more 'green' economy."*
>
> *"Agricultural multifunctionality is defined as the joint production of standard commodities (e.g., food or fiber) and 'ecological services.' Examples of the latter include increased recreational opportunities in agricultural landscapes and protection of biodiversity and water quality."*
>
> *"Multifunctional production systems can be highly valuable. The 34-million-acre Conservation Reserve Program (CRP) has been estimated to produce \$500 million/year in benefits from reduced erosion and \$737 million/year in wildlife viewing and hunting benefits at a cost of ~\$1.8 billion. If benefits such as carbon sequestration are added, CRP likely produces a net gain in many areas, if not for the entire nation."*
>
> *"Diversified grassland agroecosystems on degraded agricultural land can increase both carbon storage and net energy gain in biofuel production."*
>
> *"Research must be focused on the tradeoffs that arise, e.g., between wildlife habitat and biomass production."*

Closing Remarks

We (Vidrine et al. 2001c) began a discussion of the future of the Cajun Prairie as follows:

> *"Sadly, because no adequate pre-settlement vegetational records exist for the region, the present restoration efforts represent collections of those residual plant species, which have persisted within the small remnant strips. By the same token, it has already become evident that these residual plant species are creating a vegetational matrix, which appears to exclude the majority of invasive exotic plant species present within the region, and attracts and supports a surprisingly high diversity of native animals, both vertebrate and invertebrate. Workers are thus hopeful that over time, the entire community of organisms which once constituted the historical coastal tallgrass prairie may recreate a habitat, which very closely approximates that of the original (Vidrine et al., 1995, and Vidrine and Borsari, 1998), inspiring its human inhabitants to more ethically understand the complete circle of life, where the land itself comes to be viewed as the real community to which the people belong (Leopold, 1949 and 1999 and Jackson 1999)."*

The Cajun Prairie is a very different place today as compared to the 1600s. With the loss of vast community of plants and the animals, the introduction of cattle, horses and livestock as well as the great variety of exotic plants led to an agricultural ecosystem at the expense of the native prairie. Urbanization and the oil and gas industry further contributed to the present ecosystem. This ecosystem has provided for the livelihoods for our citizens. But is this ecosystem sustainable? Recent hurricanes and economic problems strongly suggest that the system is failing. As ecologists, we not only warn of impacts associated with change in unsustainable systems but also suggest alternatives that might lead to more sustainable choices. Prairie, the natural habitat that literally created the agricultural soils of the region, is one of those alternatives. We do have small fragments of prairie remaining, and we do have the knowledge and understanding necessary to begin recreating the prairie habitat. But do we have an incentive or even a desire to do so?

Some of us do have such a desire, and steps are being taken to begin this odyssey. The Cajun Prairie Habitat Preservation Society has provided the initial framework for collaboration and coordination in these efforts. The new Cajun Prairie will be very different, but it can be sustainable and productive, and it can help ameliorate the effects of climate change. The current predictions for climate change place the Cajun Prairie essentially back into the sea within the next 200 years—so it is essentially a fight for the life of the region.

The new version of the Cajun Prairie (a biodiversity garden) can be a beautiful place, with flowers of nearly 1000 kinds and a growing diversity of animals and soil biota. The streams can once again run clear, and the life in those streams can redevelop. The hydrology of the prairie can be restored. All of these hinge on the development of a new attitude in the people of the prairie. This prairie can be a place of solace, where a small child or an old man can once again sense nature and connect to the past by walking backwards through time in order to see Bison and prairie chickens among the grasses and forbs. If it ended there, this in itself would be enough, but there is much more.

The Cajun Prairie is connected to the marshes and the nearby forests. Restoration of the prairie is connected to restoration of the marshes and forests. The protection of the people and property from the forthcoming onslaught of hurricanes, droughts and general climatic change is in part linked to redeveloping habitat—habitat that not only survived previous onslaughts but was also derived from those very same onslaughts. The dangers are now expanded with the massive growth of the oil and chemical industries (see chapter 10 of Weisman 2007). The organisms, including man, are connected to the land (Leopold 1949); if we lose sight of this one more time, the land will literally disappear and all that is connected to it.

To close with optimism is my goal. A short story recounting a recent event will serve as my closing. In the spring of 2008 as the prairie was beginning its bloom, our local electrical supplier contracted a company from outside our immediate area to replace and fortify the electrical poles along our highway and thus along the Cajun Prairie Gardens in our front yard. The contractors took great care to lay out plywood sheets so as not to rut the neighbor's lawn, but in the prairie garden, they simply drove through and created deep ruts with abandon. When I observed the damage, I spoke with the foreman of the crew about this, took his apology and refused his generous offer to get his crew and their equipment back into the prairie to 'fix' it. He then said that they would work from the highway, as they usually do, for any additional work. So far this is not an optimistic story, but before you show disdain at my effort, the story continues. My daughter Caroline walked out to the scene of the damage. She took one look at the ruts and to my surprise, she allowed anger to well-up as she spoke more strongly than I had ever recalled. In disgust, she remarked, "How in the heck can these idiots not have seen the flowers in bloom and the sign saying the area is a garden? All the work that I have done is wasted!" I reminded her that these plants will heal the damage with just a little help from us. But I had seen for the first time the passion in my daughter that I knew so well in myself. It is this passion that moved me to do what I do. It is, in part, anger, hope, fear and a dire sense of impending doom and it is yet a source of motivation that causes me (and Caroline) to do something to conserve that which others have no concern for whatsoever. I am in the end reminded by the Kenyan Proverb, '*This land does not belong to me; it is on loan to me from my children.*'

Literature Cited

Abbott, J. C. 2005. Dragonflies and damselflies of Texas and the south-central United States. Princeton University Press (Princeton, NJ). 344 pp.

Ajilvsgi, G. 1979. Wild flowers of the Big Thicket: East Texas, and western Louisiana. Texas A. & M. University Press (College Station, TX). 361 pp.

Ajilvsgi, G. 1990. Butterfly gardening for the south. Taylor Publishing Co. (Dallas, TX). 348 pp.

Al-Dujaili, J. S., and M. F. Vidrine. 2008. Antimicrobial Activity of Cajun Prairie Herbs on the Growth of Pathogenic Bacteria in Foods. Proceedings of the 20th North American Prairie Conference 20: 165-176.

Allain, L. 2007. Coastal Prairie Restoration Information System: Version 1 (Louisiana). Data series 256. U. S. Dept. of the Interior: U. S. Geological Survey. CDrom.

Allain, L. 2010. The role of soil fertility in restoring Louisiana's coastal prairie. Proceedings of the 21st North American Prairie Conference 21: 54-62.

Allain, L. and J. B. Grace. 2001. Changes in density and height of the shrub *Baccharis halimifolia* following burning in coastal tallgrass prairie. Proceedings of the 17th North American Prairie Conference, 17: 66-72.

Allain, L. and S. Johnson. 1997. The prairies of coastal Texas and Acadiana. Wildflower (Spring): 42-45.

Allain, L., M. Vidrine, V. Grafe, C. Allen, and S. Johnson. 1999. Paradise Lost? The coastal prairie of Louisiana and Texas. U. S. Fish and Wildlife Service and U. S. Geological Survey (Lacassine National Wildlife Refuge, Lake Arthur, LA). 40 pp (full-color brochure).

Allain, L., M. Vidrine, V. Grafe, C. Allen, and S. Johnson. 2000. Paradise Lost? The Coastal Prairie of Louisiana and Texas (2nd edition). U. S. Fish and Wildlife Service and U. S. Geological Service (with Coastal Prairie Conservation Initiative, Texas). 40 pp.

Allain, L., C. Allen, P. Faulkner, M. Vidrine, J. Faust, C. C. Lockwood, G. Ross, J. Boundy, J. Thomas and R. Greco. 2003. Our vanishing wild Cajun Prairies of Louisiana. Baton Rouge Earth Day, Inc. Designed by Jane Thomas, Thomas graphics. Text by: L. Allain, C. Allen, P. Faulkner, and M. Vidrine. Photographs by: L. Allain, P. Faulkner, J. Faust, C. C. Lockwood, G. Ross, M. Vidrine, and J. Boundy. Map graphic by R. Greco. Poster.

Allain, L., R. Greco, and P. Faulkner. 2004. Remnant prairie in Louisiana: acreage by remnant. http://usgs.gov/nwrc/prairie/acres.

Literature Cited

Allain, L., L. Smith, C. Allen, M. F. Vidrine and J. B. Grace. 2006. A floristic quality assessment system for the coastal prairie of Louisiana. Proceedings of the 19th North American Prairie Conference. 19: 1-18.

Allen, C. M. 1984. A preliminary checklist of the vascular flora of Allen Parish, Louisiana. Contributions of the Herbarium of Northeast Louisiana University, 5: 1-57.

Allen, C. M. 1992. Grasses of Louisiana. Second edition. The Cajun Prairie Habitat Preservation Society (Eunice, LA). 320 pp.

Allen, C. M. 2000. Succession in the Duralde Prairie Restoration Project. Proceedings of the Louisiana Academy of Sciences 63: 58 (abstract).

Allen, C. M. 2003. Wildflowers of Louisiana. Proceedings of the 2nd Lone Star Regional Native Plant Conference, 2: 5-9.

Allen, C. M. 2006. Creating or recreating a prairie. Cajun Prairie Habitat Preservation Society Newsletter 25: 4 page insert.

Allen, C. M. 2010. An analysis of prairie species distribution at Fort Polk, Louisiana. Proceedings of the 21st North American Prairie Conference 21: 82-94.

Allen, C. M. and S. Thames. 2007. Observation on vegetation changes in Cajun Prairie, a coastal prairie flora in southwest Louisiana. Journal Botanical Research Institute of Texas 1 (2): 1141-1147.

Allen, C. M. and M. F. Vidrine. 1989. Wildflowers of the Cajun Prairie. Louisiana Conservationist 41 (3): 20-25.

Allen, C. M. and M. F. Vidrine. 1990a. Butterflies of the Cajun Prairie. Louisiana Conservationist 42 (2): 16-21.

Allen, C. M. and M. F. Vidrine. 1990b. The vegetation of St. Landry Parish, Louisiana, in the early 1800's. Proceedings of the Louisiana Academy of Sciences 53: 55.

Allen, C. M., and M. F. Vidrine. 2003. Flowering phenology of Cajun Prairie plants. Proceedings of the 18th North American Prairie Conference 18: 174-177.

Allen, C., M. Vidrine, and R. McLaughlin. 1989. Weeds of rice fields of southwestern Louisiana. Proceedings of the Louisiana Academy of Sciences 52: 8-16.

Allen, C., M. Vidrine, and B. Borsari. 1994. Analysis of the woody vegetation of a beech forest area in the Louisiana Arboretum. The Louisiana Environmental Professional 10 & 11 (1): 17-26.

Allen, C. M., M. F. Vidrine, B. Borsari, and L. Allain. 2001. Vascular flora of the Cajun Prairie of southwestern Louisiana. Proceedings of the 17th North American Prairie Conference, 17: 35-41.

Allen, C. M., D. A. Newman, and H. H. Winters. 2002. Trees, shrubs, and woody vines of Louisiana. Allen's Native Ventures (Pitkin, LA). 333 pp.

Allen, C. M., D. A. Newman, and H. H. Winters. 2004. Grasses of Louisiana, 3rd ed. Allen's Native Ventures (Pitkin, LA). 374 pp.

Allen, C. M., A. W. Allen, and H. H. Winters. 2005. Edible plants of the Gulf South. Allen's Native Ventures (Pitkin, LA). 291 pp.

Allen, C., S. Thames, L. Chance and C. Stagg. 2006. Flora of the Anacoco Prairie on Fort Polk, Louisiana. Proceedings of the 19th North American Prairie Conference, 19: 19-22.

Althans, Amy. 2002. Cajun Prairies. Louisiana Gardener (May 2002), pp. 45-46.

Ancelet, B. J., J. Edwards, and G. Pitre. 1991. Cajun County. University Press of Mississippi (Jackson, MS). 256 pp.

Anderson, B. 1997. The Louisiana Prairie. *Baton Rouge Morning Advocate,* 16 March 1997.

Ball, L. S. 2008. Ants swarm over Houston area, fouling electronics. *Yahoo News* (Wednesday, May 14, 2008). *Associated Press* news release.

Bartram, W. 1792. Travels through North and South Carolina, Georgia, and East and West Florida. Facsimile Edition. University Press of Virginia (Charlottesville, VA).

Baskin, Y. 1997. The work of nature: How the diversity of life sustains us. Island Press (Washington D. C.). 263 pp.

Bergerie, M. 1962. They tasted bayou water. Pelican Press (New Orleans, LA). 167 pp.

Bergerie, M. 2000. They tasted bayou water. Firebird Press (Gretna, LA). 196 pp.

Berry, T., P. Beasley and J. Clements, Eds. 2006. The Forgotten Expedition, 1804-1805: The Louisiana Purchase Journals of Dunbar and Hunter. Louisiana State University (LSU) Press (Baton Rouge, LA). 288 pp.

Betz, R. F. 1986. One decade of research in prairie restoration at the Fermi National Accelerator Laboratory (Fermilab), Batavia, IL. Proceedings of the 9th North American Prairie Conference, 9: 179-184.

Bick, G. W. 1957. The Odonata of Louisiana. Tulane Studies in Zoology 5 (5): 71-135.

Borsari, B. 1998. Sustainable agriculture: Concepts and educational applications. IN: B. Borsari and M. F. Vidrine (eds.). Sustainable Agriculture Seminar Proceedings. Vol. 1. Low Input Agriculture: Feasible Alternatives to Conventional Agricultural Practices. Louisiana State University at Eunice Press (Eunice, LA), pp. 14–19.

Borsari, B. 2003. The coastal prairie habitat as a model to promote sustainability in the agricultural curriculum. Proceedings of the 18th North American Prairie Conference, 18: 17-23.

Borsari, B. and V. Shirley. 1993. Preservation of natural habitats: biodiversity and farming. Annual Proceedings American Society of Environmaental

Science. pp. 181-187.

Borsari, B. and M. Vidrine. 1997. Estimates of clump size increase in clump-forming native perennials in the Cajun Prairie Restoration Project in Eunice, Louisiana. Proceedings of the Louisiana Academy of Sciences 60: 54 (abstract).

Borsari, B. and M. F. Vidrine. Eds. 1998. Low Input Agriculture: Feasible Alternatives to Conventional Agricultural Practices. Proceedings of the 1st Seminar. Louisiana State University at Eunice 30th Anniversary Celebrations. September 12-13, 1997. Louisiana State University at Eunice Press (Eunice, LA). 52 pp.

Borsari, B. and M. F. Vidrine. 2000. An Evaluation Tool for Improving Undergraduate Curricula in Agriculture. Sustainable Agriculture-Fertile Ground for Growth in the Agricultural Sciences. J. College Sci. Teaching 29(4): 235–240.

Borsari, B. and M. F. Vidrine. 2005. Agriculture Curricula in Sustainability: An Evaluation Across Borders. Journal of Sustainable Agriculture. Vol. 25 (4): 93-112.

Borsari, B., S. LaHaye and W. Sellers. 1998. Soil macrofauna and topsoil depletion: a comparative study among various cultivated fields and the prairie. IN: B. Borsari and M. F. Vidrine (eds.). Sustainable Agriculture Seminar Proceedings. Vol. 1. Low Input Agriculture: Feasible Alternatives to Conventional Agricultural Practices. Louisiana State University at Eunice Press (Eunice, LA). pp. 46-52.

Borsari, B., M. F. Vidrine and S. Doherty. 2002. Assessing students' preparedness towards sustainability in U. S. and European undergraduate agricultural curricula. American Journal of Alternative Agriculture 17 (4): 188-194.

Brady, N. C., and R. R. Weil. 2007. The nature and property of soils. 14th edition. Prentice Hall (New Jersey). 980 pp.

Brasseaux, C. A. 1987. The founding of New Acadia: The beginnings of Acadian life in Louisiana, 1765-1803. LSU Press (Baton Rouge, LA). 229 pp.

Brasseaux, C. A. 1991. Scattered to the Wind: Dispersals and Wanderings of the Acadians, 1755-1809. Louisiana Life Series, No. 6. The Center for Louisiana Studies, University of Southwestern Louisiana (Lafayette, LA). 84 pp.

Brasseaux, C. A. 2005. French, Cajun, Creole, Houma : A primer on Francophone Louisiana. LSU Press (Baton Rouge, LA). 159 pp.

Brasseaux, C. A., K. P. Fontenot, and C. F. Oubre. 1994. Creoles of color in the bayou country. University Press of Mississippi (Jackson, MS). 174 pp.

Brown, C. A. 1972. Wildflowers of Louisiana and adjoining states. LSU Press (Baton Rouge, LA). 247 pp.

Brown, C. A. 1997. Preliminary report on the isolated prairies of Louisiana. Proceedings of the Louisiana Academy of Sciences 60: 10-19.

Buchmann, S. and G. Nabhan. 1996. The forgotten pollinators. Island Press (Washington D. C.). 292 pp.

Caillet, M. and J. Mertzweiller, eds. 1988. The Louisiana Iris: The history and culture of five native species and their hybrids. Texas Gardener Press (Waco, TX). 226 pp.

Caillet, M., J. F. Campbell, K. C. Vaughn, and D. Vercher, Eds. (The Society for Louisiana Iris). 2000. The Louisiana Iris: The taming of a native American wild flower. Timber Press (Portland, OR). 254 pp.

Camel, N. 2006. The nature of things at Lake Martin: Exploring the wonder of Cypress Island Preserve in southern Louisiana. Acadian House Publishing (Lafayette, LA). 128 pp.

Chapman, H. C. and E. B. Johnson. 1986. The mosquitoes of Louisiana. Louisiana Mosquito Control Association Tech. Bull. 1 (revised), 17 pp.

Clark, H. L., G. J. Haley, E. J. Hebert, R. M. Hollier and A. J. Roy. 1962. Soil survey of Acadia Parish, Louisiana. U. S. Dept. Agric., Soil Conservation Service in cooperation with Louisiana Agricultural Experimental Station.

Collins, S. L. and L. L. Wallace (Eds.). 1990. Fire in North American Prairies. University of Oklahoma Press (Norman, OK). 176 pp.

Comeaux, M. 1983. Louisiana's Acadians: The environmental impact. IN: Conrad, G. R., Editor. The Cajuns: Essays on Their History and Culture. U.S.L. History Series, No. 11, 262 pp. Center for Louisiana Studies, University of Southwestern Louisiana (Lafayette, LA). Pp. 109-126.

Conrad, G. R., ed. 1983. The Cajuns: Essays on their history and culture. Center for Louisiana Studies, University of Southwestern Louisiana (Lafayette, LA). 262 pp.

Cottam, G. 1987. Community dynamics on an artificial prairie. Pp. 257-270 in: W. R. Jordan III, M. E. Gilpin and J. D. Aber, eds. Restoration Ecology: A synthetic approach to ecological research. Cambridge University Press (Cambridge, MA). 342 pp.

Crisler, R. M. 1968. Bayou Teche. Pp. 105-112. *In*: Davis, E. A., Editor. 1968. The rivers and bayous of Louisiana. Louisiana Education Research Association (Baton Rouge, LA).

Curry, M. G., B. Everitt, and M. F. Vidrine. 1981. Haptobenthos on shells of living freshwater clams in Louisiana. The Wasmann Journal of Biology 39 (1-2): 56-62.

Curry, J. P. 1994. Grassland Invertebrates. Chapman and Hall (London, UK). 437 pp.

Daniels, S. 1995. The wild lawn handbook: Alternatives to the traditional

front lawn. Macmillan (New York). 223 pp.
Dakin, M. E. Jr. and D. L. Felder. 1981. A record of the mayfly *Dolania americana* in Louisiana (Ephemeroptera: Behningiidae). Florida Entomologist 64 (3): 454-455.
Darby, W. 1816. Geographical description of the State of Louisiana. John Melish (Philadelphia, PA). 270 pp.
Darsie, R. F. Jr. and R. A Ward. 2005. Identification and geographical distribution of the mosquitoes of North America, north of Mexico. University Press of Florida (Gainesville, FL). 383 pp.
Darwin, C. 1881. The formation of vegetable mould, through the action of worms, with observations on their habits. Forgotten Books. 274 pp.
Davis, E. A., Editor. 1968. The rivers and bayous of Louisiana. Louisiana Education Research Association (Baton Rouge, LA). 201 pp.
Diamond, J. 2005. Collapse: How societies choose to fail or succeed. Viking (Penguin Group Inc., New York, NY). 575 pp.
Dixon, B. 2009. Last days of Last Island: The hurricane of 1856, Louisiana's first great storm. University of Louisiana at Lafayette Press (Lafayette, LA). 290 pp.
Dormon, C. 1958. Flowers native to the deep south. Mt. Pleasant Press (Harrisburg, PA). 176 pp.
Dormon, C. 1965. Natives preferred. Claitor's Book Store (Baton Rouge, LA). 217 pp.
Dormon, J. H. 1983. The people called Cajun: An introduction to an ethnohistory. The Center for Louisiana Studies, University of Southwestern Louisiana (Lafayette, LA). 98 pp.
Douglas, N. H. 1974. Freshwater fishes of Louisiana. Claitor's Publishing Division (Baton Rouge, LA). 443 pp.
Douglas, N. H. and J. J. Hoover. 2008. Fishes of the lower Red River: 1806 account of Peter Custis and present-day ichthyofauna. Bulletin of the Museum of Life Sciences (Louisiana State University in Shreveport) 14: 181-200.
Druse, K. and M. Roach. 1994. The natural habitat garden. Clarkson Potter Publ. (New York). 248 pp.
Dundee, D. S. and W. J. Harman. 1963. *Corbicula fluminea* in Louisiana. Nautilus 77: 30.
Dundee, H. A. and D. A. Rossman. 1989. The amphibians and reptiles of Louisiana. LSU Press (Baton Rouge, LA). 300 pp.
Dunkle, S. W. 1989. Dragonflies of the Florida peninsula, Bermuda, and the Bahamas. Scientific Publishers (Gainesville, FL). 155 pp.
Dunkle, S. W. 1990. Damselflies of Florida, Bermuda, and the Bahamas. Scientific Publishers (Gainesville, FL). 154 pp.
Dunkle, S. W. 2000. Dragonflies through binoculars. Oxford University

Press (New York, NY). 266 pp.

Du Pratz, Antoine Simon Le Page. 1758. History of Louisiana. Paris.

Earley, L. S. 2004. Looking for longleaf: The fall and rise of an American Forest. The University of North Carolina Press (Chapel Hill, NC). 322 pp.

Fearn, M. L. 1995. Louisiana's Cajun Prairie: Holocene History of a Southern Grassland. Ph.D. dissertation. Louisiana State University, Baton Rouge, Louisiana.

Featherman, A. 1871. Report of botanical survey of southern and central Louisiana made during the year of 1870. IN: Annual Report of the Board of Supervisors of the Louisiana State University, for the year ending December 31, 1870. New Orleans, Louisiana.

Finerty, M.W., J. D. Madden, S. E. Feagley and R. M. Grodner. 1990. Effect of environs and seasonality on metal residues in tissues of wild and pond-raised crayfish in southern Louisiana. Arch. Environ. Contam. Toxicol. 19: 94-100.

Flores, D. L. 1984. Southern counterpart to Lewis and Clark: The Freeman and Custis expedition of 1806. University of Oklahoma Press (Norman, OK). 386 pp.

Flores, D. L. 2008. Celebrating the forgotton expedition. Bulletin of the Museum of Life Sciences (Louisiana State University in Shreveport) 14: 1-6.

Fontenot, C. L. Jr. 2004. Cajun-French common names for Louisiana amphibians and reptiles. Herpetological Review 35 (4): 337-8.

Fontenot, M. A. and P. B. Freeland. 1976. Acadia Parish, Louisiana: A History to 1900. The Center for Louisiana Studies, University of Southwestern Louisiana Press (Lafayette, LA). 369 pp.

Fontenot, W. R. 1992. Native gardening in the south. A Prairie Basse Publication (Carencro, LA). 152 pp.

Fontenot, W. R. 1999. Louisiana wildflowers: A look at the state's wildflower communities. Louisiana Gardener (July/August 1999): 6-9.

Foster, S., and J. A. Duke. 1990. Eastern/Central medicinal plants. Peterson Field Guides, Houghton Mifflin Co. (New York, NY). 366 pp.

Fox, J. 2010. Gasland. International WOW Co. Video.

Freeman, T. and P. Custis. 1821. An account of the Red River in Louisiana drawn up from the returns Messrs. Freeman and Custis in the War Office of the United States, who explored the same in the year 1806. Ye Galleon Press (Fairfield, WA). 74 pp.

French, J. 1987. Louisiana prairie: Ecologists warn: Part of local nature is endangered. *The Daily World*, Sunday, July 19, 1987.

Frois, J. 2003. Around Louisiana: Louisiana Grown: Little prairie on the coast. Louisiana Life 23: 2 (summer 2003): 63.

Frois, J. 2004. Prairie Doctor. Louisiana Life 24 : 1 (spring 2004): 71.
Gahn, Robert Sr. 1972. A history of Evangeline: Its land, its men and its women who made it a beautiful place to live. Claitor's Publishing Division (Baton Rouge, LA). 148 pp. (The book cover title reads 'The Opelousas Country with a history of Evangeline Parish,' whereas the book cover reads 'A history of Evangeline Parish,' but here I list the entry on the title page of the book.)
Gandhi, K. and R. Thomas. 1989. Asteraceae of Louisiana. SIDA, Botanical Miscellany No. 4: xii + 202 pp.
George, S. G. and M. F. Vidrine. 1993. New records for freshwater mussels (Unionidae) and a snail (Pleuroceridae). Texas Journal of Science 45 (4): 363-366.
George, S., M. Mayeaux, and M. F. Vidrine. 1993. Freshwater mussels: treasures underfoot. Louisiana Conservationist 45 (2): 20-23.
Gibson, J. L. 1975. The prehistory of Acadiana. Pp. 16-40 *in*: Del Sesto, S. L. and J. L. Gibson, eds. 1975. The culture of Acadiana: Tradition and change in south Louisiana. The University of Southwestern Louisiana (Lafayette, LA). 176 pp.
Gibson, J. L. 1976a. Archaelogical survey of Mermentau River and Bayous Nezpique and Des Cannes, southwest Louisiana. The University of Southwestern Louisiana, Center for Archeological Studies, Report No. 1. 86 pp.
Gibson, J. L. 1976b. Archaelogical survey of Bayou Teche, Vermilion River, and Freshwater Bayou, south central Louisiana. The University of Southwestern Louisiana, Center for Archeological Studies, Report No. 2. 104 pp.
Glassberg, J. 1999. Butterflies through binoculars: The east. Oxford University Press (New York, NY). 400 pp.
Goins, C. R. and J. M. Caldwell. 1995. Historical Atlas of Louisiana. Univ. of Oklahoma Press (Norman, OK). Various paging.
Goldman, D. H. 1995. A new species of *Calopogon* from the Midwestern United States. Lindleyana 10(1): 37-42.
Gomez, G. M. 2008. The Louisiana coast: Guide to an American wetland. Texas A & M University (College Station, TX). 189 pp.
Gotte, G. N., D. M. Jariel and M. F. Vidrine. 2010. Cajun Prairie soils as reservoirs of water, lead and arsenic. Annual meeting of the Louisiana Academy of Sciences. Poster.
Grace, J. B. 1998. Can prescribed fire save the endangered coastal prairie ecosystem from Chinese tallow invasion? Endangered Species Update 15: 70-76. (reprinted in Wildland Weeds, Spring 1999).
Grace, J. B., L. Allain, and C. Allen. 2000a. Vegetation associations in a rare community type of coastal tallgrass prairie. Plant Ecology 147:

105-115.

Grace, J. B., L. Allain, and C. Allen. 2000b. Factors associated with plant species richness in a coastal tall-grass prairie. Journal of Vegetation Science 11: 443-452.

Grissell, E. 2001. Insects and gardens. Timber Press (Portland, OR). 345 pp.

Grossman, D. H., K. L. Goodwin, and C. L. Reuff. 1994. Rare plant communities of the coterminous United States. The Nature Conservancy, Arlington, Virginia.

Hardy, L. K. 2008. The amphibians and reptiles of the Freeman and Custis Red River Expedition: Changes to the present. Bulletin of the Museum of Life Sciences (Louisiana State University in Shreveport) 14: 201-218.

Harris, M. 2003. Botanica North America: The illustrated guide to our native plants, their botany, history, and the way they have shaped our world. Harper Resources (Harper Collins Publ. Inc., New York, NY). 665 pp.

Hebert, O. Jr. 1968. The Calcasieu. Pp. 85-93. In: E. A. Davis, Ed. The Rivers and Bayous of Louisiana. Louisiana Education Research Association (Baton Rouge, LA). 201 pp.

Hendrix, S. S., M. F. Vidrine, and R. H. Hartenstine. 1985. A list of records of freshwater aspidogastrids (Trematoda) and their hosts in North America. Proceedings of the Helminthological Society of Washington 52 (2): 289-296.

Hilgard, E. W. 1884. Report on cotton production in the United States. Part I. Mississippi valley and southwestern states. Department of the Interior, Census Office (Washington D. C.).

Hoese, H. D. 1973. Abundance of the low salinity clam, *Rangi cuneata*, in southwestern Louisiana. Proceedings of the National Shellfish Association 63: 99-106.

Hoese, H. D. and R. H. Moore. 1977. Fishes of the Gulf of Mexico: Texas, Louisiana, and adjacent waters. Texas A. and M. Press (College Station, TX). 327 pp.

Hohorst, C. Jr. and M. Bienvenu. 2009. Wings of paradise: Birds of the Louisiana wetlands. LSU Press (Baton Rouge, LA). 144 pp.

Holmes, W. C. 1990. Flore Louisiane: An Ethno-Botanical Study of French-Speaking Louisiana. Center for Louisiana Studies, University of Southwestern Louisiana (Lafayette, LA). 149 pp.

Howells, R. G., R. W. Neck, and H. D. Murray. 1996. Freshwater mussels of Texas. Texas Parks and Wildlife: Inland Fisheries Division (Austin, TX). 218 pp.

Huner, J. V. 1994. Freshwater Crayfish Aquaculture in North America,

Europe, and Australia: Families Astacidae, Cambaridae, and Parastacidae. CRC Press. 336 pp.

Huner, J. V. 1998. Vertebrate biological diversity at the University of Southwestern Louisiana Experimental Farm, St. Martin Parish, Louisiana USA. Sust. Agric. Sem. Proc. Vol. 1: 39-45.

Huner, J. V. 2000. Crawfish and water birds. American Scientist 88: 301-303.

Ingold, J. L. 2008. Birds of the Freeman and Custis Expedition, 1806 to present. Bulletin of the Museum of Life Sciences (Louisiana State University in Shreveport) 14: 219-229.

Jackson, W. 1980. New roots for agriculture. University of Nebraska Press (Lincoln, NE). 150 pp.

Jackson, W. 1987. Alters of unhewn store: Science and the earth. North Point Press (New York). 158 pp.

Jackson, W. 1994. Becoming native to this place. University Press of Kentucky (Lexington, KY). 121 pp.

Jackson, W. 1999. Natural systems agriculture: The truly radical adventure. Pp. 191-199. IN: Robert F. Sayre, ed. Rediscovering the prairie. The University of Wisconsin Press (Madison, WS).

Jariel, D. M., M. F. Vidrine, N. Bordelon, and J. Al-Dujaili. 2006. Soil chemistry properties under two different management practices: clipped Saint Augustine grass lawn and annually burned Cajun Prairie. Proceedings of the 19th North American Prairie Conference, 19: 192-199.

Jariel, D. M., M. F. Vidrine, A. A. Williams, and R. R. Fontenot. 2008.Population density of annelids and microarthropods in two differently managed soils. Proceedings of the 20th North American Prairie Conference, 20: 187-194.

Jariel, D. M., M. F. Vidrine, R. R. Fontenot, and C. E. Vidrine. 2010a. Plant and soil nutrient concentrations in a restored Cajun Prairie. Proceedings of the 21st North American Prairie Conference, 21: 95-107.

Jariel, D. M., M. F. Vidrine, and M. A. Mansfield. 2010b. Organic matter and humic substances in remnant and restored Cajun Prairie soils. Proceedings of the 21st North American Prairie Conference, 21: 116-125.

Johnson, S. 1996. Orchids of Louisiana's Cajun Prairie. North American Native Orchid Journal 2 (4): 368-372.

Johnson, S. R. 1999. Herps of the Cajun Prairie. Reptile and Amphibian Hobbyist (September), pp. 44-48.

Johnson, S. R. 2000. Louisiana Tallgrass: The scattered, surprising remnants of Cajun Prairie. The Prairie Reader 4 (3) (1999/2000 (Winter)): 4.

Jones, B. 2007. Louisiana Cowboys. Pelican Publishing Company (Gretna, LA). 208 pp.

Jordan, N., G. Boody, W. Broussard, J. D. Glover, D. Keeney, B. H. McCown, G. McIsaac, M. Muller, H. Murray, J. Neal, C. Pansing, R. E. Turner, K. Warner and D. Wyse. 2007. Sustainable development of the agriculture bio-economy. Science 316 (June 15, 2007): 1570-1571.

Kenner, R. 2009. Food, Inc. Magnolia Pictures. Video.

Kindscher, K. 1987. Edible wild plants of the prairie. University of Kansas Press (Lawrence, KS). 340 pp.

Kindscher, K. 1992. Medicinal plants of the prairie. University of Kansas Press (Lawrence, KS). 340 pp.

Kirk, R. 1995. Prairie plants of the Midwest: identification and ecology. Stipes Publishing L. L. C. (Champaign, Illinois). 137 pp.

Klemm, D. J. 1985. A guide to the freshwater Annelida (Polychaeta, Naidid and Tubificid Oligochaeta, and Hirudinea) of North America. Kendall/ Hunt Publ. Co. (Dubuque, IA). 198 pp.

Knapp, A. K., J. M. Briggs, D. C. Hartnett, and S. L. Collins, eds. 1998. Grassland dynamics: Long-term ecological research in tall grass prairie. Oxford University Press (New York, NY). 364 pp.

Kniffen, F. B. 1965. The Indians of Louisiana. Bureau of Educational Materials and Research, Louisiana State University (Baton Rouge, LA). 110 pp.

Ladd, D. (and F. Oberle). 1995. Tallgrass prairie wildflowers: a Falcon field guide. Falcon Press (Helena, MT). 262 pp.

Lambremont, E. N. 1954. The butterflies and skippers of Louisiana. Tulane Studies in Zoology 1 (10): 127-164.

Leopold, A. 1949. A Sand County Almanac, and Sketches Here and There. Oxford University Press, Inc. (New York). A special commemorative edition, 1989. 228 pp.

Leopold, A. 1999. For the health of the land: Previously unpublished essays and other writings. Edited by J. Baird Callicott and Eric T. Freyfogle. Island Press (Washington, D. C.). 243 pp.

Lockett, S. H. 1969. Louisiana As It Is: A Geographical and Topographical Description of the State. (Edited by Lauren C. Post). LSU Press (Baton Rouge, LA). (Original copy dated 1873 and resides in the Tulane University library.) 148 pp.

Lockwood, C. C. 1981. Atchafalaya: America's largest river basin swamp. Beauregard Press (Baton Rouge, LA). 105 pp.

Lockwood, C. C. 1984. The Gulf Coast: Where land meets sea. LSU Press (Baton Rouge, LA). 160 pp.

Lockwood, C. C. 1986. Discovering Louisiana. LSU Press (Baton Rouge, LA). 150 pp.

Lockwood, C. C. 1995. C. C. Lockwood's Louisiana Nature Guide. LSU Press (Baton Rouge, LA). 94 pp.

Lockwood, C. C. 2005. Marsh mission: Capturing the vanishing wetlands. LSU Press (Baton Rouge, LA). 106 pp.

Lockwood, C. C. 2007. C. C. Lockwood's Atchafalya. LSU Press (Baton Rouge, LA). 124 pp.

Loden, Michael S. 1979. A new North American species of freshwater tubificid (Oligochaeta). Proceedings of the Biological Society of Washington 92 (3): 601-605.

Logan, W. B. 1995. Dirt: The ecstatic skin of the Earth. Riverhead Books (New York, NY). 202 pp.

Longfellow, H. W. 1847. Evangeline: A tale of Acadie. Republished in 1951 by Numbus Publishing Limited (Halifax, Nova Scotia, Canada). 111 pp.

Lowery, G. H. Jr. 1974a. Louisiana birds. Third edition. LSU Press (Baton Rouge, LA). 651 pp.

Lowery, G. H. Jr. 1974b. The mammals of Louisiana and adjacent waters. LSU Press (Baton Rouge, LA). 565 pp.

Lynch, J. J. 1942. Glossary of local, English, and scientific plant names. Unpublished report to the U. S. Fish and Wildlife Service and the Louisiana Wildlife and Fisheries Commission. 5 pp.

MacRoberts, B. R. and M. H. MacRoberts. 2008a. Plant ecology and phytogeography of the West Gulf Coast Plain: An overview. Bulletin of the Museum of Life Sciences (Louisiana State University in Shreveport) 14: 7-28.

MacRoberts, M. H. and B. R. MacRoberts. 2008b. West Gulf Coastal Plain botanical ecology: Ice Age to Present. Bulletin of the Museum of Life Sciences (Louisiana State University in Shreveport) 14: 55-79.

MacRoberts, M. H., B. R. MacRoberts and D. C. Moore. 1997. Introduction and notes to Clair Brown's "Preliminary report on the isolated prairies of Louisiana." Proceedings of the Louisiana Academy of Sciences 60: 1-9.

Madden, J. D., M. W. Finerty and R. M. Grodner. 1989. Survey of persistent pesticide residues in the edible tissues of wild and pond-raised Louisiana crayfish and their habitat. Bulletin Environmental Contamination and Toxicology. 43: 779-784.

Madden, J. D., R. M. Grodner, S. E. Feagley, M. W. Finerty and L. S. Andrews. 1991. Minerals and xenobiotic residues in the edible tissues of wild and pond-raised Louisiana crayfish. Journal of Food Safety 12: 1-15.

Madson, J. 1982. Where the sky began. Houghton Mifflin (Boston, MA). 340 pp.

Madson, J. (and F. Oberle). 1993. Tallgrass prairie. Falcon Press Publ. Co. (Helena, MT). 112 pp.

Manning, R. 1995. Grassland: The history, biology, politics, and promise of the American Prairie. Viking (Penguin Books USA Inc., New York, NY). 306 pp.

Mauffray, B. 1997. The dragonflies and damselflies (Odonata) of Louisiana. Bulletin of American Odonatology 5 (1): 1-26.

McClain, W. E. 1997. Prairie establishment and landscaping. Natural Heritage Technical Publication #2, Division of Natural Heritage, Illinois Department of Natural Resources (Dpringfield, IL). 62 pp.

McLaughlin, R. E. 1984. The Riceland agroecosystem in relation to mosquito population dynamics in Acadia Parish, La. 1984. Gulf Coast Mosquito Research Laboratory, USDA, ARS (Lake Charles, LA). 44 pp.

McLaughlin, R. E. and M. F. Vidrine. 1984a. Point-source introduction system for control of *Psorophora columbiae* (Diptera: Culicidae) by *Bacillus thuringiensis* serotype H-14: Results of a large-area operational test. Environmental Entomology 13 (2): 366-370.

McLaughlin, R. E. and M. F. Vidrine. 1984b. Distribution of a flowable concentrate formulation of *Bacillus thuringiensis* serotype H-14 during irrigation of rice fields as a function of the quantity of formulation. Mosquito News 44 (3): 330-335.

McLaughlin, R. E. and M. F. Vidrine. 1984c. Comparison of three *Bacillus thuringiensis* serotype H-14 formulations against *Psorophora columbiae.* Mosquito News 44 (3): 351-356.

McLaughlin, R. E. and M. F. Vidrine. 1984d. Distribution of *Bacillus thuringiensis* (H-14) by water during flooding of rice fields. Mosquito News 44 (1): 36-42.

McLaughlin, R. E. and M. F. Vidrine. 1985. Factors affecting distribution of *Bacillus thuringiensis* serotype H-14 during flooding of rice fields. Journal of the American Mosquito Control Association 1 (3): 381-384.

McLaughlin, R. E. and M. F. Vidrine. 1986. A more accurate method for assessment of *Psorophora columbiae* larvicidal tests. Journal of the American Mosquito Control Association 2 (3): 362-363.

McLaughlin, R. E. and M. F. Vidrine. 1987. *Psorophora columbiae* larval density in southwestern Louisiana rice fields as a function of cattle density. Journal of the American Mosquito Control Association 3 (4): 633-635.

McLaughlin, R. E., M. A. Brown and M. F. Vidrine. 1987a. The sequential sampling of *Psorophora columbiae* larva (Diptera: Culicidae) in rice fields. Journal of the American Mosquito Control Association 3 (3): 423-428

McLaughlin, R. E., M. F. Vidrine and O. R. Willis. 1987b. Distribution of *Anopheles quadrimaculatus* and *An. crucians* (Diptera: Culicidae) larvae within rice field habitats in southwestern Louisiana. Journal of the American Mosquito Control Association 3 (4): 574-578.

McLaughlin, R. E., M. F. Vidrine and O. R. Willis. 1988. Incidence of patent infections of *Anopheles quadrimaculatus* (Diptera: Culicidae) larvae by *Parathelohania anophelis* (Protozoa: Microsporida) in rice fields in southwestern Louisiana. Journal of Invertebrate Pathology. 51: 172-174.

McPherson, A. B. 2008. The dynamics of Louisiana's past and present mammalian fauna in relation to the Freeman and Custis Expedition of 1806. Bulletin of the Museum of Life Sciences (Louisiana State University in Shreveport) 14: 231-245.

Merrill, E. C. 2005. Germans of Louisiana. Pelican Publishing Co. (Gretna, LA). 380 pp.

Miller, B. K., and W. R. Fontenot. 2001. Birds of the Gulf Coast. LSU Press (Baton Rouge, LA). Pp. 1-132.

Mire, P. 1989. Wildflowers of the Cajun Prairie. Attakapas Productions (Eunice, LA). 28, minute video.

Mire, P. 1992. Anything I catch: The handfishing story. Attakapas Productions (Eunice, LA). 28 minute video.

Murphy, K.E., J. K. Vidrine, D. R. McDaniel and C. L. Godfrey. 1986. Soil survey of St. Landry Parish, Louisiana. U. S. Dept. Agric., Soil Conservation Service in cooperation with Louisiana Agric. Exp. Sta. and Louisiana State Soil and Water Conservation Committee.

Nabhan, G. P. 2009. Where our food comes from: Retracing Nikolay Valvilov's quest to end famine. Island Press (Washington D. C.). 223 pp.

Nancarrow, L. and J. H. Taylor. 1998. The worm book. Ten Speed Press (Berkeley, CA). 150 pp.

Nardi, J. B. 2007. Life in the soil: A guide for naturalists and gardeners. University of Chicago Press (Chicago, IL). 293 pp.

Neck, R. W. (with G. Ajivlsgi). 1996. A field guide to butterflies of Texas. Gulf Publishing Company (Houston, TX). 323 pp.

Needham, J. G. and H. B. Heywood. 1929. A handbook of the dragonflies of North America. Charles C. Thomas (Springfield, IL). 378 pp.

Needham, J. G. and M. J. Westfall, Jr. 1955. A manual of the dragonflies of North America (Anisoptera). University of California Press (Berkeley, CA). 615 pp.

Needham, J. G., M. J. Westfall, Jr. and M. L. May. 2000. Dragonflies of North America. Scientific Publishers (Gainesville, FL). 939 pp.

Nelson, G. 2006. Atlantic Coastal Plain wildflowers: A field guide to the wildflowers of the coastal regions of Virginia, N. C., S. C., Georgia and northeastern Florida. Falcon Guide, Morris Book Publ. (Guilford, CT). 264 pp.

Newton, M. B. Jr. 1972. Atlas of Louisiana: A guide for students. The School

of Geoscience, Louisiana State University, Misc. Publ. 72-1. 196 pp.
Neyland, R. 2009. Wildflowers of the Coastal Plain: A field guide includes the lower Mississippi River valley, Gulf, and Atlantic coastal states. LSU Press (Baton Rouge, LA). 339 pp.
Onwueme, I., B. Borsari, and W. L. Filho. 2008. An analysis of some paradoxes in alternative agriculture and a vision of sustainability for future food systems. Int. J. Agricultural Resources, Governance and Ecology, 7 (3): 199-210.
Opler, P. A. and V. Malikul. 1992. A field guide to eastern butterflies. Peterson Field Guides. Houghton Mifflin Co. (New York). 396 pp.
Opler, P. A. and V. Malikul. 1998. A field guide to eastern butterflies. Peterson Field Guides. Houghton Mifflin Press (New York). 512 pp.
Opler, P. A, H. Pavulaan, and R. E. Stanford (coordinators). 1995. Butterflies of North America. Jamestown, ND: Northern Prairie Wildlife Research Center Home Page. Retrieved from: http://www.npwrc.usgs.gov/resource/distr/lepid/bflyusa/bflyusa.htm (Version 16 May 2000).
Oubre, C. 2001. Timeline: Louisiana history. Unpublished manuscript. 18 pp.
Packard, S. 1997. Interseeding. Pages 163-192. *In* S. Packard and C. F. Mutel (eds.), The Tallgrass Restoration Handbook, Island Press (Washington D. C.).
Packard, S. and C. Mutel, Eds. 1997. The tallgrass restoration handbook: For prairies, savannas, and woodlands. Island Press (Washington D. C.). 463 pp.
Parker, R. S., C. T. Hackney and M. F. Vidrine. 1984. Ecology and reproductive strategy of a south Louisiana freshwater mussel, *Glebula rotundata* (Lamarck) (Unionidae: Lampsilini). Freshwater Invertebrate Biology. 3 (2): 53-58.
Pastorek, M. 2003. Prairie restoration. Proceedings of the 2nd Lone Star Regional Native Plant Conference, 2: 81-82.
Pastorek, M. 2007. Meadow Makers: Fireproof wildflowers for the meadow and garden. Seed catalog. 24 pp.
Pastorek, M. 2008. A dozen dirty reasons to grow prairies and savannahs or Twelve Little Known Facts about Prairie Gardening. Cajun Prairie Habitat Preservation Society Newsletter 32 (November): 2-3.
Pilkey, O. H., and L. Pilkey-Jarvis. 2007. Useless arithmetic: Why environmental scientists can't predict the future. Columbia University Press (New York, NY). 230 pp.
Post, L. C. 1962. Cajun Sketches From the Prairies of Southwest Louisiana. LSU Press (Baton Rouge, LA). 215 pp.
Post, L. C., Ed. 1969. Louisiana as it is: A geographical and topographical description of the state 1873, orignal by Samuel H. Lockett. LSU Press

(Baton Rouge, LA). 148 pp.

Price, P., T. Lewinsohn, G. W. Fernandes, and W. Benson. Eds. 1991. Pant-animal interactions: Evolutionary ecology in tropical and temperate regions. John Wiley and Sons, Inc. (New York). 639 pp.

Raeburn, P. 1995. The last harvest: The genetic gamble that threatens to destroy American agriculture. Simon and Schuster (New York, NY). 269 pp.

Ramsay, J. C. Jr. 1996. Jean Laffite: Prince of pirates. Eakin Press (Austin, TX). 209 pp.

Reed, A. B. and V. Reed. 2009. Images of America: Eunice. Arcadia Publishing (Charleston, SC). 127 pp.

Rees, G. 1976. The Dauterive Compact: The foundation of Acadian cattle industry. *Attakapas (LA) Gazette* II: 91.

Reese, W. D. 1992. Acadiana flora. Center for Louisiana Studies, University of Southwestern Louisiana (Lafayette, LA). 125 pp.

Reese. W. D. and C. M. Allen. 2004. Mamou, Acadian folklore, natural history, and botany of the Mamou plant, *Erythrina herbacea L* (Fabaceae). University of Louisiana at Lafayette Press (Lafayette, LA). 65 pp.

Roback, S. S., D. J. Bereza and M. F. Vidrine. 1980. Description of an *Ablabesmyia* (Diptera: Chironomidae: Tanypodinae) symbiont of unionid fresh-water mussels (Mollusca: Bivalvia: Unionacea), with notes on its biology and zoogeography. Transactions of the American Entomological Society 105: 577-620.

Robin, C. C. 2000. *Voyages dans l'intérieure de la Louisiane, de la Floride occidéntale et dans les islés de la Martinique et de Saint-Domigue pendant les années 1802, 1804, 1805 et 1806 ...En outre, contenant ce qui s'est passé de plus interèssant, relativement a l'etablissément des Anglo-Americains á la Louisiane*. Three volumes. Paris, France: F. Buisson. Original dated 1807. Voyage to Louisiana. (Translated by Stuart O. Landry, Jr.) Pelican Publishing Company (Gretna, LA). 272 pp.

Ross, G. 1994. Gardening for Butterflies in Louisiana. Louisiana Dept. of Wildlife and Fisheries, Natural Heritage Program (Baton Rouge, LA). 41 pp.

Ross, G. N. and E. N. Lambremont. 1963. An annotated supplement to the state list of Louisiana butterflies and skippers. Journal of Lepidopterist Society 17: 148-158.

Roth, S. 1997. Natural landscaping: Gardening with nature to create a backyard paradise. Rodale Press Inc. (Emmaus, PA). 255 pp.

Sachs, J. D. 2008. Common Wealth: Economics for a crowded planet. Penguin Press (New York, NY). 388 pp.

Saunders, J. W., R. D. Mandel, R. T. Saucier, E. T. Allen, C. T. Hallmark, J. K. Johnson, E. H. Jackson, C. M. Allen, G. L. Stringer, D. S. Frink, J. K.

Feathers, S. Williams, K. J. Gremillion, M. F. Vidrine, and R. Jones. 1997. A mound complex in Louisiana at 5400-5000 years before the Present. Science 277 (19 September 1997): 1796-1799.

Saunders, J. W., R. D. Mandel, C. G. Sampson, C. M. Allen, E. T. Allen, D. A. Bush, J. K. Feathers, K. J. Gremillion, C. T. Hallmark, H. E. Jackson, J. K. Johnson, R. Jones, R. T. Saucier, G. L. Stringer, and M. F. Vidrine. 2005. Watson Brake, a Middle Archaic Mound Complex in northeast Louisiana. American Antiquity 70 (4): 631-668.

Savage, C. 2004. Prairie: A natural history. Greystone Books (Berkeley, CA). 308 pp.

Savoy, A. A. 1984. Cajun Music: A reflection of a people. Volume I. Bluebird Press (Eunice, LA). 424 pp.

Schramm, P. 1990. Prairie restoration: A twenty-five year perspective on establishment and management. Proceedings of the 12th North American Prairie Conference, 12: 169-178.

Semar, M. and M. F. Vidrine. 2000. Creating a postage stamp prairie with Cajun Prairie plants. Proceedings of the Louisiana Academy of Sciencies 63: 58. (abstract).

Shirley, S. 1994. Restoring the tallgrass prairie: An illustrated manual for Iowa and the upper midwest. Univ. of Iowa Press (Iowa City, IA). 330 pp.

Shively, S. H. and M. F. Vidrine. 1984. Fresh-water mollusks in the alimentary tract of a Mississippi map turtle. Proceedings of the Louisiana Academy of Sciences 47: 27-29.

Smeins, F. E., D. D. Diamond, and C. W. Hanselka. 1992. Coastal prairie. Pages 269-290. *In* R. T. Coupland (ed.) Ecosystems of the World 8A, Natural Grasslands, Introduction and Western Hemisphere, Elsevier, NY.

Smith, A. 1996. Big bluestem: Journey into the tall grass. Council Oak Books (Tulsa, OK). 288 pp.

Smith, D. G. 2001. Pennak's Freshwater Invertebrates of the United States: Porifera to Crustacea. John Wiley and Sons, Inc. (New York). 648 pp.

Smith, L. M. 1995. State rare plant communities in Louisiana. Louisiana Natural Heritage, Baton Rouge, LA.

Soule, J. D. and J. K. Piper. 1992. Farming in nature's image: An ecological approach to agriculture. Island Press (Washington D. C.). 286 pp.

Stewart, A. 2004. The earth moved: On the remarkable achievements of earthworms. Algonquin Book of Chapel Hill (Chapel Hill, NC). 223 pp.

Stones, M. 1991. Flora of Louisiana: water color drawings by Margaret Stones with botanical descriptions by Lowell Urbatsch. LSU Press (Baton Rouge, LA). 220 pp.

Strayer, D. L., J. A. Downing, W. R. Haag, T. L. King, J. B. Layzer, T. J.

Newton, and S. J. Nichols. 2004. Changing perspectives on pearly mussels, North America's most imperiled animals. BioScience 54 (5): 429-439.

Strecker, J. K. 1926. The extension of range of the nine-banded armadillo. Journal of Mammology 7: 206-210.

Strickland, G. and J. Strickland. 2008. Website with dragonflies. http://public.fotki.com/gstrick3/

Stubbendieck, J., S. Hatch, and C. Butterfield. 1997. North American range plants. Fifth Ed. University of Nebraska Press (Lincoln, NE). 501 pp.

Suarez, R. 1968. The Sabine. Pp. 77-84. In: E. A. Davis, Ed. The Rivers and Bayous of Louisiana. Louisiana Education Research Association (Baton Rouge, LA). 201 pp.

Taber, S. W. 2000. Fire ants. Texas A. and M. Press (College Station, TX). 308 pp.

Thomas, R. D., and C. M. Allen. 1993. Atlas of the vascular flora of Louisiana, Vol. 1: Ferns and fern allies, conifers, and monocotyledons. Louisiana Department of Wildlife and Fisheries (Baton Rouge, LA). 217 pp.

Thomas, R. D., and C. M. Allen. 1996. Atlas of the vascular flora of Louisiana, Vol. 2: Dicotyledons: Acanthaceae-Euphorbiaceae. Louisiana Department of Wildlife and Fisheries (Baton Rouge, LA). 213 pp.

Thomas, R. D., and C. M. Allen. 1998. Atlas of the vascular flora of Louisiana, Vol. 3: Dicotyledons: Fabaceae-Zygophyllaceae. Louisiana Department of Wildlife and Fisheries (Baton Rouge, LA). 248 pp.

Thompson, J. R. 1992. Prairies, forests, and wetlands: The restoration of natural landscape communities in Iowa. University of Iowa Press (Iowa City, IA). 139 pp.

Thorp, J. H. and A. P. Covich, Eds. 2001. Ecology and Classification of North American Freshwater Invertebrates. Second Edition. Academic Press (New York). 911 pp.

Thorp, R. W. 1991. Bowie knife. Phillips Publications (Williamstown, NJ). 167 pp.

Timm, T. 2006. A Nearctic tubificid *Varichaetadrilus harmani* (Loden) n. comb. in a Dutch wetland, with remarks on *Tubifex tubifex* (Müller)(Annelida: Oligochaeta). Zootaxa 1281: 21-39.

Tompkins, P. and C. Bird. 1998. Secrets of the soil: New solutions for restoring our Planet. Earthpulse Press Inc. (Anchorage, AK). 422 pp.

Touchet, B. A., H. L. Clark, C. M. Redlich, A. J. Roy, B. L. Griffis, S. J. Zaunbrecher and R. M. Hollier. 1974. Soil survey of Evangeline Parish, Louisiana. U. S. Dept. Agric., Soil Conservation Service in cooperation with Louisiana Agric. Exp. Sta. and Louisiana State Soil and Water

Conservation Committee.
Touchstone, S. 1983. Herbal and folk medicine of Louisiana and adjacent states. Folk-Life Books (Princeton, LA). 175 pp.
Tveten, J. and G. Tveten. 1993. Wildflowers of Houston. Rice University Press (Houston, TX). 309 pp.
Tveten J. and G. Tveten. 1996. Butterflies of Houston and southeast Texas. University of Texas Press (Austin, TX). 304 pp.
USDA. NRCS (Natural Resources Conservation Service). 2007. PLANTS database. U. S. Department of Agriculture, accessed in August 2007 at: http://plants.usda.gov/ .
Valvilov, N. I. 1992. Origin and Geography of Cultivated Plants. Translated by Doris Löve. Cambridge University Press (Cambridge, UK). 484 pp.
Valvilov, N. I. 1997. Five continents. Edited by L. E. Rodin, translated by Doris Löve. Rome: IPGRI (International Plant Genetics Research Institute) (Biodiversity International); St. Petersburg: VIR. 198 pp.
Venter, J. C. 2007. A life decoded: My genome: My life. Penguin Group, Inc. (New York, NY). 390 pp.
Vidrine, C. E. and S. R. Hazelton-Robichaux. 2010. Anhydrobiosis in colonial rotifers of the Cajun Prairie ecosystem: a 2009 URSI project. Annual meeting of the Louisiana Academy of Sciences. Poster.
Vidrine, E. 1924. The Life of Euzebe Vidrine. Privately published (original in Louisiana Room, LSU, Baton Rouge, LA). 22 pp.
Vidrine, J. O. 1981. Vidrine – Védrines: Volume 1: 1600-1750: Our Védrines in France. Acadiana Press (University of Southwestern Louisiana, Lafayette, LA). 244 pp.Vidrine, M. F. 1973. Fresh-water mussels (Bivalvia: Unionidae) from Evangeline Parish, Louisiana, parasitized by water-mites (Acarina: Trombidiformes: Unionicolidae) and aspidogastrid trematodes (Trematoda: Aspidogasteridae). Proceedings of the Louisiana Academy of Sciences 36: 53.
Vidrine, M. F. 1985. Fresh-water mussels (Unionacea) of Louisiana; a zoogeographical checklist of post-1890 records. The Louisiana Environmental Professional 2 (1): 50-59.
Vidrine, M. F. 1989. A summary of the mollusk-mite associations of Louisiana and adjacent waters. The Louisiana Environmental Professional 6 (1): 30-63.
Vidrine, M. F. 1990. Fresh-water mussel-mite and mussel-*Ablabesmyia* associations in Village Creek, Hardin County, Texas. Proceedings of the Louisiana Academy of Sciences 53: 1-4.
Vidrine, M. F. 1992. Identification of benthic samples. Contract No. 24400-92-24. 21 December 1992. Submitted to: Jan Boydstun, Water Quality Management Division, Department of Environmental Quality, P. O. Box

82215, Baton Rouge, LA 70884. 27 pp.

Vidrine, M. F. 1993. The historical distributions of freshwater mussels in Louisiana. Gail Q. Vidrine Collectibles (Eunice, LA). 225 pp.

Vidrine, M. F. 1996a. North American *Najadicola* and *Unionicola*: Collections and Communities. Gail Q. Vidrine Collectables (Eunice, LA). 259 pp.

Vidrine, M. F. 1996b. North American *Najadicola* and *Unionicola*: Diagnoses and Distributions. Gail Q. Vidrine Collectables (Eunice, LA). 355 pp.

Vidrine, M. F. 2008. The Mollusca of the Freeman and Custis Expedition of 1806: Mollusks (emphasis on mussels and associated parasites) of the Red River drainages. Bulletin of the Museum of Life Sciences (Louisiana State University in Shreveport) 14: 111-146.

Vidrine, M. F. and C. M. Allen. 1993. Dragonflies of the Cajun Prairie. Louisiana Conservationist 45 (1): 10-13.

Vidrine, M. F. and B. Borsari. 1994. Freshwater mussel-mite associations in Calcasieu River, Rapides Parish, Louisiana. Louisiana Environmental Professional 10 & 11: 11-16.

Vidrine, M. F. and B. Borsari. 1998. Coastal prairie: A plausible model for sustainable agriculture in southwestern Louisiana. pp. 27-38 In: Borsari, B. and M. F. Vidrine. Editors. 1998. Low Input Agriculture: Feasible Alternatives to Conventional Agricultural Practices. Proceedings of the 1st Seminar. Louisiana State University at Eunice 30th Anniversary Celebrations. September 12-13, 1997. Eunice, LA: Louisiana State University at Eunice. 52 pp.

Vidrine, M. F. and B. Borsari. 1999. Restoring native habitats as part of IPM in southwestern Louisiana, USA. Annals of ANPP (Association Nationale de Protection des Plantes) Fifth International Conference on Pests in Agriculture (AGRO-Montpellier), Tome III: 853-860.

Vidrine, M. F. and R. P. Hatler. 1995. *Hemidactylus turcicus* (Mediterranean Gecko). New Parish Record. Herpetological Review 26 (3): 155.

Vidrine, M. F. and S. R. Hazelton-Robichaux. 2003. Butterflies and dragonflies of the Cajun Prairie: Landscaping for native insects. Proceedings of the 2nd Lone Star Regional Native Plant Conference, 2: 94-103.

Vidrine, M. F. and S. R. Hazelton-Robichaux. 2008. Free-swimming Colonial Rotifers in Prairie Habitats in Arkansas, Louisiana, and Texas: Native or Exotic? Proceedings of the 20th North American Prairie Conference 20: 195-202.

Vidrine, M. and T. Hillman. 2008. Remnant status: Morbidity and mortality. Cajun Prairie Habitat Preservation Society Newsletter, No. 31 (2 page insert).

Vidrine, M. F. and G. J. Quillman-Vidrine. 1994. Freshwater mussel-mite associations in Bayou Wauksha, St. Landry Parish, Louisiana. Louisiana Environmental Professional 10 & 11: 1-5.

Vidrine, M. F. and M. F. Vidrine II. 1987. Macroinvertebrate benthic community structure in Louisiana Irrigation Canal. The Louisiana Environmental Professional 4 (1); 66-74.

Vidrine, M. F., R. E. McLaughlin and O. R. Willis. 1985. Free-swimming colonial rotifers (Monogononta: Flosculariacea: Flosculariidae) in southwestern Louisiana rice fields. Freshwater Invertebrate Biology 4: 187-193.

Vidrine, M. F., R. E. McLaughlin and O. R. Willis. 1987a. Aquatic snails (Mollusca: Gastropoda) of southwestern Louisiana rice fields. Proceedings of the Louisiana Academy of Sciences 50: 16-17.

Vidrine, M. F., R. E. McLaughlin and O. R. Willis. 1987b. Report of water mites (Acari: Hydrachnellae) in southwestern Louisiana rice fields. Proceedings of the Louisiana Academy of Sciences 50: 18-20.

Vidrine, M. F., S. L. Sissom and R. E. McLaughlin. 1987c. *Eulimnadia texana* Packard (Conchostraca: Limnadiidae) in rice fields in southwestern Louisiana. Southwestern Naturalist 32 (1): 1-4.

Vidrine, M. F., C. M. Allen and H. D. Guillory. 1992a. List of parish distribution records of Odonata in Louisiana. The Louisiana Environmental Professional 9 (1): 19-39.

Vidrine, M. F., C. M. Allen and H. D. Guillory. 1992b. Flight records of Odonata in Southwestern Louisiana. The Louisiana Environmental Professional 9 (1): 40-53.

Vidrine, M. F., C. M. Allen and H. D. Guillory. 1992c. Flight records of Papilionoidea of southwestern Louisiana. The Louisiana Environmental Professional 9 (1): 54-66.

Vidrine, M. F., C. M. Allen, and W. R. Fontenot. 1995. A Cajun Prairie Restoration Journal: 1988-1995. Gail Q. Vidrine Collectables (Eunice, Louisiana). 183 pp.

Vidrine, M. F., B. Borsari, J. E. Cordes, M. S. McClendon, and C. M. Allen. 1999 (2001)a. Landscaping with native prairie in southwestern Louisiana: Part I. Cajun Prairie and the natural landscaping paradigm. Proceedings of the Louisiana Academy of Sciences 62: 49.

Vidrine, M. F., B. Borsari, J. E. Cordes, M. S. McClendon, and C. M. Allen. 1999 (2001)b. Landscaping with native prairie in southwestern Louisiana: Part II. The Cajun Prairie natural landscape models. Proceedings of the Louisiana Academy of Sciences 62: 49-50.

Vidrine, M. F., B. Borsari, J. E. Cordes, M. S. McClendon, and C. M. Allen. 1999 (2001)c. Landscaping with native prairie in southwestern Louisiana: Part III. Benefits and problems with implementing the Cajun Prairie natural landscaping models. Proceedings of the Louisiana Academy of Sciences 62: 50.

Vidrine, M. F., C. M. Allen, and B. Borsari. 2000. The Cajun Prairie Restoration Project: A classroom for restoration ecology. Proceedings of the Louisiana Academy of Sciences 63: 59. (abstract).

Vidrine, M. F., C. M. Allen, B. Borsari, L. Allain, and S. R. Johnson. 2001a. The Cajun Prairie Restoration Project. Proceedings of the 17th North American Prairie Conference, 17: 151-154.

Vidrine, M. F., C. M. Allen, B. Borsari, and L. Allain. 2001b. Lepidopteran and Odonate Communities in the Cajun Prairie Ecosystem in Southwestern Louisiana. Proceedings of the 17th North American Prairie Conference, 17: 206-214.

Vidrine, M. F., W. R. Fontenot, C. M. Allen, B. Borsari, and L. Allain. 2001c. Prairie Cajuns and the Cajun Prairie: A History. Proceedings of the 17th North American Prairie Conference, 17: 220-224.

Vidrine, M. F., G. J. Quillman-Vidrine, D. J. Vidrine, and C. E. Vidrine. 2003. The Cajun Prairie Gardens: Demonstrations of sustainable landscaping in southwestern Louisiana. Proceedings of the 18th North American Prairie Conference, 18: 95-99.

Vidrine, M. F., G. J. Quillman-Vidrine, M. F. Vidrine II, D. J. Vidrine, and C. E. Vidrine. 2006. Freshwater mussels (Bivalvia : Unionidae) in the Cajun Prairie ecosystem in southwestern Louisiana. Proceedings of the 19th North American Prairie Conference, 19: 133-136.

Vidrine, M. F., J. S. Al-Dujaili, K. D. Joubert, and C. E. Vidrine. 2008a. Meat, Salad and Microbes: Antimicrobial Plants of the Cajun Prairie. Proceedings of the 4th Lone Star Regional Native Plant Conference, 4: 71-80.

Vidrine, M. F., B. Borsari, S. L. H. Fuller, and D. J. Bereza. 2010. Mites in mollusks in the upper Mississippi River and their community structure. Proceedings of the 21st North American Prairie Conference 21: 171-178.

Vidrine, M. F., B. Borsari, D. M. Jariel, and C. M. Allen. 2008b. The Cajun Prairie Gardens: Analysis of Vegetational Patterns in a Small Prairie Restoration. Proceedings of the 20th North American Prairie Conference 20: 233-240.

Vidrine, O. and C. Venzon. 2006. How not to celebrate a birthday. Good Old Days Specials (July 2006), Drg Publishing (Berne, IN). Pp. 38-39.

Villeré, Sidney Louis. 1971. The Canary Islands Migration to Louisiana 1778-1783: The history and passenger lists of the Islenos volunteer recruits and their families. The Genealogicical Research Society of New Orleans (New Orleans, LA). 94 pp.

Voorhies, J. K. 1983. The Acadians: The search for the promised land. IN: Conrad, G. R., Editor. The Cajuns: Essays on Their History and Culture. U.S.L. History Series, No. 11, 262 pp. Center for Louisiana

Studies, University of Southwestern Louisiana (Lafayette, LA). Pp. 79-94.

Wade, M. G. 1995. Sugar dynasty: M. A. Patout & Son, Ltd., 1791-1993. Center for Louisiana Studies (University Siana). 405 pp.

Walls, J. G. 2009. Crawfishes of Louisiana. LSU Press (Baton Rouge, LA). 240 pp.

Walls, J. G., and J. B. Black. 2008. A new crayfish, *Procambarus (Pennides) pentastylus*, from southwestern Louisiana (Crustacea: Decapoda: Cambaridae), with a key to western species of the subgenus. Proceedings of the Biological Society of Washington 121 (1): 49-61.

Walsh, G. P., E. E. Storrs, H. P. Burchfield, E. H. Cottrell, M. F. Vidrine and C. H. Binford. 1975. Leprosy-like disease occurring naturally in armadillos. Journal of the Reticuloendothelial Society 18 (6): 347-351.

Wasowski, S. and A. Wasowski. 1988. Native Texas Plants. Landscaping region by region. Gulf Publishing Co. (Houston, TX). 406 pp.

Wasowski, S. and A. Wasowski. 1992. Requiem for a lawnmower. Taylor Publishing Co. (Dallas, TX). 182 pp.

Wasowski, A. and S. Wasowski. 2000. The Landscaping Revolution: Garden with mother nature, not against her. Contemporary Books (Chicago, IL). 166 pp.

Weaver, J. E. 1954. North American prairie. Johnson Publ. Co. (Lincoln, NE). 348 pp.

Weaver, J. E. 1968. Prairie plants and their environment: A fifty-year study in the Midwest. Univ. of Nebraska Press (Lincoln, NE). 276 pp.

Weaver, J. E. and T. J. Fitzpatrick. 1934. The prairie. Reprinted in 1980 by Prairie-Plains Institute of Nebraska (Aurora, NE).

Weisman, A. 2007. The world without us. Picador (New York). 416 pp.

Westfall, M. J. Jr., and M. L. May. 1996. Damselflies of North America. Scientific Publishers (Gainesville, FL). 502 pp.

Williams, A. A. and J. Trahan, III. 2009. Faunistics of a Restored Tallgrass Prairie. Proceedings of the Louisiana Academy of Sciences. (abstract, in press).

Wilson, J. 1992. Landscaping with wildflowers: An environmental approach to gardening. Houghton Mifflin Company (New York). 244 pp.

Wright, H. A. and A. W. Bailey. 1982. Fire ecology: United States and Canada. John Wiley and Sons, Inc. (New York). 501 pp.

Zohrer, J. J. 2001. Prairie/wetland complex restorations in the Prairie Pothole Region of Iowa. Proceedings of the 17th North American Prairie Conference, 17: 136-137.

Appendixes

Appendix 1
Timeline for Cajun Prairie and Louisiana Territory (in part adapted from Oubre 2001)

1539-42	Spanish expeditions left livestock in the southern states
1600-1700's	Attakapas and Opelousas Indians occupied region and lived along bayous; Coureurs du Bois and other traders explored region
1713-15	Natchitoches established; first settlement in Louisiana Territory; Native Americans trade cattle and horses
1718-21	New Orleans established
1735-45	Trading post established by Blanpain and LeKintrek; first in Cajun Prairie
1755	Grand Dérangement—Cajuns leave Acadie (Nova Scotia)
1762	French lose French and Indian War; Opelousas country and Cajun Prairie ceded to the Spanish by the French
1765	Poste des Opelousas officially established
1765 & 1785	Cajuns (approximately 3000) arrived in Louisiana
1778	Isleños (Canary Islanders) arrive and settle Bayou Lafourche (Donaldsonville) and Malaga settle Bayou Teche at Poste de Attakapas (New Iberia)
1787 (April 3)	Spanish land grant to Pierre Vidrine (one mile north of Washington on Bayou Courtableu)
1799	Mermentau colony established along Mermentau River
1800	Spanish returned Louisiana Territory to the French
1803	French transferred Louisiana Territory to the United States; the Cajun Prairie was claimed by both the United States and Spain until treaty in 1819; C. C. Robin visited Opelousas; land survey of St. Landry Parish and Cajun Prairie
1805	William Darby visited Cajun Prairie
1812	Jim Bowie moved to Louisiana; he was legendary in both in Louisiana and Texas. He died at the Alamo in 1836
1860-65	Civil War; Union troops round up cattle in the Cajun Prairie; many Cajuns served in the Confederate Army; Cajun Prairie entered 80 years of depression
1869-72	Col. Samuel Lockett mapped Louisiana and the Cajun Prairie
1872-73	German and Midwesterners settled the open Cajun Prairie: Fabacher community settled—railroad built from Vermilionville to Lake Charles and to Texas by 1880
1880-81	First commercial rice crop (Fabacher) on the Faquetaique

	Prairie
1887	Crowley established along the southern railroad near Robert's Cove and Rayne
1890s	Massive canal systems built to control rice agroecosystem; rice cultivation expanded throughout the Cajun Prairie
1894	Eunice established at the intersection of the east-west and north-south railroad lines; Germans, Midwesterners, Cajuns, etc. moved out onto the Cajun Prairie
1901	Oil is discovered near Jennings
1929-1950	The Great Depression, World War II, and massive building of roads and bridges in the Cajun Prairie
1986	Several authors proclaimed that the natural Cajun Prairie habitat was lost, but remnants along railroad rights-of-way persisted
1988	Cajun Prairie restoration begins (Cajun Prairie Habitat Preservation Society)

Appendix 2
Genealogy of Malcolm Francis Vidrine

The preparation of this book gave me a chance to compile my genealogy. Fortunately a number of people had diligently pursued this as they sought out their own genealogies (see list at the end of this appendix). Like so much else in this book, I am more a scribe than a participant in the actual research.

My mother's lineage:

Josef Hidalgo[1] (born (b.) ca. 1745) married (m.) Isabel Sambrana Morales (b. ca. 1743-45)

|

Joseph "PePe" Antonio Hidalgo (b. Jan. 6, 1784 in Donaldsonville, LA) m. (Oct. 19, 1807) Felipa Rosalia Rodriguez

|

Joseph Vincent Hidalgo (b. Jan. 21, 1810, died in Lafayette, LA Feb. 4, 1852) m. (April 19, 1830 in Plattenville, LA) Marie Dominguez

|

Joseph Luc (Luke) Hidalgo (b. Dec. 3, 1833 in Lafayette, LA) m. (May 31, 1858 in Church Point, LA) Louise Martin[11]

|

Joseph Hidalgo m. Orphese Matt(e) (daughter of Olin Matt(e) m. Modess Savoy)

|

Felix Hidalgo Sr. m. Angelina Lee[2] (15 children; only one died as a child)

|

Marie Odile Hidalgo (b. Feb. 25, 1921) m. (Nov. 16, 1946) Numa Lee Vidrine[3]

|

Malcolm Francis Vidrine (b. June 23, 1949)

1. The original couple arrived in the port of New Orleans after boarding the San Ignacio de Loyola on October 8, 1778, leaving the Canary Islands (Isleños). They are possibly from Point Hidalgo area on Tenerife Island (northeast corner) (see Villeré 1971). He is listed as a soldier on the ship (Armed Personnel). The couple had 3 children on the ship's manifest: Gregoro, Francesca, and Juan. Joseph was born in Donaldsonville in 1784 in the Spanish colony of Louisiana. Donaldsonville is near junction of Bayou Lafourche and the Mississippi River, i. e. south of Baton Rouge. Joseph Vincent moved to Lafayette, where Joseph Luc was born in 1833, which indicates that part of the family has moved into the Cajun Prairie region.

2. George Lee[15] m. Harriet Goss[13]
|
Robert Edmund Lee

Edmund Thibodeaux[14] m. Josephine Rayo[16]
|
Aristella (Aristillo) Thibodeaux

Robert Edmund Lee m. Aristella (Aristillo) Thibodeaux
|
Angelina Lee

---

3. My father's lineage :

François Védrines (1599-1657) m. Marguerite Roche
|
Jean Védrines (1635?-1706) m. Françoise Cautier
|
Jean Védrines (1678?-?) m. Marie Raymond
|
Jean Baptiste Lapaise Védrines[4] (1712-1788)
m. Elizabeth Tisserand de Moncharvaux[5]
|
Jean Baptiste Pierre Lapaise Vidrine[6]
and Etienne Lapaise Vedrine[7]

4. Jacqueline O. Vidrine (1981) traced the family lineage in France until the dispersal to Louisiana. Jean Baptiste Lapaise Védrines (from the area called Lapaise in Bordeaux, France) came as a member of the King's military to the Louisiana territory ca. 1744 (date uncertain—could have been as early as 1739). Jean Baptiste Lapaise Védrines left 3 siblings in France: one brother entered the priesthood, a sister, and Pierre Vedrines (an older brother) who has descendents still living in France. Jean Baptiste Lapaise Védrines apparently married Elizabeth after his retirement from the military.

5. Elizabeth was quarter-blood Kaskaskian; her great grandfather was a Chieftan

of the Kaskaskia, a member tribe of the Illinois (Illiniwek Confederacy) Nation. Elizabeth de Monchervaux was the daughter of Marie Agnes Chessin (b. unk. d. October 24, 1758) and Michael Chessin (a French soldier reported to be an armorer (person who repaired firearms)). Marie Agnes Chessin was 1/2 Kaskaskian. Her mother was Agnes Chessin (pure Kaskaskian) who was the daughter of the Kaskaskian Chief. (credit: John R. (Randy) Vidrine and Winston DeVille.). Elizabeth's father (Jean François Tisserand was commander at the Kaskaskia Poste in Illinois country (south of St. Louis, Missouri on the Mississippi River; also south of the large Cahokia mound village), where he apparently married Elizabeth's mother. Elizabeth married Vidrine in Illinois in 1758. They had two sons (Pierre and Etienne).

Elisabeth's Family Tree:

Kaskaskian Chief
|
Agnes Phillippe
m. Michael Chessin
|
Marie Agnes Chessin
m. Jean François Tisserand de Moncharvaux
|
Elizabeth Tisserand de Moncharvaux

6. Jean Baptiste PierreVidrine[8] m. (1788) Marie Josephe Brignac
|
Pierre Vidrine (b. January 3, 1790)
m. Adelaide Larose Fontenot
|
Pierre Vidrine (b. April 24, 1824)
m. Celesie L. J. Fontenot
|
Onezime Vidrine m. Amelie Fontenot
|
Numa Onezime Vidrine[9] (b. Sept. 5, 1887)
m. Marie Zoe Vidrine (b. March 1, 1891)
|
Numa Lee Vidrine[18]
m. Marie Odile Hidalgo[17]
|
Malcolm Francis Vidrine

7. Etienne Lapaise Vedrine
m. (1795) Victoire Soileau
|
Etienne Vidrine m. Domilise Guillory
|
Aurelie Vidrine[10] m.(July 3, 1865) Zoe Fuselier
|
Marie Zoe Vidrine
m. Numa Onezime Vidrine (This couple has 12 children)
|
Numa Lee Vidrine
m. Marie Odile Hidalgo (This couple has 5 sons)
|
Malcolm Francis Vidrine[12]

8. I have a copy of the map (Appendix 3) of his Spanish land grant dated April 3, 1787 for a triangular plot of land fronting on the Bayou Cocodrie ca. 1 mile north of Washington, Louisiana, and adjacent to the Thistlewaite Natural Area. It is currently transected by I-49. It is labeled "Spanish Patent: Pierre Vidrine: April 3, 1787" with adjacent grants for Gregoire Guillory, John Gradenigo, Jacques Courtableau, Francois Lemele, and Michael Brignac. The grant was apparently made to Jean Baptiste Lapaise Védrines, the first Vidrine (J. O. Vidrine, Pers. Comm., 2008). He died in 1788 at this property. Elizabeth died in Baton Rouge at her son Pierre's residence later. The two sons had land grants north of Ville Platte (J. O. Vidrine, Pers. Comm., 2008).

9. Numa O. had two siblings:
Sister: Loritta Vidrine (b. September 29, 1890)
Brother: Rene Vidrine (b. January 12, 1896)

10. Aurelie (referred to as Aurelis in Gahn (1972)) was born on February 10, 1843. He served in the Confederate Army in Company K, 28th Louisiana Infantry Regiment (also known as Thomas's 28th Infantry Regiment) and fought in the Vicksburg Campaign, where he was captured. Upon release, he married Zoe Fuselier. They settled in south central Evangeline Parish near the current town of Vidrine. They had 10 children; Julius Caesar Vidrine became an MD and is the founder of the town of Vidrine.

11. Joseph Luc and Louise Martin Hidalgo had 8 children: Marie Alexandrine, Joseph, Jean Baptiste, Marie Eve Sidonia, Paul Philibert, Pierre Gilbert, and Annie. Louise Martin is likely a descendent of Andre Martin, a Cajun *vacher* (rancher) with herds (over a thousands) and one of the first settlers of Vermilionville (now

Lafayette) and is mentioned several times in Acadia Parish history (Fontenot and Freeland 1976).

12. My brothers and I sat down and counted first cousins one afternoon—to the best of my recollection, more than 100 first cousins were in my generation. I would recount but I can't remember them all.

13. The Goss family is German/Midwestern and is mentioned in Acadia Parish.

14. The Thibodeaux family is very large and considered as 'Cajun" as "Cajun" can be. Apparently members of the family arrived both in 1865 and 1885 as exiles from Acadie and France. This family is fairly well studied. Pierre Thibodeau is considered the forefather of all Thibodeauxs in the U. S. from France. Pierre Thibodeau from Beau Sejour, Acadia, Nova Scotia, is mentioned several times in Acadia Parish history (Fontenot and Freeland 1976).

15. The Lee family is probably Midwestern.

16. The Rayo, Matt, Savoy, Soileau, Guillory and Fuselier families are all historical families of the Cajun Prairie region.

17. Marie Odile Hidalgo Vidrine (most knew her as 'Odile') was born February 25, 1921, in Opelousas. She is a descendant from her mom in the Thibodeaux, Goss, Martin and Lee lines; her mother was Angelina Lee. Her father was Felix Hidalgo Sr. from the Canary Island Hidalgo line. She typified the Cajun heritage in that the Thibodeaux and Martin lines originated from the Grand Derangement of the Acadian migration from Nova Scotia to the Cajun Prairie. She also represented the German and Midwestern lines with her grandmothers in the Goss and Lee lines. Mom made it through the 8th grade and then settled down to being a seamstress in Opelousas. She married Dad at the age of 26. She finished high school with a GED when I was in high school. Her pride regarding this accomplishment never ceased. When I first brought her plants from the prairie, I could see in her eyes a 'familiarity.' When I told her about 'pock-pock' plants and gave her a couple to grow, she used them the next Easter to dye her eggs for 'pocking.' Although I ask her many times about the prairie, she had not formed such images in her mind as a young person. Call it serendipity, people observe that which they are trained to observe. As such Mom saw the prairie, hoed these plants down, grew cotton, vegetables, chickens, and only knew prairie plants as plants that 'had been.' She cooperated with the entire 'civilization' in eliminating the prairie in order to grow crops and animals. Unfortunately, we continue in this effort in that we are destroying the diversity of the natural world with our daily activities. This destruction is deemed necessary for survival, profit, success, and

mere ‘living.’ Conserving this habitat was not even a concept. But she caught on quickly when I started to discuss the prairie and to save the prairie. Her surprise that it had survived in those remnants was evident. She would look at the plants in my gardens for an hour and just delight in their variety—often, I think, she was searching her memories for these plants—it was as if she knew their names and had forgotten them. Mom passed Aug 21, 2005. She had 14 siblings.

18. Numa Lee Vidrine (most knew him as ‘NL’ pronounced ‘en-el’) was born June 27, 1922, in Ville Platte. He is a descendant from the French militia first arriving in Louisiana in the 1740s (J. O. Vidrine 1982). The Jean Batiste Vidrine (Spanish spelling, originally Védrines) married a Native American descendent and had two sons. Numa Onezime Vidrine and Zoe Vidrine, hid dad and mom, were descendents of the two sons’ lines in the Vidrine family. Zoe was the sister of Julius, a physician for whom the town of Vidrine was named, and daughter of Aurelie, a Confederate veteran of the Civil War. Dad went to Vidrine High School, where he completed high school in 1940. He grew up with his family as tenant farmers (share-croppers) on his Uncle Julius’s land—the square mile northwest of the intersection in Vidrine at the high school. Dad drove me around the property one wintry day and told me of this legacy. Dad was a carpenter, attended trade school at night after returning from the war, and developed a contracting business in 1967—when I graduated from high school and received a National Merit Scholarship as a Southern Region semifinalist. Dad never said that he enjoyed plants—he simply walked around the prairie restoration and quietly observed. I can’t remember a single word—just a laugh! when he heard people complaining about the ‘tall weeds.’ He too would develop a pride for what I was doing—he liked starting up conversations with people about my projects, I think, just to see what they would say. If they spoke negatively, he would slip into the conversation that his son was deeply involved in the effort. That always had a dramatic effect—and maybe it was for dramatic effect as my dad was reported to be a great French joke-teller—he always hit the punch-line. He passed Feb 26, 1995. He and my mom married November 16, 1946, after Numa returned from his tour of duty in the European Theatre of WWII. They moved to Eunice. They had 5 sons; I am their second son—the others are Reynold Patrick, Kenneth Paul, Blake Jude and Blaine Joseph. Dad had 11 siblings.

Credits for information in this genealogy:

Jackie Caton
Joann Hidalgo Duncan
Robert Gahn (1972)
James Oakley St. Cyr
Jacqueline O. Vidrine (1981)

John R. "Randy" Vidrine
Marie Odile Hidalgo Vidrine
Sidney Villeré (1971)

Appendix 3
Spanish land grant to Pierre Vidrine (April 3, 1787)

The triangular-shaped long lot (top center) assigned to Pierre Vidrine borders on Bayou Coutableu to the west at approximately one mile north of Washington, Louisiana. It also borders on Thistlewaite Natural Area to the east. It is currently traversed by I-49 just north of the Washington exit. The land was granted to the original Vidrine several years before his death at the property (J. O. Vidrine, personal communication 2008).

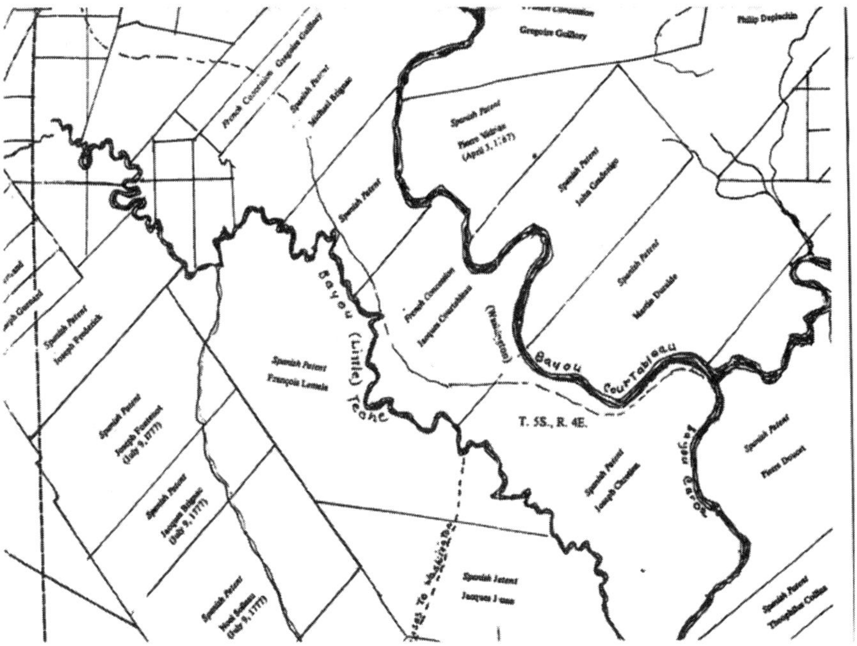

Figure 2. Spanish land grant to Pierre Vidrine (April 3, 1787). Copy provided by J. R. Guillory.

Appendix 4
Autobiographical Comments

I was born in 1949 in Eunice in the Cajun Prairie. My family moved to the open prairie within 3 months and settled on the Mamou Prairie just north of Bayou des Cannes near Fenris in a region/community called Duralde. Duralde is a rural region of southwestern Evangeline Parish just north of Eunice. Twenty acres of farmland and 10 acres of forest were our estate. By the time I can remember, we are growing cotton, cattle, pigs, chickens, horses, and lots of ruits and vegetables.

A day/year of my life in the 1955-65 period can be summarized as follows. Barefoot was the order of the day unless I was going to school or to church. Boots in winter were used to get to the barn where a foot-deep of mud and/or manure was found everywhere during the cool-wet season.

My four brothers and I routinely fed animals and milked cows every morning before school and every evening. In the spring, we manured the cropland—digging up the floor of the barn and literally carrying it and spreading it over the fields. Fences were repaired. Equipment for planting was repaired and readied for spring planting.

April and May were spent plowing and planting. Remaking rows for cotton planting was routine. In the early 1960s, pre-emergent herbicides changed the entire cotton planting and hoeing regimen. School ended in late May, and we began chopping and hoeing our 10 acres of cotton. Planting a garden of corn, beans, squash and melons was routine. We also grew lettuce, cabbage, okra, tomatoes, cucumbers, peanuts, turnips and a variety of peas. We grew much of what we ate in the way of vegetables, fruit and seed. A nearby family that was skill in growing vegetables often provided additional food.

Neighbors planted rice all around us. We were in the most western prairie area for planting cotton in the Cajun Prairie. June and July involved intense hoeing of cotton as weeds were ferocious; it also meant spraying cotton for bugs and pests. For extra money, we cut levees in rice fields to drain the fields for harvesting. I remember well the 50 cents an hour pay that came after a long morning in the rice field. Pulling coffeeweeds was another of the rice field chores—I didn't know *Sesbania* or any other scientific names at this time.

In August, we began hauling hay for our own cattle and for our neighbors' cattle for additional spending money. We still were hoeing cotton, but the cotton was in full bloom and forming bolls. We also were still spraying for pests. The days were hot and long. No air-conditioning was available! A cool shade created by a tree was the best place to look for us. I spent hours reading in the shade of our Chinaberry trees.

September was the beginning of the cotton harvest. Picking cotton was an enormous and seemingly endless job. From early in the morning until dark, we pulled sacks along the rows and hauled in as much as 100 pounds of cotton. My mother and other pickers could pick nearly 200 pounds in a day. We hired locals to help pick at the peak of the season. The 10 acres of cotton was usually picked at least 5 times each year (multiple pickings were required as the bolls ripened over time; today, the field is treated in order to ripen all the bolls at once for machine picking). One year, we picked 17 bales of cotton, and we were picking until Thanksgiving weekend. Fingers especially cuticles are ripped to shreds while picking cotton. The dry carpals on the bolls of cotton have distinctive hard points that rip and tear hurried hands. Many used gloves for cotton picking—a luxury we often could not afford. Rain got us out of the cotton fields; we had to put pans out in the house to catch the rain falling through the roof. In fact if rain was heavy, we had to make sure the cotton field would not hold water by digging additional drainage.

The smell of the oils of the cotton pervaded everything—even your body released these oils in sweat. The scent persisted on the skin. I recall my fellow classmates from the town making comments about that smell. It made labeling countryfolk rather easy. With no air-conditioning, sweating exacerbated the odor. The same was said of the odor of cattle and raw milk.

We lived on land measured in arpents as the old land deeds had laid it out. Land grants were strips that ran from the bayou to the prairie. We lived on approximately 20 acres of the prairie. We also owned 10 acres of woodland in the bottomland of the gallery forest on the bayou (Bayou des Cannes). The intervening 10 acres had been sold to a rice farmer. We used the 10 acres close to our home to grow cotton and vegetables. A barn served for storing hay and keeping cows, pigs, chickens, sheep, horses, and more. The lower 10 acres were used as pasture—we dug a pond for watering our animals—it filled in during my lifetime. I loved exploring the nearby lanes and woodland, and I was entertained simply watching Dung beetles roll the cattle dung in the pasture. We used the woodland to graze cattle, the prairie for pasture and growing cotton and vegetables. We picked blackberries, dug Mamou (Coralbean) and cut our Christmas tree (a pine tree) in the forests. We also played in the forest.

We returned to school in early to mid September. While we never missed school to pick cotton; however, many of our friends did. My Mom just wouldn't have it—nor did she allow my dad to work us in the fields on Sunday. There were rules! Sunday afternoon was for visiting—often for lunch—we routinely visited my mom and dad's brothers and sisters. The cousins played together.

Several summers we had cousins stay with us for several weeks. Such visits were occasioned when mom's sisters were sick or had some other business to tend to. I remember little of this—the fun of having one or more visitors was just overwhelming—any change in the routine made for a holiday of sorts.

Hurricanes were important. We all remembered Hurricane Audry in June 27, 1957, but most hurricanes occurred in the September-October period. My mother was especially affected by the storms; she wrote an article about a couple of storms (Vidrine and Venzon 2006). Several destroyed the cotton crop and some of our animals.

During the heat of the afternoon, we occasionally headed to the bayou to play around in the water or to the local water pump that filled the canals to flood the rice. The water pumps pumped ice cold water (~70^0 F), and it was fun to cool off there on summer afternoons. Later, a resort swimming pond (Lake View Park) was opened just north of Eunice—we went there often after 1964.

We played with cousins and anyone else available. We had some neighbors that came over. Reading the Ancelet et al. (1991) essay about the many games that Cajun children played brought back a lot of memories—we had played all these games. We got a television when I was 6 years old and at about the same time that we got indoor plumbing and toilets (1954-55). We were convinced that we had joined modern society with these changes, but we had a way to go. After we had the television, we played baseball, softball, football and the more common school games. Many days we just ran around, rode horseback and read in the shade of a tree.

Endrows and lanes had prairie plants even in my childhood. By 1960, a massive movement to plant up to the edge of the road led to the disappearance of remaining prairie plants. I only vaguely remember seeing some of the prairie plants, and I certainly never imagined that I would someday try to recall them with such passion.

Winters kept us inside the house, in the barn or at school. Animals had to be taken care of and the trotlines in the bayou had to be set and run. Later, we would spend much of the winter working with my father, who was a carpenter and later a building contractor. Winter, however, was punctuated by the *boucheries* every other weekend—we gathered with the neighbors to slaughter a pig, calf or other large animals and shared the meat. The entire day was filled with varied activities that made for extraordinary memories.

School was the center of my life. My parents wanted me and my brothers to treat school like a job, and they thought us how to work on the farm. I was the first in two generations to go to college—my great-uncle Julius Vidrine had been the last. His medical practice and the establishment of the town of Vidrine were oft-mentioned stories in our home. My parents knew what an education might bring, and they insisted upon it. Mom had finished high school at night during my junior year of high school—she was thus adamant about taking school seriously.

Appendix 5
An Essay on The Cajun Prairie of the 1600s

From William Darby's 1805 visit to the Cajun Prairie as detailed in his book (1816) and repeated here from Chapter 3 in the present book:
Darby was immensely impressed by what he saw.

> *"The face of the earth exhibits an expanse of grass, interrupted only by an occasional clump of oak or pine trees...The winds breathe over the pathless waste of savanna. The wild fowl is seen flitting or the deer skimming over the plain."*

In describing the Attakapas prairie between Bayous Teche and Vermilion and the Opelousas prairie between the Teche and the Mermentau River, Darby noted:

> *"Here you behold those vast herds of cattle which afford sustenance to the natives, and the inhabitants of New Orleans. It is certainly one of the most agreeable views in nature, to behold from a point of elevation, thousands of horses and cows, of all sizes, scattered over the interminable mead, intermingled in wild confusion. The mind feels a glow of corresponding innocent enjoyment, with those useful and inoffensive animals grazing in a sea of plenty. If the active horsemen who guard them, would keep their distance, fancy would transport us backwards into the pastoral ages. When we estimate the extent of ground that must forever remain covered with grass, it is no extravagant declaration to call this one of the meadows of America."*

One of the goals of this book is to provide a view of the Cajun Prairie in the 1600s, the time before the arrival of Europeans, cattle, horses, pigs and sheep. No accounts of this period exist. We here attempt to create one.

'A sea of grasses and forbs' is the simple description of the prairie. In more detail, it was a quilt-work mosaic of clumps of grasses and forbs, each group seasonally obvious by their blooms and their growth habits. For example, Big bluestem would form large circles that shot up to 6-8 feet in height in September and remained so until the winds of late summer or autumn knocked the stems down forming an entangled mass just above the ground. The height is recorded as tall enough to obscure a group of horseback riders from one another except for their hats. The largest circle that Charles and I found in the remnant prairies was 33 steps (approximately 30 yards) in diameter—impressive and suggestive of what things may have been like. This circle was apparently all one plant—at our 22 year old restoration site, circles of Big bluestem started from single rhizomes are at best 2 yards in diameter. My conclusion is that the largest circle was easily 100 years of age. Most of the obvious prairie grasses form clumps, including the large grasses like Eastern gama grass, Switchgrass, and Yellow Indian grass. Some grasses like little bluestem form clumps that seem to achieve a maximum diameter of less than 2 feet and as such are numerous in the Eunice Prairie. Little bluestem and other shorter grasses, usually 4-5 feet in height, covered large areas of the remnant prairies and the Eunice restoration site and probably the Cajun Prairie of the 1600s. Although evidence is lacking, there is a possibility that large areas once covered with Big bluestem were heavily grazed and gave way to Little bluestem; thus, the Cajun Prairie as it appeared in the 1800s and today is mainly Little bluestem dominated. In areas with Big bluestem and taller forbs, the Cajun Prairie appears to grow in height as a community, such that: in March-April, it is 1 foot tall; in May-June, it is 2-4 feet tall; in July-August, it is 5-6 feet tall; and at the end of the growing season in September-October, it is 7-9 feet tall. The height increases at roughly 1 foot per month through the growing season from March to October.

The forbs, or showy flowering plants, were of great variety and numerous; each species bloomed in a specific habitat at a specific time of the year. The Plates 2-28 represent photographs of remnant prairies along railroad rights-of-way as they appeared from the mid-1980s to the mid-1990s. These remnants were burned routinely by local farmers for at least the last 100 years. The result of this simple management regimen was strips of highly diverse, often forb-dominated prairies. Each remnant was different in that it had distinctive dominant species, but different remnants of the same larger prairie often resembled one another. Remnants that were burned most often apparently had a higher 'species evenness' in that grasses that readily responded favorably to fire were clearly dominant and covered most of the soil surface. In wetter sites, sites that were somewhat disturbed or sites that were not subjected to fire as often, forbs commonly appeared to be dominant and 'species richness' was higher. However, in either case, fire was essential in order to maintain the diversity and the prairie habitat itself. The seasonal blooming of specific varieties of forbs provided a panorama of flower color and variety, usually site specific. In all the prairie remnants, high plant biodiversity was evident. High plant diversity (species richness) signaled high animal, bacterial, fungal and protist diversity. Thus, the diversity evident by visual inspection of the flora of these remnants is indicative of the diversity of other forms of life. By viewing the Plates of these remnants, the reader should conjure up an image of the Cajun Prairie flora—a mosaic of dramatic and colorful species varying in time and space. This virtual garden was once a reality that flourished across more than 2 million acres in southwestern Louisiana.

Large animals, including deer, Bison, birds, varied predators and man, were present and did impact the ecosystem, but those impacts were decidedly different. Man's impact centered on burning the prairies, Three Sisters' agriculture, and hunting and fishing local game/non-game species. Native Americans apparently had a vast trading economy with regional centers that are now viewed as 'mound cities.' The Cajun Prairie is only moderately distant to known trade centers with the closest one in Marksville, Louisiana.

In addition to larger animals, a tremendous variety of smaller birds, reptiles, amphibians, fish, insects and mussels were present in the ecosystem. Bison, Monarch butterflies and migratory birds were transient or seasonal in their visits, while most other animals were residents. Insects were abundant and diverse, and to some extent remain so. Butterflies, however, are now very rare in general, and even the most modest effort to grow plants for them is rewarded with numerous butterflies as most species remain sporadically in the gallery forests or in the nearby habitats.

The Cajun Prairie was maintained to a large extent by fire, both natural and man-made. Thousands of years of fire maintenance, a tradition among Native Americans throughout the Midwest, selected for fire-tolerant plant and animal species. Fire was not only used in the prairie but also in the savanna and marsh habitats, thus the entire region was subject to fire with the probable exception of the bottomland hardwood forests, and they may even have partially burned on occasion.

Large swaths of grassland were burned by the Native Americans or by lightning-started fires. The underground prairie was still alive in spite of the seeming death of all aboveground organisms. In fact, the underground biota comprises more than 70% of the biomass of the prairie, and it is much more diverse in terms of root variety and 'species richness' than the aboveground prairie. These underground parts of the plants may well have been hundreds of years old. Spring following a burn was absolutely splendiferous. Week after week, plants would literally break from the soil and explode into bloom. Each prairie had a different assemblage based upon moisture, natural residents, disturbance, time of the burn, etc. Following a winter fire, the plants and animals responded.

In spring and early summer, the drier areas splashed blooms of wild indigoes, rosinweeds, tickseed, spiderworts, Black-eyed Susans, orchids, milkweeds, coneflowers, gauras, winecups, bergamonts, mountain-mints, Coralbeans and obedient plants, while the wetter areas would bloom with irises, mallows, Indian plantains, Texas brown-eyed Susans, milkweeds and orchids. Late summer into fall, the drier areas bloomed with blazing stars, sunflowers, goldenrods, Compass plants and a mass of grasses from Little bluestem to Big bluestem, while in the wetter areas, ironweeds, Switchgrass and beggarticks would bloom. Winter turned golden-yellow, *Paille jaune*, and swayed to the winter winds. This prairie, as the rainy season began, would become nearly impassible by foot or wagon. The people could tell the time of the year by the plants in bloom, and they certainly came to depend on some of these plants for food for themselves and their animals.

Physiographically and topographically, the Cajun Prairie was divided into prairies by the tributaries of the Mermentau, Vermilion and Calcasieu Rivers. Each prairie was different, with the phenology of blooms (bloom seasons) of grasses and varied forbs making these differences apparent. The prairies had distinctive pimple mounds and wet areas, and the vegetation associated with each of these hydrologic features (runoff verses runon) was distinctive. The prairies also intergraded with surrounding habitats without distinctively changing appearance, for example, the prairie and marsh intergraded by simply increasing the number of wet sites until the entire area was wet. And for another example, the prairie intergraded with the savanna by at first having a few pine trees greatly interspersed but becoming ever more common until the savanna-type habitat was evident. These kinds of intergrades are still present today, but very uncommon. The gallery forest and the prairie intergraded by forming coves, where prairie habitat was almost completely surrounded by encroaching forest near the river or vice versa. This prairie and its associated habitats were dynamic and ever-changing; every fire would move the line of intergradations in one direction or the other. Large animals also modified the prairie habitat, often leaving a seemingly devastated mud-hole or damaged vegetation in their trail.

The broad expanse of grassland with Bison trails would have been similar at first glance to the Midwestern prairie. The climate would have been very different in that seasons would not have been nearly as discrete as the Midwest in the Cajun Prairie. However, Bison would have moved as herds across the prairie literally chewing through the tallgrass and beating down a trail. The Bison may have gathered at the wetter sites, but they would not have lingered for days as cattle do. The Bison chew the tops of the grasses and forbs and minimally damage the remaining parts of the plants, whereas the cattle chew the entire plant and usually, given the opportunity, pull up the plant with its roots. Thus prairies usually spring back after a Bison encounter, but an encounter with cattle is more devastating and in fact may ultimately destroy a patch of prairie.

This view of the prairie may be idyllic, but the whole of it must have been beautiful and at the same time near maddening. When burned, the new growth was splendid and easy to traverse, but late in the year or in the absence of a burn and in the presence of rain, the prairie turned into a marsh that was nearly impassable. Native Americans lived near the rivers and ventured out onto the prairie during specific times of the year. Early settlers did the same—the open prairie was not a safe place to be—a true wilderness. Later when the cattle arrived, the prairie would become even less safe. The trails multiplied, thus the repeated request of '*bon chemin*?' from all who traveled across the prairie. The surprise encounter with feral cattle might have been less of an adventure and more of a contest. The cattle promoted the development of wet and impassible trails. The vegetation changed in association with the grazing and habits of the cattle, hogs, and other feral livestock. The number of ticks and flies and other pests associated with cattle, horses and hogs multiplied—stories of mass tick infestations were common. It is only with the introduction of Fire Ants (and maybe the mirex treatments) that the numbers of ticks crashed in the Cajun Prairie and its surroundings.

We can reconstruct a flora and a fauna for the Cajun Prairie, and in fact, we can reconstruct a habitat that looks like our vision of the prairie. However, the essence of the prairie is gone. The large expanses of Cajun Prairie are mere legends now, and the prairies, if recreated, will be functionally biodiversity gardens or biomass-production farms (for ethanol or biogasoline production). The prairie chicken may have an opportunity to use this habitat, but the efforts to maintain populations of native birds and companion pets, dogs and cats, in the same vicinity are not optimistic. Opportunities for butterfly and hummingbird habitat and maybe even Whooping Crane habitat reconstruction and maintenance are more likely, but the use of herbicides and insecticides and the habit of rearing purple martins make the opportunity for butterflies less optimistic.

The dynamic habitat known as Cajun Prairie now has a basic description, a foundation for future study and an opportunity for reconstruction on several levels. The habitat is however forever changed, but the message of the habitat can be written short in the form of a garden. This message, like that of many other habitats and ecosystems, is one of 'singularity' of man and environment. The message compels humans to 'gather close' and sense the nature that surrounds them, because the secret to success on this planet is 'sustainability.' This secret of sustainability is the message hidden in the habitats that have survived for eons of time. The grasslands are the source of human food and may be a primary source of fuel in the future. If humans do not understand the message, no longer a secret, they are doomed to go the way of the dinosaurs.

Appendix 6
Flora of the Cajun Prairie

A partial list of the scientific names (left) and USDA PLANTS database common names (right) of the plant species found in the Cajun Prairie with Coefficients of Conservatism (C) values 4-10 and selected examples of plants with lower C-values. The list begins with plants with a C = 10 (the most conservative prairie species) (Allain et al. 2006). Older common names used in the literature are mentioned in parentheses.

C = 10

Manfreda virginica .... American aloe agave
Calopogon oklahomensis .... Bearded grass-pink
Platanthera nivea .... Snowy orchid
Pteroglossapsis ecristata .... Giant orchid (Wild coco)
Schizacharium tenerum .... Slender bluestem
Sorghastrum nutans .... Yellow Indiangrass
Sporobolus compositus var. drummondii .. Meadow dropseed
Sporobolus junceus .... Pineywoods dropseed
Sporobolus silveanus .... Silven's dropseed
Tridens ambiguus .... Pinebarrens tridens
Ruellia humilus .... Prairie petunia
Polytaenia nuttalliana .... Nuttall's prairie parsley
Echinacea pallida .... Purple cone flower (Pale coneflower)
Gaillardia aestivalis var. aestivalis .... Lanceleaf blanketflower (Indian blanket)
Helianthus mollis .... Ashy sunflower (Wooly or Hairy sunflower)
Liatris elegans .... Pinkscale gayfeather (White-bract blazing-star)
Liatris spicata .... Dense blazing-star (Spiked blazing-star)
Liatris squarrosa .... Scaly blazing-star
Ratibida pinnata .... Pinnate prairie coneflower
Silphium laciniatum .... Compass plant
Solidago odora .... Sweet goldenrod
Solidago rugosa .... Wrinkled leaf goldenrod
Euphorbia corollata .... Flowering spurge
Tragia betonicifolia .... Betonyleaf noseburn

Tephrosia onybrychoides............................Multibloom-hoarypea
Monarda lindheimeri................................Lindheimer's bergamont (Bee-balm)
Salvia azurea var. grandiflora....................Blue sage
Anemone caroliniana................................Carolina anemone
Hedyotis nigricans....................................Diamondflowers

29 species

C = 9

Cyperus cephalanthus...............................Buttonbush flatsedge
Scleria pauciflora.......................................Carolina whipgrass
Habertia lahue ssp. caerulea.......................Prairienymph
Andropogon gerardii.................................Big bluestem (Turkeyfoot grass)
Coelorachis cylindrica...............................Cylindrical jointgrass
Gymnopogon brevifolius............................Shortleaf skeletongrass
Paspalum praecox.....................................Early paspalum
Tripsacum dactyloides...............................Eastern gama grass
Eryngium yuccifolium..............................Button snakeroot
Arnoglossum plantagineum......................Indian plantain
Bigelowia virgata......................................Rayless goldenrod
Liatris pycnostachya.................................Kansas gayfeather (Prairie blazing-star)
Pityopsis graminifolia var. graminifolia Narrowleaf silkgrass (Grass-leaf golden- aster)
Rudbeckia texana.......................................Texas brown-eyed Susan
Silphium gracile..Slender rosinweed
Symphyotrichum oolentangiense..............Sky blue aster
Symphyotrichum pratense.........................Barrens silky aster
Dalea candida..White prairie-clover
Scutellaria integrifolia..............................Helmet flower (Skullcap)
Polygala incarnata....................................Processionflower
Buchnera americana..................................American bluehearts
Penstamon digitalis..................................Talus slope penstemon (Beard-tongue)

22 species

C = 8

Carex triangularis.....................................Caric sedge
Rhynchospora harveyi...............................Harvey's beaksedge
Aletris aurea..Golden colic-root
Aletris farinosa..White colic-root

Hypoxis hirsuta .... Eastern yellow stargrass
Aristida purpurascens .... Arrowfeather three-awn grass
Ctenium aromaticum .... Toothache grass
Dicanthelium oligosanthes .... Scribner's rosette-grass
Muhlenbergia capillaris .... Gulf coast muhly
Paspalum floridanum .... Florida paspalum
Asclepias viridiflora .... Green-flowered milkweed (Green milkweed)
Chrysopsis mariana .... Maryland golden-aster
Eupatorium leucolepis .... Justiceweed
Liatris acedota .... Slender gayfeather (Blazing-star)
Packera tomentosa .... Wooly ragwort
Rudbeckia grandifolia var. alismifolia .... Rough coneflower
Symphyotrichum patens .... Late purple aster
Evolvulus sericeus .... Silky evolvulus
Stylisma aquaticus .... Water dawnflower
Desmosdium sessilifolium .... Sessile leaf ticktrefoil
Galactea volubilis .... Downy milkpea
Lespedeza capitata .... Round-head lespedeza
Lespedeza virginica .... Slender lespedeza
Mimosa hystricina (not microphylla) .... Sensitive brier
Neptunia pubescens .... Tropical puff
Orbexilum pedunculatum var. psoraliodes .... Sampson's snakeroot
Orbexium simplex .... Single-stem scurfpea (Snakeroot)
Stylosanthes biflora .... Sidebeak pencil-flower
Sabatia gentianoides .... Pine-woods rose-gentian
Monarda fistulosa .... Bergamont (Bee-balm)
Callirhoe papaver .... Woods poppymallow (Wine-cups)
Oenothera pilosella ssp. sessilis .... Meadow evening primrose
Phlox pilosa ssp. pilosa .... Downy phlox (Prairie phlox)
Polygala ramosa .... Low pinebarren milkwort
Pedicularis canadensis .... Canadian louse-wort
Penstamon laxiflorus .... Nodding beardtongue

36 species

C = 7

Carex microdonta .... Little tooth sedge
Rhynchospora cephalantha .... Branched beaksedge
Rhynchospora chalarocephala .... Loose-head beaksedge
Rhynchospora elliotii .... Elliot's beaksedge

Rhynchospora microcarpa..........................Southern beakrush
Rhynchospora pusilla................................Fairy beakrush
Rhynchospora rariflora ....Few-flowered beakrush
Scleria ciliata..Fringed nutrush
Scleria reticularis......................................Netted nutrush
Scleria verticillata ....Low nutrush
Allium canadense var. mobilense ....Meadow garlic
Andropogon gyrans var. gyrans...............Elliot's beardgrass
Anthaenanthia rufa...................................Purple silky-scale
Coelorachis rugosa ....Wrinkled jointgrass
Dichanthelium acuminatum
var. acuminatum ....Tapered rosette-grass
Dichanthelium ovale.................................Eggleaf rosette-grass
Digitaria cognata.......................................Fall witch-grass
Schizacharium scoparium..........................Little bluestem
Xyris difformis var. difformis ....Common yellow-eyed-grass
Xyris laxifolia var. laxifolia ....Iris-leaf yellow-eyed-grass
Asclepias longifolia ....Long-leaf milkweed
Asclepias obovata.......................................Pineland milkweed
Asclepias tuberosa.....................................Butterfly-weed
Asclepias verticillata.................................Whorled milkweed
Coreopsis gladiata ....Coastalplain tickseed
Coreopsis tripteris......................................Tall tickseed
Eupatorium rotundifolium ....Roundleaf thoroughwort
Eurybia hemisphaerica..............................Showy aster
Helenium drummondii ....Fringed sneezeweed
Oligoneuron nitidum.................................Shiny golden-rod
Lobelia appendiculata................................Pale lobelia
Lobelia puberula var. puberula ....Downy lobelia
Hypericum cistifolium ....Round-pod St. John's wort
Hypericum nudiflorum.............................Early St. John's wort
Crotolaria sagittalis ....Arrowleaf rattlebox
Erythrina herbacea....................................Coralbean (Mamou)
Physostegia digitalis ....Finger false dragonhead (Obedient plant)
Pycnanthemum muticum ....Clustered mountain-mint
Pycnanthemum tenuifolium......................Narrowleaf mountain-mint
Scutellaria parvula....................................Small skullcap
Pinguicula pumila ....Small butterwort
Rhexia mariana..Maryland meadow beauty
Ludwigia linearis ....Narrow-leaf primrosewillow
Oenothera spachiana.................................Spach's evening primrose

Polygala cruciata..Drumheads
Claytonia virginica....................................Narrow-leaf springbeauty
Caenothus americanus................................New Jersey tea
Viola lanceolata..Lance-leaf violet
Viola sagittata..Arrow-leaf violet

49 species

C = 6

Pteridium aquilinum.................................Bracken fern
Tradescantia hirsutiflora............................Hairyflower spiderwort
Tradescantia virginiana.............................Virginia spiderwort
Cyperus croceus...Baldwin's flatsedge
Fuirena pumila..Dwarf umbrellasedge
Rhynchospora caduca.................................Anglestem sedge
Rhynchospora globularis............................Globe beaksedge
Rhynchospora glomerata............................Clustered beaksedge
Cooperia drummondii................................Evening rain lily
Spiranthes vernalis...................................Spring ladies'-tresses (Spiral orchids)
Andropogon ternarius var. ternarius........Splitbeard bluestem
Dichanthelium aciculare............................Needleleaf rosette-grass
Eragrostis elliottii.....................................Field lovegrass
Panicum virgatum.....................................Switchgrass
Paspalum plicatulum.................................Brown-seed paspalum
Spartina patens..Marshhay cordgrass
Xyris torta...Twisted yellow-eyed-grass
Ruellia pedunculata...................................Stalked wild-petunia
Amsomia tabernaemontana.......................Eastern blue-star
Asclepias lanceolata...................................Fewflower milkweed (Red milkweed)
Boltonia diffusa..Smallhead doll's daisy
Coreopsis lanceolata...................................Lance-leaf tickseed
Helenium flexuosum..................................Purple-head sneezeweed
Pluchea foetida...Marsh fleabane
Solidago semipervirens var. mexicana.......Seaside goldenrod
Vernonia texana...Texas ironweed
Lechea mucronata......................................Hairy pinweed
Lechea tenuifolia..Narrowleaf pinweed
Hypericum crux-andreae...........................St. Peter's-wort
Hypericum hypericoides ssp. hypericoides.St. Andrew's-cross
Ipomoea sagittata......................................Saltmarsh morning-glory

Drosera brevifolia........Dwarf sundew
Baptisia alba var. macrophylla........Largeleaf wild-indigo (White false indigo)
Baptisia bracteata var. laevicaulis........Long-bract wild-indigo (False indigo)
Baptisia bracteata var. leucophea........Nodding wild-indigo (False indigo)
Baptisia nuttalliana........Nuttall's wild indigo (False indigo)
Baptisia sphaerocarpa........Yellow wild-indigo (Yellow false indigo)
Centrosema virginianum........Spurred butterfly pea
Desmanthus illinoensis........Illinois bundleflower
Desmanthus ciliare........Hairy small leaf ticktrefoil
Neptunia lutea........Yellow-puff
Strophostyles umbellata........Pink fuzzybean
Castanea pumila var. pumila........Allegheny chinkapin
Lycopus americanus........American bugleweed
Physostegia intermedia........Slender false dragonhead (Obedient plant)
Physostegia virginiana spp. praemorsa........Obedient plant
Pycnanthemum albescens........Whiteleaf mountain-mint
Lythrum alatum var. lanceolatum........Winged lythrum
Oenothera linifolia........Threadleaf evening primrose
Orobanche uniflora........Oneflowered broomrape
Polygala lepticaulis........Swamp milkwort
Polygala mariana........Maryland milkwort
Polygala nana........Candyroot
Polygala sanguinea........Blood milkwort
Ranunculus fascicularis........Prairie butter-cup
Solanum dimidiatum........Western horsenettle
Styrax americanus........American snowbell

57 species

C = 5

Cyperus acuminatus........Tapertip flatsedge
Cyperus echinatus........Globe flatsedge
Rhynchospora colorata........White-top-sedge
Iris brevicaulis........Zigzag iris (Louisiana iris)
Iris virginica........Southern-blue-flag
Sisyrinchium angustifolium........Narrow blue-eyed-grass

Sisyrinchium atlanticum .... Eastern blue-eyed-grass
Sisyrinchium langloisii .... Roadside blue-eyed-grass
Juncus elliottii .... Bog rush
Hymenocallis liriosme .... Fragrant spider-lily
Nothoscordum bivalve .... Crowpoison (False garlic)
Dichanthelium sphaerocarpon .... Roundseed panicum
Eragrostis hirsuta .... Big-top lovegrass
Eragrostis lugens .... Mourning lovegrass
Eragrostis refracta .... Coastal lovegrass
Eragrostis spectabilis .... Purple lovegrass
Panicum brachyanthum .... Prairie panic grass
Paspalum laeve .... Field paspalum
Paspalum setaceum .... Thin paspalum
Saccharum giganteum .... Sugarcane plumegrass
Spartina spartinae .... Gulf cordgrass
Steinchisma hians .... Gaping panicum
Centella erecta .... Erect centella
Eryngium integrifolium .... Blue-flower eryngo
Boltonia asteroides .... Large-flowered doll's daisy
Chromolaena ivifolia .... Ivyleaf thoroughwort
Coreopsis pubescens .... Star tickseed
Erigeron strigosus .... Prairie fleabane
Eupatorium hyssopifolium .... Hyssopleaf thoroughwort
Eupatorium semiserratum .... Small-flower thoroughwort
Euthamia leptocephala .... Bushy goldentop (Flat-top goldenrod)
Euthamia tenuifolia .... Slender goldentop (Flat-top goldenrod)
Helianthus angustifolius .... Swamp narrowleaf sunflower
Krigia dandelion .... Potato dwarf-dandelion
Pseudognaphalium obtusifolium .... Rabbittobacco
Pterocaulon virgatum .... Wand black root
Rudbeckia hirta .... Black-eyed Susan
Vernonia gigantea .... Giant ironweed
Hypericum galioides .... Bedstraw St. John's wort
Hypericum gentianoides .... Pineweed
Hypericum gymnanthum .... Clasping-leaf St. John's wort
Cuscuta indecora .... Pretty dodder
Croton willdenowii .... Willdenow's croton
Desmodium paniculatum .... Narrow-leaf ticktrefoil
Lespedeza repens .... Creeping lespedeza
Sabatia brachiata .... Narrow-leaf rose-gentian

Hyptis alata... Cluster bushmint
Utricularia subulata .... Zigzag bladderwort
Linum medium var. texanum.................... Stiff yellow flax
Linum sulcatum... Grooved flax
Hibiscus moscheutos var. lasiocarpos........ Crimsoneyed rose-mallow
Gaura lindheimeri..................................... Lindheimer's beeblossum (Butterfly flower)
Gaura longiflora.. Longflowered beeblossum
Passiflora incarnata .... Purple passionflower (Maypop)
Polygala verticillata .... Whorled milkwort
Galium virgatum .... Southwestern bedstraw
Agalinis oligophylla .... Ridge stem false-foxglove
Agalinis skinneriana.................................. Skinner's false-foxglove
Agalinis viridis .... Green false-foxglove
Mecardonia acuminata .... Axilflower

60 species

C = 4

Asplenium platyneuron .... Ebony spleenwort
Commelina erecta....................................... Erect dayflower
Bulbostylis capillaris................................. Threadleaf beakseed
Carex alata .... Broadwing sedge
Carex albolutescens................................... Greenish-white sedge
Carex vulpinoidea .... Fox sedge
Carex mariscus var. jamaicense................ Jamaica sawgrass
Eleocharis microcarpa .... Smallfruit spikerush
Eleocharis montana................................... Dombey's spikerush
Eleocharis obtusa....................................... Blunt spikerush
Eleocharis tuberculosa .... Long-tubercle spikerush
Fimbristylis puberula................................ Vahl's hairy fimbry
Kyllinga odorata... Fragrant spikesedge
Psilocarya nitens.. Shortbeak beaksedge
Juncus dichotomus..................................... Forked rush
Juncus polycephalus.................................. Many-head rush
Juncus validus.. Round-head rush
Agrostis hyemalis....................................... Winter bentgrass
Aristida oligantha .... Prairie three-awn grass
Dicanthelium dichotomum .... Cypress panic grass
Dicanthelium scoparium............................ Velvet panic grass
Digitaria filiformis .... Slender crabgrass
Elymus virginicus...................................... Virginia wildrye

Panicum hemitomon .... Maiden-cane
Panicum rigidulum var. rigidulum .... Red-top panic grass
Setaria parviflora .... Knotroot bristlegrass
Tridens strictus .... Long-spike tridens
Ruellia caroliniensis ssp. Caroliniensis .... Carolina wild-petunia
Cicuta maculata .... Water hemlock
Cynosciadium digitatum .... Finger dogshade
Ptilimnium costatum .... Ribbed mock bishop-weed
Asclepias viridis .... Green milkweed
Conoclinium coelestinum .... Blue mistflower
Epatorium perfoliatum .... Common boneset
Pluchea rosea .... Rosy camphor-weed
Hypericum drummondii .... Nits and lice
Diospyros virginiana .... Common persimmon
Croton monanthogynus .... Prairie tea
Apios americana .... Groundnut
Chamaecrista fasciculata .... Partridge pea
Sabatia campestris .... Prairie rose-gentian
Sabatia stellaris .... Rose of Plymouth
Hydrolea ovata .... Hairy hydrolea (Blue waterleaf)
Hedeoma hispida .... Rough falsepennyroyal
Monarda punctata .... Spotted beebalm (Horse-mint)
Teucrium canadense .... American germander
Mitreola petiolata .... Lax hornpod
Mitreola sessilifolia .... Swamp hornpod
Lythrum lineare .... Wand lythrum
Ludwigia hirtella .... Spindleroot
Oenothera biennis .... Common evening primrose
Oxalis violacea .... Violet wood-sorrel
Polygonum hydropiperoides .... Swamp smartweed
Galium tinctorium .... Stiff marsh bedstraw
Bacopa caroliniana .... Lemon bacopa
Vitis cinerea .... Graybark grape

56 species

C < 4 (mentioned in text) (C-value in parentheses)

Lygodium japonicum .... Japanese climbing fern (-2)
Pinus palustris .... Longleaf pine (3)
Pinus taeda .... Loblolly pine (2)
Taxodium distichum .... Bald cypress (1)
Andropogon glomeratus .... Bushy bluestem (3)

Andropogon virginicus	Broom-sedge bluestem (2)
Axonopus fissifolius	Southern carpet grass (0)
Briza minor	Little quaking grass (-1)
Paspalum notatum	Bahia grass (-2)
Rottboellia cochinchinensis	Itch grass (Raoul grass) (-1)
Sorghum halepense	Johnson grass (-2)
Paspalum urvillei	Vasey grass (-2)
Pontedaria cordata	Pickerelweed (1)
Typha latifolia	Broad-leaf cattail (1)
Baccharis halimifolia	Eastern baccharis (2)
Bidens aristosa	Bearded beggar's-ticks (3)
Cirsium horridulum	Bull thistle (0)
Coreopsis tinctoria	Plains coreopsis (Painted coreopsis)(3)
Erigeron annus	Eastern daisy fleabane (2)
Erigeron philadelphicus	Philadelphia daisy fleabane (0)
Eupatorium serotinum	Fall boneset (2)
Mikania scandens	Climbing hempweed (1)
Pakera glabella	Butterweed (0)
Pyrrhopappus carolinianus	False dandelion (1)
Solidago canadensis	Canada goldenrod (1)
Symphyotrichum dumosum	Rice button aster (2)
Triodanus perfoliata	Venus' looking-glass (3)
Lonicera japonica	Japanese honeysuckle (-1)
Croton capitatus	Wooly croton (3)
Triadica sebiferum	Chinese tallow (Chicken-tree) (-3)
Prunella vulgaris	Heal-all (2)
Salvia lyrata	Lyreleaf sage (2)
Stachys floridana	Florida hedgenettle (1)
Melia azedarach	Chinaberry (-1) (Chinaball tree)
Morella cerifera	Wax-myrtle (3)
Oenothera speciosa	Showey evening primrose (2)
Oxalis stricta	Yellow wood-sorrel (0)
Rubus argutus	Southern blackberry (3)
Rubus trivialis	Southern dewberry (3)
Sambucus nigra ssp. Canadensis	American elderberry (2)
Diodia teres	Poorjoe (2)
Houstonia micrantha	Southern bluets (3)
Agalinis fasciculata	Beach false-foxglove (3)
Nuttallanthus canadensis	Old field toadflax (3)
Verbena braziliensis	Brazilian vervain (-2)

45 species

Appendix 7
French/Native American Names of Some Plants and Animals

Woody Cajun Prairie ecosystem plants with English and Cajun French common names and remarks regarding uses by the Cajuns (adapted from Holmes 1990).

Botanical name	English	Cajun French	Remarks
Arundinaria gigantea	Bamboo	Cannes	common along bayous
Ascyrum hypericoides	St. Andrew's cross	Hérbe a chien	translates as "dog herb"
Carya illinoiensis	Pecan	Pacanier	pecan is the fruit
Celtis laevigata	Hackberry	Bois inconnú	"unknown tree"
Cephalanthus occidentalis	Buttonbush	Bois de marais	swamp tree
Cornus spp.	Dogwoods	Bois de fléche	stems as arrowshafts
Crataegus opaca	Mayhaw	Pommiér	apple tree
Diospyros virginiana	Persimmon	Plaquéminier	Native American name
Erythrina herbacea	Coral bean	Mamou	Native American name
Liquidambar styraciflua	Sweetgum	Copal	resin (Spanish)
Maclura pomifera	Osage orange	Bois d'arc	bow wood
Magnolia grandiflora	Magnolia	Magnolia	palm (Palm Sunday)
Magnolia virginica	Sweet bay	Lauriér doux	sweet laurel
Myrica cerifera	Wax myrtle	Ciriér	wax producer
Nyssa aquatica	Tupelo	Oliviér	olive tree
Persea palustris	Red bay	Tit lauriér	*petite* (little) laurel
Pinus spp.	Pines	Pin	pine (French)
Planera aquatica	Water elm	Chataignier	chestnut tree
Prunus serotina	Black cherry	Merisier	cherry tree (French)
Quercus spp.	Oaks	Chêne	oak (French)
Rubus spp	Blackberry	Éronce	adapted from French
Sabal minor	Palmetto	Lataniér	little palm (Easter palm)
Salix spp.	Willow	Saulé	willow (French)
Sassafras albidum	Sassafras	Filé	gumbo seasoning
Taxodium distichum	Bald cypress	Cypré	cypress (French)
Vitis rotundifolia	Muscadine	Liane de socó	grape vine
Zanthoxylum clava-herculis	Prickly ash	Fréne piquant	toothache treatment

Paille jaune or yellow straw—general French term for prairie grass in winter.

Many areas named for animals (Native American names and French/ Spanish names): (adapted from Fontenot 2006)

Animals:

Calcasieu .... 'crying eagle'
Carencro .... vulture (carrion crow)
Chaoui .... raccoon
Cheveuril .... deer
Choc .... blackbird
Choupique .... swamp fish
Cocodrie .... 'crocodile or aligator'
Faiquetaique .... 'female turkey'
Maringouin .... mosquito
Moiselle .... dragonfly
Ouaouaron .... bullfrog
Patassa .... 'flat fish'
Queue du Tortue .... 'tail of the turtle'
Teche or Serpent .... 'snake'

Other terms in this book are French terms (I took the liberty of not italicizing prairie throughout the text):

Marais .... (little) marsh
Paille jeune .... yellow straw
Petit habitant .... small farm with polyculture
Platin .... pond
Prairie .... meadow
Vache .... cow (cattle)
Vacher .... cattleman (rancher)
Vacherie .... ranch

Appendix 8
Flora of Prairies by Clair Brown

Clair A. Brown in his *Wildflowers of Louisiana and Adjoining States* (1972) delineated the Cajun Prairie very well and lists 53 plant species as 'prairie' species. Some of his species names have changed, and many of the common names have changed. However, this provided a 'working flora' for the prairies in Louisiana including the Cajun Prairie.

Page	Species	Common name
14	*Aletris aurea*	colic root
25	*Cooperia drummondii*	Drumond's rain lily
26	*Hypoxis hirsuta*	yellow star grass
29	*Alophia drummondii*	Hebertia
34	*Sisyrinchium atlanticum*	blue-eyed grass
63	*Drosera intermedia*	sundew
74	*Baptisia leucantha*	white indigo
74	*Baptisia leucophaea*	nodding indigo
75	*Baptisia sphaerocarpa*	
75	*Baptisia nuttaliana*	Nuttall indigo
84	*Neptunia lutea*	yellow sensitive plant
85	*Petalostemum candidum*	white prairie clover
87	*Psoralea psoralioides*	Sampson snake root
88	*Stylosanthes biflora*	pencil flower
91	*Trifolium reflexum*	buffalo clover
107	*Caenothus americanus*	New Jersey tea
108	*Callirhoe papaver*	poppy-mallow
113	*Hypericum cistifolium*	St. John's-wort
121	*Gaura lindheimeri*	
123	*Eryngium integrifolium*	eryngo
124	*Eryngium yuccifolium*	button snakeroot
125	*Ptilimnium costatum*	mock bishop's-weed
126	*Zizia aurea*	golden alexanders
139	*Sabatia brachiata*	rose-gentian
140	*Sabatia gentianoides*	rose-gentian
142	*Amsonia tabernaemontana*	blue-star
145	*Asclepias viridis*	spider milkweed
149	*Ipomoea sagittata*	morning-glory
150	*Ipomoea hederacea*	morning-glory
152	*Hydrolea ovata*	blue waterleaf
159	*Physostegia digitalis*	obedient plant

Page	Species	Common name
161	*Salvia azurea*	blue sage
170	*Penstamon laxiflorus*	beardtongue
173	*Pinguicula caerulea*	butterwort
182	*Lobelia appendiculata*	pale lobelia
189	*Bidens aristosa*	beggar ticks
192	*Cacalia plantaginea*	Indian plantain
196	*Coreopsis lanceolata*	tickseed
198	*Echinacea pallida*	purple coneflower
203	*Helianthus annuus*	sunflower
204	*Helianthus mollis*	sunflower
204	*Helianthus cucumerifolius*	cucumber leaf sunflower
209	*Krigia dandelion*	potato dandelion
209	*Liatris elegans*	blazing-star
210	*Liatris squarrulosa*	blazing-star
211	*Liatris squarrosa*	blazing-star
213	*Ratibida columnaris*	cone-flower
214	*Rudbeckia hirta*	black-eyed susan
215	*Rudbeckia maxima*	giant cone-flower
215	*Rudbeckia fulgida*	bracted cone-flower
217	*Silphium laciniatum*	compass-plant
217	*Silphium gracile*	rosin weed
226	*Vernonia missurica*	ironweed

Appendix 9

Phenology: Blooming times of plants are listed in the order of first blooms in Eunice, Louisiana. Plant scientific names and photos are at PLANTS database (USDA) online. Plants with Coefficient of Conservatism (C) < 5 designated with an asterisk (*).

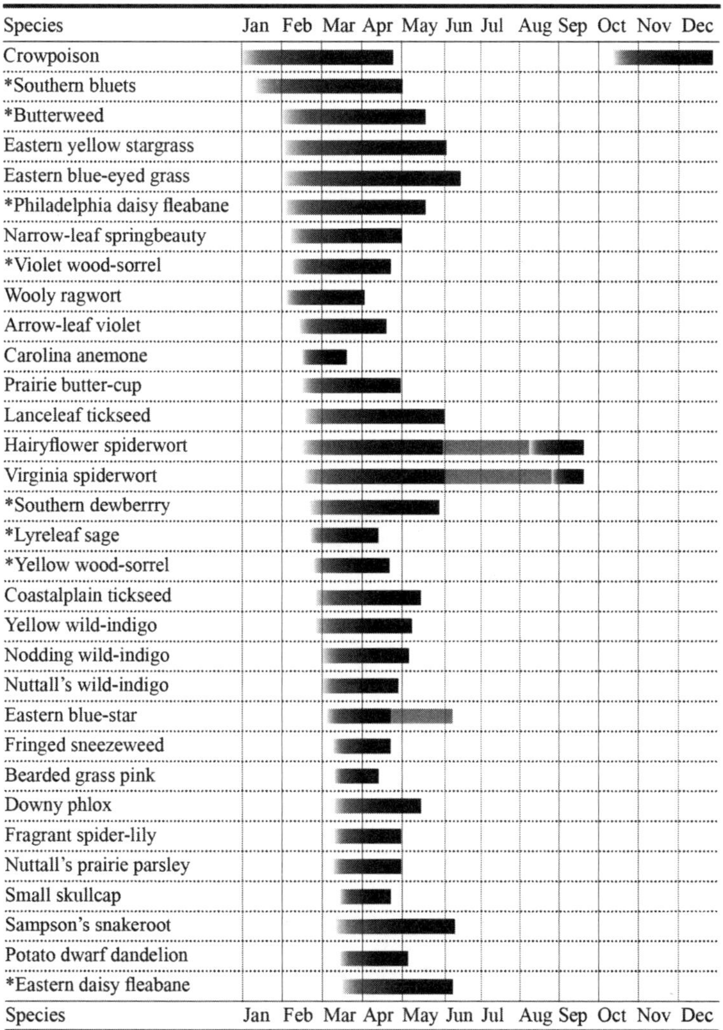

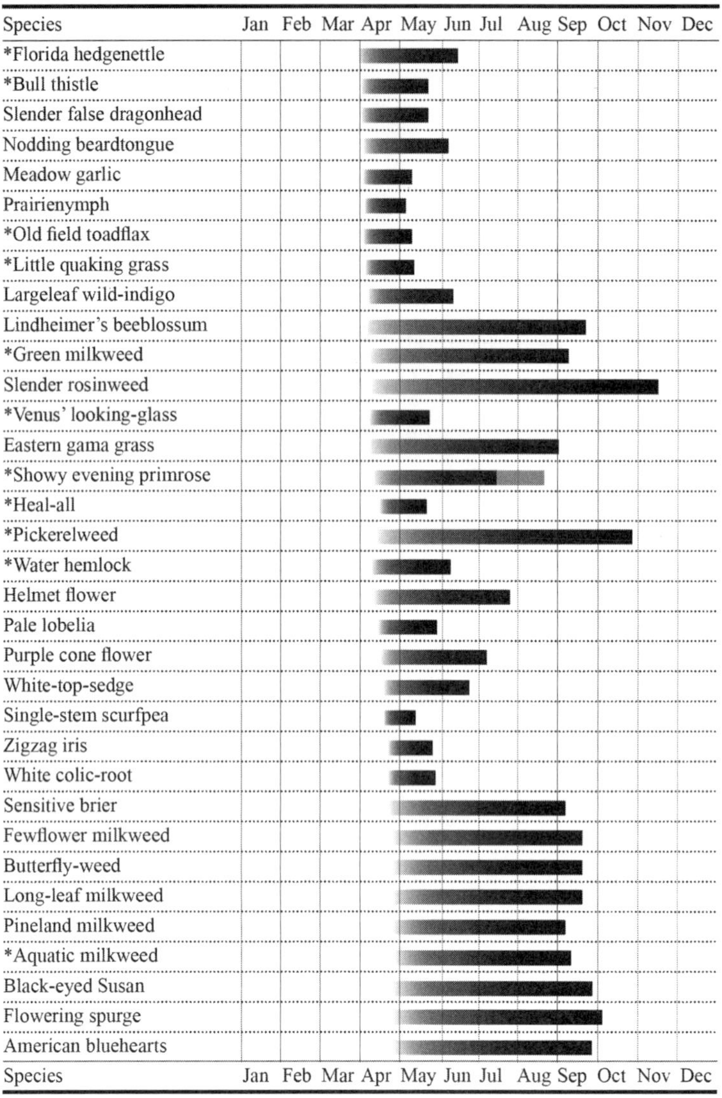
Species
Jan Feb Mar Apr May Jun Jul Aug Sep Oct Nov Dec
*Florida hedgenettle
*Bull thistle
Slender false dragonhead
Nodding beardtongue
Meadow garlic
Prairienymph
*Old field toadflax
*Little quaking grass
Largeleaf wild-indigo
Lindheimer's beeblossum
*Green milkweed
Slender rosinweed
*Venus' looking-glass
Eastern gama grass
*Showy evening primrose
*Heal-all
*Pickerelweed
*Water hemlock
Helmet flower
Pale lobelia
Purple cone flower
White-top-sedge
Single-stem scurfpea
Zigzag iris
White colic-root
Sensitive brier
Fewflower milkweed
Butterfly-weed
Long-leaf milkweed
Pineland milkweed
*Aquatic milkweed
Black-eyed Susan
Flowering spurge
American bluehearts
Species
Jan Feb Mar Apr May Jun Jul Aug Sep Oct Nov Dec

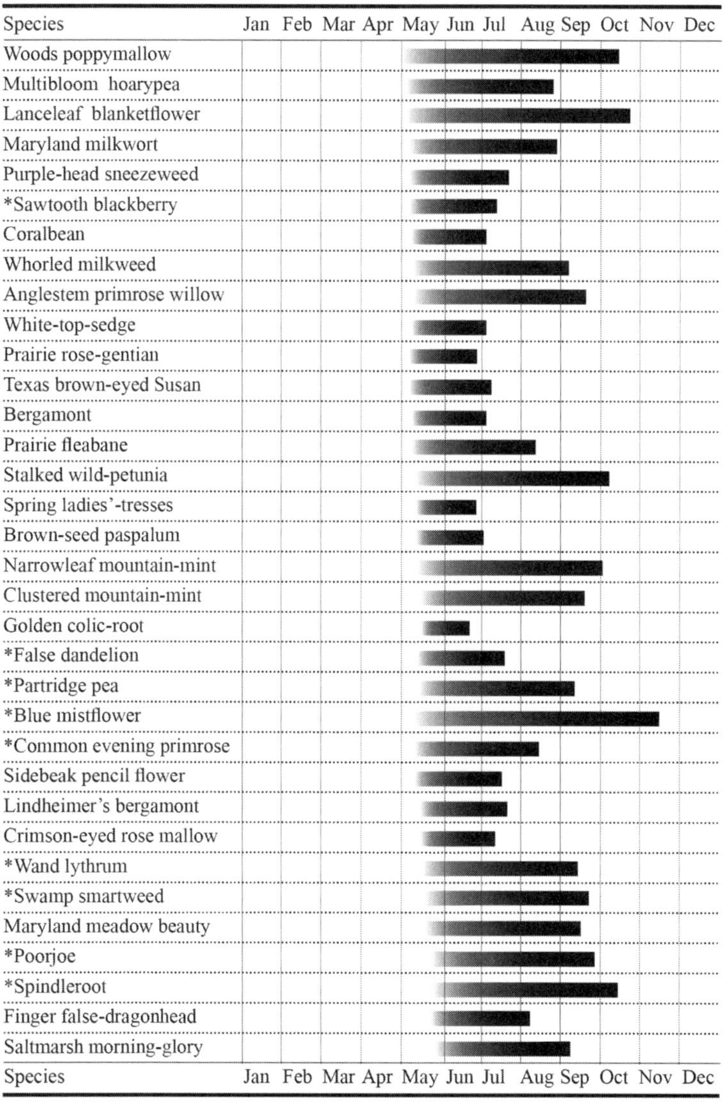
Species
Jan Feb Mar Apr May Jun Jul Aug Sep Oct Nov Dec
Woods poppymallow
Multibloom hoarypea
Lanceleaf blanketflower
Maryland milkwort
Purple-head sneezeweed
*Sawtooth blackberry
Coralbean
Whorled milkweed
Anglestem primrose willow
White-top-sedge
Prairie rose-gentian
Texas brown-eyed Susan
Bergamont
Prairie fleabane
Stalked wild-petunia
Spring ladies'-tresses
Brown-seed paspalum
Narrowleaf mountain-mint
Clustered mountain-mint
Golden colic-root
*False dandelion
*Partridge pea
*Blue mistflower
*Common evening primrose
Sidebeak pencil flower
Lindheimer's bergamont
Crimson-eyed rose mallow
*Wand lythrum
*Swamp smartweed
Maryland meadow beauty
*Poorjoe
*Spindleroot
Finger false-dragonhead
Saltmarsh morning-glory
Species
Jan Feb Mar Apr May Jun Jul Aug Sep Oct Nov Dec

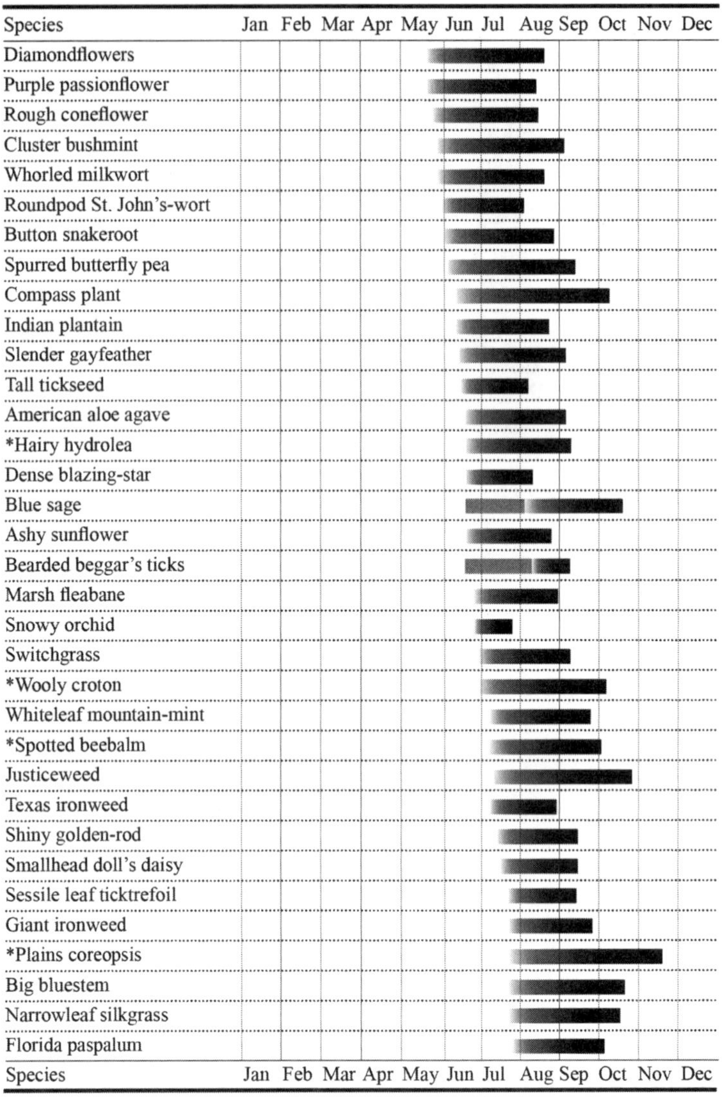
Species
Jan Feb Mar Apr May Jun Jul Aug Sep Oct Nov Dec
Diamondflowers
Purple passionflower
Rough coneflower
Cluster bushmint
Whorled milkwort
Roundpod St. John's-wort
Button snakeroot
Spurred butterfly pea
Compass plant
Indian plantain
Slender gayfeather
Tall tickseed
American aloe agave
*Hairy hydrolea
Dense blazing-star
Blue sage
Ashy sunflower
Bearded beggar's ticks
Marsh fleabane
Snowy orchid
Switchgrass
*Wooly croton
Whiteleaf mountain-mint
*Spotted beebalm
Justiceweed
Texas ironweed
Shiny golden-rod
Smallhead doll's daisy
Sessile leaf ticktrefoil
Giant ironweed
*Plains coreopsis
Big bluestem
Narrowleaf silkgrass
Florida paspalum
Species
Jan Feb Mar Apr May Jun Jul Aug Sep Oct Nov Dec

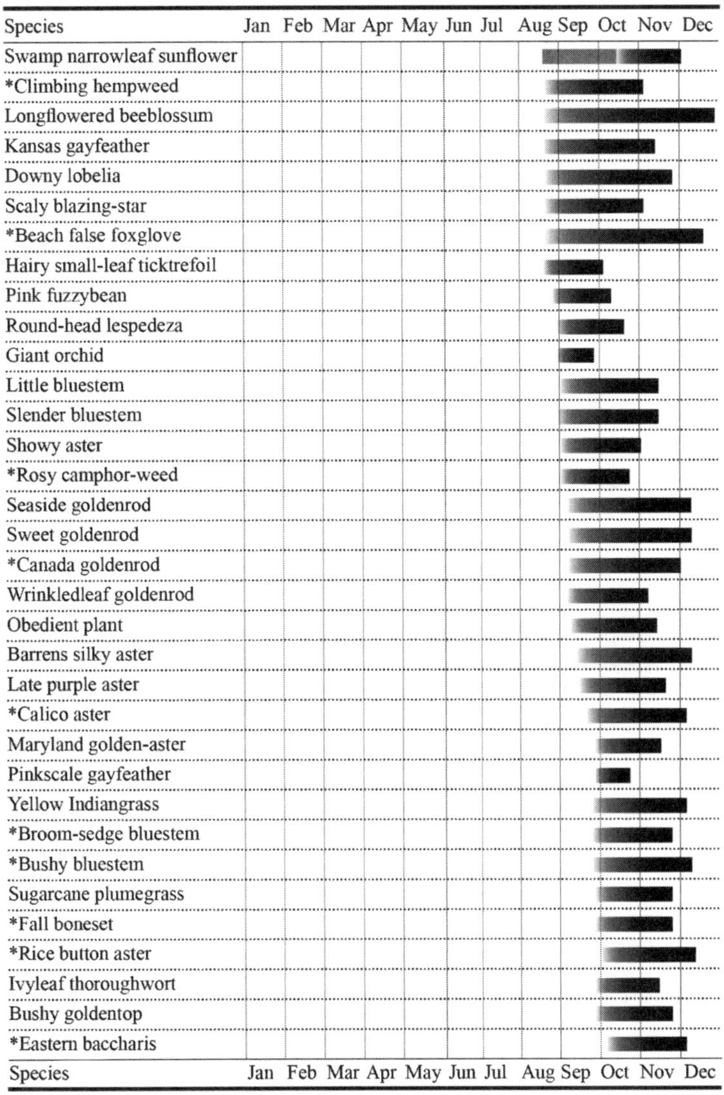
Species
Jan Feb Mar Apr May Jun Jul Aug Sep Oct Nov Dec
Swamp narrowleaf sunflower
*Climbing hempweed
Longflowered beeblossum
Kansas gayfeather
Downy lobelia
Scaly blazing-star
*Beach false foxglove
Hairy small-leaf ticktrefoil
Pink fuzzybean
Round-head lespedeza
Giant orchid
Little bluestem
Slender bluestem
Showy aster
*Rosy camphor-weed
Seaside goldenrod
Sweet goldenrod
*Canada goldenrod
Wrinkledleaf goldenrod
Obedient plant
Barrens silky aster
Late purple aster
*Calico aster
Maryland golden-aster
Pinkscale gayfeather
Yellow Indiangrass
*Broom-sedge bluestem
*Bushy bluestem
Sugarcane plumegrass
*Fall boneset
*Rice button aster
Ivyleaf thoroughwort
Bushy goldentop
*Eastern baccharis
Species
Jan Feb Mar Apr May Jun Jul Aug Sep Oct Nov Dec

Appendix 10
Factors affecting blooming time (onset and length of blooming period) in various species and varieties of plants.

1. The latitude and elevation where the plant grows dramatically affect blooming season. Cajun Prairie plants collectively bloom usually 11 months of the year, whereas upper Midwestern prairie plants bloom usually no more than 6 months of the year.
2. The amount of rain or moisture regimen strikingly affects not only growth but also blooming season.
3. The geology and moisture gradient of the soil controls plant growth and affects bloom season.
4. The burning time of year and the number of burns or mows per year. Any variation in burning or mowing regimen will impact blooming.
5. The genetics of the plant and the genetic varieties that are planted. A clone planting may very well all bloom at once.
6. The age of plants is important in that young plants may bloom once and produce few flowers whereas an older establish plant or planting may produce not only more blooms but also an extended blooming season.
7. The companion plants may impact bloom by both mechanical and chemical means. Legumes may enrich soil with nitrogen and reduce blooming, while fungi (mycorrhizae) may enhance phosphorus levels and exacerbate blooming.
8. Shade has a dramatic impact on bloom. The impact is so significant that plants are categorized as shade-lovers or sun-lovers.
9. The health of plants dramatically affect the bloom season. However, some plants bloom extraordinarily when in poor health as if to attempt one last batch of offspring.
10. Temperature, even minor changes, dramatically affects blooming.
11. Unusual weather extremes may cause plants to bloom at unusual time, e.g., a warm period in winter may result in a blooming event.
12. In habitat restoration and in natural succession, the maturity of the restoration or the succession sere dramatically affects not only which plants are growing but also their blooming seasons.
13. Stress (fertilizer, mowing, grazing, enrichment, etc.) may significantly alter blooming strategies and seasons of plants.

Appendix 11
Animals of the Cajun Prairie Ecosystem

This list includes the invertebrate animals that I have had the opportunity to do research upon. The sheer number of kinds of these organisms is nearly astronomical; unfortunately many are unknown. Groups that contain many species that I have not worked upon include spiders, skippers, moths, bees, wasps, grasshoppers, mantids, crayfish, etc. Walls (2009) provided a list of crayfish. Birds and other vertebrates will be treated elsewhere (Appendix 12).

Porifera .... sponges
  Spongillidae .... freshwater sponges
  Dosilia radiospiculata (Mills 1888)
  Ephydatia fluviatilis (Linnaeus 1758)
  Trochospongilla leidyi (Bowerbank 1863)
  Trochospongilla horrida (Weltner 1893)
  Trochospongilla pennsylvanica (Potts 1882)
Cnidaria
  Clavidae
  Cordylophora lacustris Allman (= *Cordylophora caspia* (Pallas 1771))
Entoprocta (Urnatellidae):
  Urnatella gracilis Leidy 1851
Ectoprocta (Paludicellidae).. moss animals
  Pottsiella erecta (Potts 1884)
  Fredericella sultana (Blumenbach 1779)
  Plumatella repens (Linnaeus 1758)
  Pectinatella magnifica (Leidy 1851)
Platyhelminthes.. flatworms
  Trematoda ....flukes
  Aspidogastridae
    Aspidogaster conchicola von Baer 1826
    Cotylaspis insignis Leidy 1857
Rotifera ....rotifers (wheel animals)
  Conochilopsis causeyae (Vidrine, McLaughlin, and Willis 1985)
  Lacinularia flosculosa (Mueller 1773)
  Lacinularia ismailoviensis (Poggenpol 1872)
  Lacinularia elliptica Shepard 1897
  Sinantherina socialis (Linnaeus 1758)
  Sinantherina semibullata (Thorpe 1889)
  Sinantherina spinosa (Thorpe 1893)

Annelida
- Polychaeta .... polychetes
  - *Streblospio benedicti* Webster
  - *Hypaniola florida* (Hartman)
  - *Nemalycastis abiuma* (Müller)
  - *Nereis succinea* (Frey and Leuckart)
  - *Parandalia filiformis* (Claparede)
- Oligochaeta .... earthworms
  - Lumbricidae
- Lumbriculidae
  - *Lumbriculus variegatus* (Müller)
- Naididae
  - *Dero* sp.
- Tubificidae .... aquatic earthworms
  - *Branchiura sowerbyi* Beddard
  - *Varichaetadrilus harmani* (Loden)
  - *Spirosperma ferox* (Eisen)
  - *Limnodrilus cervix* Brinkhurst
  - *Limnodrilus claparedianus* Ratzel
  - *Limnodrilus hoffmeisteri* Claparede
  - *Limnodrilus maumeensis* Brinkhurst and Cook
  - *Limnodrilus rubipenis* Loden
  - *Limnodrilus udekemianus* Claparede
- Hirudinea .... leeches
  - Glossiphoniidae
    - *Placobdella montifera* Moore 1906
    - *Helobdella stagnalis* (Linnaeus 1858)

Mollusca
- Gastropoda .... snails
  - *Viviparus subpurpureus* (Say 1829) .... Olive mysterysnail
  - *Campeloma decisum* (Say 1816) .... Pointed campeloma
  - *Amnicola* sp.
  - *Gyraulis* sp.
  - *Physa* sp.
  - *Ferissa* sp.
- Bivalvia .... clams and mussels
    - *Musculium transversum* (Say) .... Long fingernailclam
    - *Eupera cubensis* (Prime) .... Mottled fingernailclam
    - *Rangia cuneata* Gray .... Atlantic rangia
    - *Pyganodon grandis* (Say 1829) .... Giant floater
    - *Utterbackia imbecillis* (Say 1829) .... Paper pondshell
    - *Arcidens confragosus* (Say 1829) .... Rock pocketbook

Strophitus undulatus (Say 1817) .... Creeper
Amblema plicata (Say 1817) .... Threeridge
Megalonaias nervosa (Rafinesque 1820) .... Washboard
Plectomerus dombeyanus (Valenciennes 1827) .... Bankclimber
Quadrula mortoni (Conrad 1835) .... Western pimpleback
Quadrula apiculata (Say 1829) .... Southern mapleleaf
Quadrula quadrula (Rafinesque 1820) .... Mapleleaf
Tritogonia verrucosa (Rafinesque 1820) .... Pistolgrip
Pleurobema riddelli (Lea 1861) .... Louisiana pigtoe
Fusconaia askewi (Marsh 1896) .... Texas pigtoe
Fusconaia flava (Rafinesque 1820) .... Wabash pigtoe
Uniomerus tetralasmus (Say 1831) .... Pondhorn
Uniomerus declivus (Say 1831) .... Tapered pondhorn
Glebula rotundata (Lamarck 1819) .... Round pearlshell
Lampsilis satura (Lea 1852) .... Sandbank pocketbook
Lampsilis hydiana (Lea 1838) .... Louisiana fatmucket
Lampsilis teres (Rafinesque 1820) .... Yellow sandshell
Ligumia subrostrata (Say 1831) .... Pondmussel
Leptodea fragilis (Rafinesque 1820) .... Fragile papershell
Potamilus purpuratus (Lamarck 1819) .... Bleufer
Toxolasmus parvus (Barnes 1823) .... Lilliput
Toxolasmus texasensis (Lea 1857) .... Texas lilliput
Obliquaria reflexa (Rafinesque 1820) .... Threehorn wartyback
Obovaria jacksoniana (Frierson 1912) .... Southern hickorynut
Truncilla donaciformis (Lea 1828) .... Fawnsfoot
Truncilla truncata (Rafinesque 1820) .... Deertoe
Villosa lienosa (Conrad 1834) .... Little spectaclecase
Corbicula fluminea (Mueller) .... Asian clam

Arthropoda
Insecta .... insects
Collembola
Ephemeroptera
Hexagenia sp. .... mayflies
Diptera
Chironomidae .... midges
Chironomus sp.
Cryptochironomus sp.
Stenochironomus sp.
Procladius sp.
Brillia sp.
Ablabesmyia sp.
Paratendipes sp.

Pentaneura sp.
Culicidae.......... mosquitoes
Psorophora ferox.......... Woodland mosquito
Psorophora columbiae.......... Dark ricefield mosquito
Psorophora ciliata
Culex salinarius
Culex pipiens quinquefasciatus.......... Southern house mosquito
Culex resturans
Anopheles crucians
Anopheles quadrimaculatus.......... Malaria mosquito
Ochlerotatus solicitans.......... Tan salt marsh mosquito
Aedes vexans.......... Floodwater mosquito
Aedes aegypti.......... Yellow fever mosquito
Aedes albopictus.......... Asian tiger mosquito
Culiseta inornata.......... Winter mosquito
Uranotaenia spp.
Toxorhynchites sp.
Chaoboridae
Chaoborus sp.
Megaloptera
Sialis sp.
Neuroptera
Climacia areolaris (Hagen).......... Sponge fly
Odonata
Zygoptera.......... damselflies
Calopterygidae
Calopteryx dimidiata Burmeister.......... Sparkling jewelwing
Calopteryx maculata (Beauvois).......... Ebony jewelwing
Hetaerina titia (Drury).......... Smoky rubyspot
Lestidae
Lestes disjunctus australis Walker.......... Common spreadwing
Lestes vigilax Hagen in Selys.......... Swamp spreadwing
Coenagrionidae
Argia apicalis (Say).......... Blue-fronted dancer
Argia fumipennis violacea (Hagen).......... Variable dancer
Argia moesta (Hagen).......... Powdered dancer
Argia tibialis (Rambur).......... Blue-tipped dancer
Enallagma civile (Hagen).......... Familiar bluet
Enallagma divagans Selys.......... Turquoise bluet
Enallagma dubium Root.......... Burgandy bluet
Enallagma durum (Hagen).......... Big bluet
Enallagma exsulans (Hagen).......... Stream bluet

Enallagma geminatum Kellicott .... Skimming bluet
Enallagma signatum (Hagen) .... Orange bluet
Enallagma traviatum westfalli Donnelly .... Slender bluet
Ischnura (Anomalagrion) hastata (Say) .... Citrine forktail
Ischnura posita Hagen... Fragile forktail
Ischnura prognata Hagen... Furtive forktail
Ischnura ramburii (Selys) .... Rambur's forktail
Nehalennia integrollis Calvert .... Southern sprite
Telebasis byersi Westfall.. Duckweed firetail

Anisoptera .... dragonflies

Petaluridae

Tachopteryx thoreyi (Hagen) .... Gray petaltail

Aeshnidae

Anax junius (Drury) .... Common green darner
Basiaeschna janata (Say) .... Springtime darner
Boyeria vinosa (Say)... Fawn darner
Coryphaeschna ingens (Rambur) .... Royal darner
Epiaeschna heros (Fabricius) .... Swamp darner
Gomphaeschna antilope (Hagen)......................... Taper-tailed darner
Gomphaeschna furcillata (Say) .... Harlequin darner
Nasiaeschna pentacantha (Rambur)........................... Cyrano darner

Gomphidae

Aphylla angustifolia Garrison .... Broad-striped forceptail
Aphylla williamsoni (Gloyd).......................... Two-striped forceptail
Arigomphus maxwelli (Ferguson)............................... Bayou clubtail
Arigomphus submedianus (Williamson) .... Jade clubtail
Dromogomphus armatus Selys .... Southeastern spineyleg
Dromogomphus spinosus (Selys).......... Black-shouldered spineyleg
Dromogomphus spoliatus (Hagen) .... Flag-tailed spineyleg
Gomphus (Gomphorus) hybridus Williamson .... Cocoa clubtail
Gomphus (Gomphus) lividus Selys .... Ashy clubtail
Gomphus (Gomphus) oklahomensis Pritchard .... Oklahoma clubtail
Hagenius brevistylis Selys .... Dragonhunter
Progomphus obscurus (Rambur)..................... Common sanddragon
Stylurus laurae (Williamson).................................. Laura's clubtail
Stylurus plagiatus (Selys) .... Russet-tipped clubtail

Cordulegastridae

Cordulegaster obliqua fasciata Rambur............ Arrowhead spiketail

Libellulidae

Macromiinae

Didymops transversa (Say)... Stream cruiser
Macromia illinoiensis georgina (Selys)............. Illinois river cruiser

Macromia taeniolata Rambur .... Royal river cruiser

Corduliinae

Epicordulia princeps Hagen .... Prince baskettail
Helocordulia selysii (Hagen) .... Selys' sundragon
Neurocordulia alabamensis Hodges .... Alabama shadowdragon
Neurocordulia molesta (Walsh).................... Smoky shadowdragon
Neurocordulia obsoleta (Say) .... Umber shadowdragon
Somatochlora filosa (Hagen)............................... Fine-lined emerald
Somatochlora georgiana Walker .... Coppery emerald
Somatochlora linearis (Hagen) .... Mocha emerald
Tetragoneuria cynosura (Say)........................... Common baskettail

Libellulinae

Brachymesia gravida (Calvert)....................... Four-spotted pennant
Celithemis amanda (Hagen) .... Amanda's pennant
Celithemis bertha Williamson .... Red-veined pennant
Celithemis elisa (Hagen)... Calico pennant
Celithemis eponina (Drury) .... Holloween pennant
Celithemis fasciata Kirby .... Banded pennant
Celithemis ornata (Rambur)...................................... Faded pennant
Celithemis verna Pritchard .... Double-ringed pennant
Erythemis simplicicollis (Say)............................ Eastern pondhawk
Erythrodiplax berenice berenice (Drury).............. Seaside dragonlet
Erythrodiplax miniscula (Rambur)................. Little blue dragonlet
Ladona deplanata Rambur.. Blue corporal
Libellula auripennis Burmeister............... Golden-winged skimmer
Libellula axilena Westwood........................... Bar-winged skimmer
Libellula cyanea Fabricius .... Spangled skimmer
Libellula flavida Rambur .... Yellow-sided skimmer
Libellula incesta Hagen .... Slaty skimmer
Libellula luctuosa Burmeister .... Widow skimmer
Libellula lydia (Drury)...................................... Common whitetail
Libellula needhami Westfall .... Needham's skimmer
Libellula semifasciata Burmeister.......................... Painted skimmer
Libellula vibrans Fabricius .... Great blue skimmer
Macrodiplax balteata (Hagen)..................................... Marl pennant
Miathyria marcella (Selys in Sagra) .... Hyacinth glider
Orthemis ferruginea (Fabricius) .... Roseate skimmer
Pachydiplax longipennis (Burmeister).......................... Blue dasher
Pantala flavescens (Fabricius).............................. Wandering glider
Pantala hymenaea (Say).................................... Spot-winged glider
Perithemis tenera (Say)..................................... Eastern amberwing
Sympetrum corruptum (Hagen) .... Blue-faced meadowhawk

Tramea carolina (Linnaeus) .... Carolina saddlebags
Tramea lacerata Hagen .... Black saddlebags
Tramea onusta Hagen .... Red-mantled saddlebags

Lepidoptera — butterflies and moths

Papilionidae

Papilioninae

Battus philenor (Linnaeus) .... Pipevine swallowtail
Eurytides marcellus (Cramer) .... Zebra swallowtail
Papilio polyxenes Stoll .... Black swallowtail
Papilio cresphontes Cramer .... Giant swallowtail
Papilio glaucus (Linnaeus) .... Eastern tiger swallowtail
Papilio troilus Smith .... Spicebush swallowtail
Papilio palamedes Drury .... Palamedes swallowtail

Pierinae

Anthocharis midea (Huebner) .... Falcate orangetip
Pontia protodice (Boisduval and Leconte) .... Checkered white

Coliadinae

Colias eurytheme Boisduval .... Orange sulphur
Colias cesonia (Stoll) .... Southern dogface
Eurema daira (Latrielle) .... Barred yellow
Eurema lisa (Boisduval and Leconte) .... Little sulphur
Eurema nicippe (Cramer) .... Sleepy orange
Phoebis sennae (Linnaeus) .... Cloudless sulphur

Lycaenidae

Theclinae

Atlides halesus (Cramer) .... Great purple hairstreak
Callophrys niphon (Huebner) .... Eastern pine elfin
Parrhasius m-album (Boisduval and Leconte) .. White-M hairstreak
Satyrium calanus (Huebner) .... Banded hairstreak
Satyrium liparops (Leconte) .... Striped hairstreak
Strymon melinus (Huebner) .... Gray hairstreak

Polyommatinae

Celastrina ladon (Cramer) .... Spring azure
Celastrina neglecta Opler and Krized .... Summer azure
Everes comyntas (Godart) .... Eastern tailed-blue

Riodinidae

Calephelis virginiensis (Guerlin) .... Little metalmark

Nymphalidae

Libytheinae

Libytheana carinenta (Cramer) .... American snout

Heliconiinae

Agraulis vanillae (Linnaeus) .... Gulf fritillary

Dryas julia (Fabricius) .... Julia
Euptoieta claudia (Cramer) .... Variegated fritillary
Heliconius charithonius (Linnaeus) .... Zebra

Nymphalinae

Nymphalis antiopa (Linnaeus 1758) .... Mourning cloak
Chlosyne gorgone (Huebner) .... Gorgone checkerspot
Junonia coenia (Huebner) .... Common buckeye
Phycoides phaon (Edwards) .... Phaon crescent
Phycoides texana (Edwards) .... Texan crescent
Phycoides tharos (Drury) .... Pearl crescent
Polygonia interrogationis (Fabricius) .... Question mark
Polygonia comma (Harris) .... Eastern comma
Vanessa atalanta (Linnaeus) .... Red admiral
Vanessa cardui (Linnaeus) .... Painted lady
Vanessa virginiensis (Drury) .... American lady

Limenitidinae

Limenitis archippus (dos Passos) .... Viceroy
Limenitis arthemis (Fabricius) .... Red-spotted purple

Charaxinae

Anaea andria (Scudder) .... Goatweed leafwing

Apaturinae

Asterocampa celtis (Edwards) .... Hackberry emperor
Asterocampa clyton (Boisduval and Leconte) .... Tawny emperor

Satyrinae

Cercyonis pegala (Fabricius) .... Common wood-nymph
Cyllopsis gemma (Huebner) .... Gemmed satyr
Enodia creola (Skinner) .... Creole pearly-eye
Enodia portlandica (Fabricius) .... Southern pearly-eye
Hermeuptychia sosybius (Fabricius) .... Carolina satyr
Megisto cymela (Cramer) .... Little wood satyr
Neonympha areolata (Smith) .... Georgia satyr

Danainae

Danaus gilippus (Cramer) .... Queen
Danaus plexippus (Linnaeus) .... Monarch

Crustacea

Amphipoda

Gammarus sp.

Isopoda

Asellus sp.

Copepoda

Cyclops sp.
Leptocaris sp.

Mysidacea
Taphromysis louisianae Banner
Mysidopsis almyra Bowman
Decapoda
Paleomonetes kadiakensis Rathbun....................................Grass shrimp
Conchostraca (clam shrimps)
Eulimnadia texana Packard.. Clam shrimp
Arachnida .... spiders and mites
Acari .... mites
Pionidae
Najadicola ingens (Koenike 1890)
Unionicolidae
Unionicola campelomaicola Marshall 1935
Unionicola viviparaicola Vidrine 1985
Unionicola aculeata (Koenike 1890)
Unionicola laurentiana Crowell and Davids 1979
Unionicola hoesei Vidrine 1986
Unionicola stricta (Wolcott 1898)
Unionicola vamana Mitchell and Wilson 1965
Unionicola amandita Mitchell and Wilson 1965
Unionicola tupara Mitchell and Wilson 1965
Unionicola vikitra Mitchell and Wilson 1965
Unionicola vikitrella Vidrine 1987
Unionicola gailae Vidrine 1987
Unionicola parkeri Vidrine 1987
Unionicola kavanaghi Vidrine 1987
Unionicola serrata (Wolcott 1898)
Unionicola megachela Vidrine 1985
Unionicola abnormipes (Wolcott 1898)
Unionicola australindistincta Vidrine 1985
Unionicola causeyae Vidrine 1985
Unionicola dimocki Vidrine 1986
Unionicola formosa (Dana and Whelpley 1836)
Unionicola foili Edwards and Vidrine 1994
Unionicola tumida (Wolcott 1898)
Unionicola belli Vidrine 1986
Unionicola fulleri Vidrine 1986
Unionicola poirrieri Vidrine 1984
Unionicola tenuis (Lundblad 1935)
Unionicola clarki Vidrine 1986
Unionicola mitchelli Conroy 1982

Appendix 12
Vertebrates of the Cajun Prairie (Huner's List)

A rather complete list of vertebrates known from the Cajun Prairie can be constructed by refering to the lists and maps of several major books on specific groups of vertebrates: fish (Douglas 1974), amphibians and reptiles (Dundee and Rossman 1989), birds (Lowery 1974a) and mammals (Lowery 1974b). Several more recent articles and books add species and portray them in a more ecological setting, e. g., Miller and Fontenot 2001. Rather than recreate these as lists, I have chosen to use Jay Huner's list of vertebrates from the Cade Farm in the southeastern part of the Cajun Prairie.

Jay Huner (1998) described his project as follows: "*The USL* (now University of Louisiana Lafayette (ULL)) *College of Applied Life Sciences has developed its 600 acre (240 ha) Experimental Farm in St. Martin Parish - south-central Louisiana - with the realization that the maintenance and enhancement of* Biological Diversity *must be an important component of any agricultural research program. This unit is situated astride the natural loessal terrace and adjacent alluvial lowland about 3 miles (5 km) west of Bayou Teche. Subunits include managed (aquaculture) and forested semi-natural, short-hydroperiod wetlands as well as livestock (beef, dairy, sheep and swine) pasture, organic waste lagoons, hay fields, and crop lands."*

"Special attention has been paid to vertebrate biodiversity on this farm since the university acquired it in the 1980s. To date, July 1998, 186 North American bird, 30 mammal, 27 reptile, 13 amphibian, and 26 fish species have been documented on the farm. Additional species will be documented as they are encountered. For example, there are a number of fish, amphibian, reptile, bird and mammal species that should appear based on their presence in rural areas elsewhere in the surrounding environs. Some of these animals do interfere with production of food and fiber and research is underway to reduce conflict with man with minimal impact on the animal resources on the farm."

The farm is commonly called the Cade Farm—a sustainable agriculture farm. This presentation of Huner's list as an appendix provides the opportunity to view the extensive 'species richness' of a small area in the southeastern Cajun Prairie. The diversity of vertebrates in a small region of the Cajun Prairie is not to be considered as a complete list by any means. For example, I have chosen to illustrate two vertebrates in Plate 32 (an Eastern box turtle and a Rough green snake) that were not on Huner's list but rather found in the remnant prairies in Acadia and Jefferson Davis Parish. His list included common names—not an uncommon practice with vertebrates as they are usually large enough in size to be well known to many people by their common names. Avery Williams kindly provided updated names for some of the amphibians and reptilies in Huner's list (in parentheses). There remains a need for a book that specifically deals with the vertebrates of the Cajun Prairie.

Huner's list included both natives and exotics and follows:

Birds:

Acadian Flycatcher
American Bittern
American Coot
American Crow
American Goldfinch
American Kestrel
American Pipit
American Redstart
American Robin
American White Pelican
American Woodcock
Anhinga
Barn Owl
Barn Swallow
Barred Owl
Belted Kingfisher
Bewick's Wren
Black Vulture
Black-and-white Warbler
Black-bellied Plover
Blackburnian Warbler
Black-crowned Night Heron
Black-throated Green Warbler
Blue Jay
Blue-gray Gnatcatcher
Blue Grosbeak
Blue-headed Vireo
Blue-winged Teal
Boat-tailed Grackle
Bobolink
Bonaparte's Gull
Brown-headed Cowbird
Brown Thrasher
Bufflehead
Canada Goose
Canada Warbler

Carolina Chickadee
Carolina Wren
Caspian Tern
Cattle Egret
Cedar Waxwing
Chimney Swift
Chipping Sparrow
Cinnamon Teal
Common Grackle
Common Ground Dove
Common Moorhen
Common Nighthawk
Common Snipe
Common Yellowthroat
Cooper's Hawk
Dickcissel
Double-crested Cormorant
Downy Woodpecker
Eastern Bluebird
Eastern Kingbird
Eastern Meadowlark
Eastern Phoebe
Eastern Screech Owl
Eastern Towhee
Eastern Wood-pewee
European Starling
Fish Crow
Forester's Tern
Fulvous Whistling Duck
Gadwall
Glossy Ibis
Gray Catbird
Great Blue Heron
Great-crested Flycatcher
Great Horned Owl
Greater Scaup
Greater White-fronted Goose
Greater Yellowlegs Great Egret
Green Heron
Green-winged Teal
Golden-crowned Kinglet
Hairy Woodpecker

Hermit Thrush
Herring Gull
Hooded Merganser
House Sparrow
House Wren
Indigo Bunting
King Rail
Laughing Gull
Least Bittern
Least Sandpiper
Lesser Scaup
Lesser Yellowlegs
Little Blue Heron
Loggerhead Shrike
Long-eared Owl
Magnificent Frigatebird
Magnolia Warbler
Mallard
Marsh Wren
Merlin
Mississippi Kite
Mottled Duck
Mourning Dove
Neotropic Cormorant
Northern Bobwhite
Northern Cardinal
Northern Flicker
Northern Harrier
Northern Mockingbird
Northern Parula
Northern Pintail
Northern Rough-winged Swallow
Northern Shoveler
Orange-crowned Warbler
Orchard Oriole
Osprey
Painted Bunting
Palm Warbler
Pectoral Sandpiper
Philadelphia Vireo
Pied-billed Grebe
Prothonotary Warbler

Purple Gallinule
Purple Martin
Red-bellied Woodpecker
Red-breasted Merganser
Reddish Egret
Red-eyed Vireo
Redhead
Red-headed Woodpecker
Red-shouldered Hawk
Red-tailed Hawk
Red-winged Blackbird
Ring-billed Gull
Ring-necked Duck
Roseate Spoonbill
Rock Dove
Rose-breasted Grosbeak
Ruby-crowned Kinglet
Ruby-throated Hummingbird
Ruddy Duck
Rusty Blackbird
Savannah Sparrow
Scarlet Tanager
Scissor-tailed Flycatcher
Sedge Wren
Sharp-shinned Hawk
Short-billed Dowitcher
Short-eared Owl
Snow Goose
Snowy Egret
Solitary Sandpiper
Song Sparrow
Sora
Spotted Sandpiper
Stilt Sandpiper
Summer Tanager
Swainson's Warbler
Swallow-tailed Kite
Swamp Sparrow
Tennessee Warbler
Tree Swallow
Tricolored Heron
Tufted Titmouse

Turkey Vulture
Vermillion Flycatcher
Virginia Rail
Western Sandpiper
Whip-poor-will
White Ibis
White-eyed Vireo
White-faced Ibis
White-rumped Sandpiper
White-throated Sparrow
Wilson's Warbler
Wood Duck
Wood Stork
Yellow-bellied Sapsucker
Yellow-billed Cuckoo
Yellow-crowned Night Heron
Yellow-headed Blackbird
Yellow-rumped Warbler
Yellow-throated Vireo
Yellow-throated Warbler

Mammals:

American Beaver
Bobcat
Common Muskrat
Cougar
Coyote
Eastern Cottontail
Eastern Mole
Eastern Wood Rat
Fox Squirrel
Fulvous Harvest Mouse
Gray Fox
Gray Squirrel
Hispid Cotton rat
House Mouse
Least Shrew
Marsh Rice Rat
Nearctic River Otter
Nine-banded Armadillo
North American Mink

Northern Raccoon
Norway Rat
Nutria
Red Fox
Southern Flying Squirrel
Striped Skunk
Swamp Rabbit
Vespertilionid Bats
Virginia Opossum
White-footed Mouse
White-tailed Deer

Amphibians:

Bullfrog
Eastern Narrow-mouthed Toad (Eastern Narrowmouth Toad)
Eastern Spadefoot Toad
Green Treefrog
Gulf Coast Toad (Coastal Plain Toad)
Northern Cricket Frog
Pig Frog
Southern Leopard Frog (Florida Leopard Frog)
Spring Peeper
Striped Chorus Frog (Cajun Chorus Frog)
Three-toed Amphiuma
Western Lesser Siren (Lesser Siren)
Woodhouse‘s Toad

Reptiles:

American Alligator
Common Kingsnake
Common Garter Snake
Cooter Turtle (Florida Cooter)
Copperhead
Corn Snake (Eastern Corn Snake)
Cottonmouth
Diamond-backed Water Snake (Diamondback Water Snake)
Eastern Hog-Nosed Snake (Eastern Hognose Snake)
Graham's Crayfish Snake
Green Anole
Hieroglyphic River Cooter Turtle (Eastern River Cooter)

Mud Snake
Painted Turtle (Northern Painted Turtle)
Plain-bellied Water Snake (Plainbelly Water Snake)
Racer (Eastern Racer)
Rat Snake (Western Rat Snake)
Razor-backed Musk Turtle (Razorback Musk Turtle)
Slider Turtle
Snapping Turtle (Common Snapping Turtle)
Southern Water Snake
Spiny Softshell Turtle
Stinkpot Turtle (Common Musk Turtle)
Western Green Water Snake (Mississippi Green Water Snake)
Western Ribbon Snake

Fishes:

Bantam Sunfish
Black Bullhead
Black Crappie
Brown Bullhead
Bowfin
Channel Catfish
Common Carp
Flyer
Gizzard Shad
Golden Shiner
Green Sunfish
Least Killifish
Mosquitofish
Northern Largemouth Bass
Pirate Perch
Pygmy Sunfish
Redear Sunfish
Sailfin Molly
Smallmouth Buffalo
Spotted Gar
Spotted Sunfish
Thread Fin Shad
Yellow Bullhead
Warmouth
White Crappie

Appendix 13
Flight Seasons of Dragonflies and Butterflies in the Cajun Prairie (Adapted from Vidrine et al. (2001b))

Flight records of dragonflies (and damselflies) from the Cajun Prairie ecosystem in southwestern Louisiana. The records are sight records collected during 1986-1991 (Vidrine et al. 1992b). A total of 86 species made up 73,742 sightings at 2,487 stations. Totals of monthly sightings of twenty-five of the more abundant species are here presented. Many dragonflies fly in restricted flight seasons and may be seen only during a single month—these seldom had a large number of observations.

Number of adult individuals observed

Dragonflies	Months											
	Jan	Feb	Mar	Apr	May	Jun	Jul	Aug	Sep	Oct	Nov	Dec
Blue dasher	0	0	81	1465	4161	2579	2995	2476	759	424	21	1
Eastern pondhawk	0	0	28	787	2858	2393	2032	1188	753	488	29	0
Halloween pennant	0	0	0	6	942	2227	1734	786	64	4	0	0
Common green darner	18	4	70	283	478	392	692	1174	905	739	184	89
Four-spotted pennant	0	0	0	0	445	1521	931	568	147	7	0	0
Blue-tipped dancer	0	0	30	1065	1358	630	134	235	105	11	4	0
Wandering glider	0	1	0	0	3	67	941	1137	648	486	112	20
*Golden-winged skimmer	0	0	0	4	921	1253	652	387	3	0	0	0
Eastern amberwing	0	0	0	104	523	479	612	379	378	104	11	0
Spot-winged glider	0	0	1	53	273	1190	542	186	5	0	0	0
Citrine forktail	0	69	260	541	236	232	119	272	26	42	30	1
Common whitetail	0	0	30	226	259	221	568	376	124	34	1	0
Common baskettail	0	0	574	781	100	0	0	0	0	0	0	0
Great blue skimmer	0	0	0	8	248	374	495	276	18	0	0	0
**Black saddlebags	0	0	2	26	123	72	293	334	248	133	6	4
Slaty skimmer	0	0	0	3	100	299	440	308	37	1	0	0
Seaside dragonlet	0	0	0	0	106	342	105	515	6	1	0	0
Roseate skimmer	7	0	0	0	2	18	206	244	141	165	101	44
Ebony jewelwing	0	0	2	26	379	180	136	56	35	28	0	0
Powdered dancer	0	0	0	0	87	37	79	276	76	8	0	0
Rambur's forktail	1	11	30	103	75	89	57	89	29	47	13	4
Carolina saddlebags	0	0	0	2	27	46	138	126	46	16	0	0
Royal darner	0	0	1	34	113	228	19	3	2	0	0	0
Little blue dragonlet	0	0	2	24	121	33	25	84	12	14	10	0
Broad-striped forceptail	0	0	0	0	12	88	101	34	3	0	0	0

**may be confused with Needham's skimmer*

***may be confused with Red-mantled saddlebags*

Flight records of butterflies from the Cajun Prairie ecosystem in southwestern Louisiana. The records are sight records collected during 1986-1991 (Vidrine et al. 1992c). A total of 47 species made up 8,517 sightings at 2,487 stations. Totals of monthly sightings of twenty-five of the more abundant species are here presented. Butterfly sightings are greatly affected by nearby plantings of nectiferous plants and host plants for their larvae.

	Jan	Feb	Mar	Apr	May	Jun	Jul	Aug	Sep	Oct	Nov	Dec
Gulf fritillary	2	0	1	3	5	14	90	529	667	222	74	11
Common buckeye	2	2	16	46	109	193	93	129	122	72	132	11
Pearl crescent	0	0	60	71	75	96	60	56	144	106	145	5
Variegated fritillary	0	0	0	2	18	62	128	213	87	24	19	2
Cloudless sulphur	12	2	24	10	10	24	56	207	107	64	3	12
Little sulphur	0	0	6	7	10	20	51	197	125	84	21	2
Carolina satyr	0	1	12	80	31	53	30	109	45	33	2	0
Orange sulphur	29	15	27	78	54	1	13	8	17	71	63	17
Viceroy	0	0	0	22	23	6	19	42	83	77	10	0
Sleepy orange	3	0	0	2	8	2	5	117	137	3	1	0
Monarch	1	0	46	49	23	0	1	7	10	86	9	1
Little wood satyr	0	0	82	33	62	21	0	0	0	0	0	0
Question mark	0	3	2	65	77	13	8	6	9	2	1	0
Georgia satyr	0	0	0	1	41	54	11	15	22	32	0	0
Pipevine swallowtail	0	1	23	7	30	20	32	26	10	2	0	0
Spicebush swallowtail	0	0	28	10	12	20	16	44	9	1	0	0
Black swallowtail	0	0	26	6	15	39	25	17	8	0	0	0
American snout	0	0	0	14	101	0	2	3	1	1	0	1
Palamedes swallowtail	0	0	8	14	8	5	22	12	15	11	0	0
American lady	0	0	2	24	54	8	1	3	1	0	0	0
Eastern tiger swallowtail	0	0	12	11	5	6	21	17	6	4	0	0
Gorgone checkerspot	0	0	8	19	8	3	18	0	1	0	0	0
Red admiral	0	0	4	7	14	4	8	6	4	3	3	0
Painted lady	2	1	0	2	3	0	0	1	3	39	0	1
Giant swallowtail	0	0	1	6	0	21	10	8	5	1	0	0

Appendix 14
The Cajun Prairie Habitat Preservation Society

A non-profit society (501 (C) (3)) was established in 1988 in order to preserve and restore prairie in southwestern Louisiana. Charles Allen was president, and Malcolm Vidrine was vice-president. Barbara Savoy was secretary, and Shirley Vige Jr. was treasurer. Kenneth Pitre served as our legal resource. Harland Guillory, Janet Seale and Philip Bourgeois were also board members. The society is a non-profit, tax-exempt society. The primary purposes of the society are to protect and preserve the remaining prairies, but more importantly, to restore the Cajun Prairie habitat. This latter purpose is celebrated in the Cajun Prairie Restoration Project in Eunice, Louisiana—a 10 acre park under the management of the society. I like to think of the park as a publication, one of my finest publications, where I am one of more than 100 coauthors. The park depicts the prairie vista, but it is circled by a paved trail connecting to a parking lot with a covered rest area. The park is the topic of a 1995 book entitled *A Cajun Prairie Restoration Journal: 1988-1995* by Malcolm Vidrine, Charles Allen and William (Bill) Fontenot. More than 1000 visitors from all over the world have personally seen the park during one of the society meetings or on private tours. Annual tours by classes from ULL and LSU as well as other organizations are led by society members. The society was reinvigorated in 1997 with new and additional officers, a vibrant membership and a newsletter.

Presidents of the society:
Charles Allen (1988-1997)
Bruno Borsari (1997-1999)
Larry Allain (1999-2001)
Marc Pastorek (2001-2003)
Peter Loos (2003-2005)
Sara Thames (2005-2007)
Jim Foret (2007-2009)
Ken Wilson (2009-present)

Current Officers:
President: Ken Wilson
Vice President: Jackie Duncan
Treasurer: Patricia Lewis
Secretary: Malcolm Vidrine
Publicity Chair: Margaret Frey
Web Master: Doug Miller
Newsletter Editor: Charles Allen

Past President: Jim Foret
Education Coordinator: Margaret Frey

The society now owns 14 acres, and the members are in the process of restoring prairie on these properties. The society also manages the restoration process at the Duralde Prairie for the Lacassine National Refuge in Duralde, Louisiana. The website www.cajunprairie.org provides a calendar of events and other resources. Newsletters in digital form are available online and from the secretary of the society.

The society has some extraordinary members, whom I refer to as 'shakers and movers' for a variety of reasons. The one thing that they all have in common is a wonderful sense of humor—something that is essential when you are in the middle of a revolution of ideas. Here is an annotated list of them and their major accomplishments to my knowledge.

Charles M. Allen is a retired professor of botany from University of Louisiana at Monroe. He is however still working. He is currently serving as botanist for the military base at Fort Polk under the auspices of Colorado State University. He and his wife, Susan, currently host visitors to their bed and breakfast and gardens, Allen's Acres Inc., in Pitkin, Louisiana. As a botanist, Charles collects plants and people. He rediscovered the Cajun Prairie, studied it in great detail and set out to restore it in as many places as he could. He began the society, served as president for a decade and continues to serve as newsletter editor. His real specialty is agrostology—the study of grasses. He is always looking for the 'Easter egg' find—a rare plant or animal. He excels at leading field trips and making them an adventure. He is the leader in any group that he joins. He inspired a legion of prairie enthusiasts with his excitement about the prairie and plants in general. Charles took the leadership role in organizing, collecting, preparing and shipping seeds in the society's seed sales. His record for obtaining grants for the society is notable, where his connections seem limitless. He is currently working on a new book, 'Wildflowers of Louisiana,' with Ken Wilson and Lowell Urbatsch. Charles is a great joke-teller, and he does tell them frequently. He is involved in everything 'Cajun Prairie.' However, Charles is a naturalized Cajun!

Larry Allain is a botanist with the Wetland Center and United States Geological Service in Lafayette, Louisiana. As a one-time commercial landscaper, he brings both the botanist and the business perspectives to a problem. He served as president of the society, but his activities center more with restorations around the Lafayette area. He is prolific in his writings on the Cajun Prairie and Texas coastal prairies. He led the effort on the Paradise Lost brochure, an Earth Day poster, a CD on prairie plants and several major papers on the botany of the Cajun Prairie. As a Cajun, he has a deep interest in the history of the area as well. His soft-spoken, task-oriented nature provided several major opportunities for the society. He lives in Lafayette with his wife, Becky, and works extensively with Jim

Grace and Stephen Johnson on coastal prairies. Larry is now playing a leadership role in the Coastal Prairie Conservation Initiative—the initiative is a cooperative effort undertaken by several agencies, organizations, and private landowners to conserve, restore, and enhance prairie in 19 counties along the Texas coast.

Bruno Borsari is currently a professor of biology education at Winona State University in Winona, Minnesota. Bruno is an agriculturist with a laurea doctorate in animal science from the University of Bologna in Italy. He is a naturalized Cajun who married Julie Chaisson—a Cajun from Lafayette. He met her in Africa during their Peace Corps-type service. Bruno arrived in Eunice just as Charles Allen was leaving for Monroe, and as such he was conscripted into service as a Cajun Prairie scientist. He did this with an energy that was, to say the least, inspiring and energizing. Bruno completed a PhD in education from the University of New Orleans and moved with his family to Pennsylvania. His deep interest in agriculture has taken the prairie in a different direction. His mentor, Wes Jackson, inspired him to write papers and books on the integration of prairie and agriculture leading to a sustainable lifestyle. He served as president of the society when Charles stepped down and helped to reinvigorate the society.

William R. (Bill) Fontenot is a zoologist and botanist and an all-around natural landscape engineer. He is retired from the Nature Center for the Lafayette Natural History Museum and operates Prairie Basse Nursery in Carencro with his wife Lydia. Bill is a French descendent from a family that is closely related to me. He is an avid historian and has a deep interest in anything natural. He writes a piece about nature in south Louisiana for the Lafayette Advertiser every Sunday. He worked with Pat Mire, Charles and me on the video 'Wildflowers of the Cajun Prairie.' He has written books on natural landscaping and birds of the gulf coast. He coauthored a book with Charles and me on the restoration project in Eunice.

Marc Pastorek is a landscape engineer from the New Orleans area. He and his wife, Candi, operate MeadowMakers Inc.—a landscaping initiative—in Covington. He is closely associated with the landscaping school at Louisiana State University in Baton Rouge and other local universities. He is restoring a major prairie in the New Orleans area proper. He served as president for the society during the period when the society was attempting to buy the restoration site in Eunice. Marc is ever present and leading one effort after another. He was heavily involved in the planning for a parking lot and paved trails for the restoration site in Eunice. With Charles, he and Peter Loos obtained Trails grants from the State of Louisiana to accomplish these tasks.

Peter Loos is a landscape engineer from the northeastern Texas area. He is closely associated with the Stephen F. Austin State University in Nacogdoches, Texas. He and David Creech host the biannual meeting of the Lone Star Regional Native Plant Conference. He and his wife, Cass, are restoring prairie habitat in east Texas near Nacogdoches. He also served as president for the society and succeeded in acquiring the restoration site in Eunice for the society. Peter led the society members in restoring several new projects in and around Eunice. Contributions of the Peter Loos/Marc Pastorek team have made both the restoration site and the society make tremendous gains during and after their terms of office. Both are always searching for new varieties of prairie plants that may serve new uses in the landscape.

Sara Thames (botanist at Fort Polk and working with Charles Allen) and **Jim Foret** (professor at the University of Louisiana Lafayette sustainable agriculture department) served as presidents of the society during the building of the parking lot and paved trails. They led the society in a massive activity to restore the rest of the Eunice site. **Ken Wilson** (avid nature photographer and financial advisor) is currently president. He is continuing this effort. He and Becky, his wife, were heavily involved in the auction to purchase the property. **Jackie Duncan** (a forester with the United States Department of Forestry) is currently the vice-president of the society, and if all goes as expected, she will become the next president, since the vice-president serves as a president-elect. She already has led the fight to eliminate Chinese tallow trees from the Eunice site.

Patricia Lewis is the treasurer for the society. She took over after my tenure in the early 2000s and succeeded into organizing a financially sound and forward-looking organization. She and her husband, David, are mycologists (fungi specialists), and they have a tremendous track-record on finding 'Easter egg' fungi. They publish and present papers, and they give field trips an added dimension, since they literally see what is going on below the soil. She is a key player in every decision that the society undertakes.

The backbone of the society is its membership. **Sara Simmonds**, **Margaret Frey**, **Celeste Gomez**, **Tom Hillman**, **Dawn Allen Newman**, **Andy Allen** and **Doug Miller** are members who have shared each their special skills to make the society a much better organization. There are other members that we can count on for not only attending meetings but getting involved in our functions; these include **Kent and Sonie Milton**, **Tyrone Foreman**, **George Broussard**, **Paul Fontenot**, **Firman Guillory**, **Chris Naquin** and Charles Allen's students and coworkers. Members of the LSUE faculty including **Renee Robichaux**, **Domingo Jariel**, **Jameel Al-dujaili** and **Myriam Stanford** have also recruited students in their endeavors to study Cajun Prairie. The trouble with any list is that you forget someone. Many of our members are part of family memberships. On numerous occasions since the inception of the society, students from local schools and colleges have volunteered to help in restoration activities—these are all honorary members of society. We have lost two notable members, **Sue Forest** and **Corrine Roberie**, who passed away recently. We have also recently lost a former student **Miranda Doucette** who worked with antimicrobial properties of prairie mints. We have a number of new members and members who are involved in a variety of projects for the society. I would like to acknowledge and celebrate all of their activities, but I will leave that to another book. The future of the society will hinge upon their very efforts. Without the original board members, the society would have not gotten off the ground, but the new and vital membership permits the society to take on many new challenges.

Appendix 15
The Cajun Prairie Gardens

The gardens are owned and operated by Malcolm and Gail Vidrine. They were created in 1996 as demonstration gardens. The gardens illustrate a variety of prairie plants that are horticulturally useful in the landscape and the various ways that the plants can be arranged in the landscape. Prairie and marsh habitats are in various stages of restoration. Butterfly gardening is a major focus. Postage stamp gardens and a large Louisiana iris marsh are among the focuses of the garden. These landscaping projects have been discussed in two papers (Vidrine et al. 2002 and 2006). A number of photographs of the gardens have been published in a variety of magazines, journals and webpages. Numerous tours of the gardens occur annually. Plate 31 depicts the gardens.

This appendix is literally my dedication of the gardens to my wife and my children. We have built these gardens from scratch and literally watched each plant grow. We have noted each butterfly and insect. The children built their insect collections in the gardens for their class projects. They helped me build the tower that was knocked down by Hurricane Lili. Daniel and Caroline transplanted and seeded the plants with me. The garden is a place of refuge and relaxation today. In times of stress, I have seen Gail, Daniel and Caroline wandering down its trails and focusing on their concerns.

The list of possible women who would have permitted a prairie garden to be the first thing you see when you arrive at their home is short, very short. Gail is that exception for me. She accepted the garden and made it her own from day one. She took criticism from neighbors and 'drivers-by' while tending the garden. She accepted in stride the attitudes of neighbors and the garbage they threw into the garden. She felt deeply when neighbors or others stopped and dug plants from the garden. The garden for me is more than just my effort to restore prairie—it is a celebration of life and my gift to all, but first, my wife.

Appendix 16
Creating or Recreating a Prairie

(Charles Allen's (2006) insert on prairie restoration in the CPHPS Newsletter is here reprinted with permission of the author.)

By Charles Allen through grants from U. S. Fish and Wildlife Service, Lacassine Refuge

The following "How To" for the development of prairies is mainly written for the coastal prairies of Louisiana and Texas but is applicable elsewhere especially in the southeastern United States. Thanks to the many people who have added to my knowledge of prairies especially Malcolm Vidrine, Peter Loos, and Marc Pastorek. Plans call for plant lists, seed and plant sources etc. in future newsletters.

Prairies are usually defined as areas dominated by herbaceous perennials, in particular grass species and few woody plants, especially trees. Both woody plants and herbaceous perennials live for more than two years but the above ground stems of woody plants do not die back to the ground during the Winter so the above ground growth of stems for woody plants is cumulative. The above ground stems of herbaceous perennials die back to the ground yearly so new above ground stem growth begins fresh every Spring. The underground growth of herbaceous perennials is cumulative as this material accumulates from year to year. Other terms that have applied to prairies are meadow, field, plain, clearing, pasture, steppe, and alp. Artificial prairies have been created by humans throughout the world. Pastures, golf courses, parks, lawns, graveyards, railroad right of ways, and roadside right of ways are dominated by herbaceous perennials and are treeless or have areas that are treeless. Many areas that were cleared for agriculture and now abandoned often are allowed to develop into prairies or meadows.

Prairies are (or in some cases we have to replace "are" with "were" as the prairies are now destroyed) formed in many parts of the world under several environmental conditions. Prairies are created by elements that restrict the growth of trees and other woody plants and allow the growth of herbaceous perennials. Factors reported to restrict woody plants and thus promote prairie creation are:

(1) low annual rainfall—the prairies of the Midwest United States are reported to be mainly created by total annual rainfall of less than 30 inches;

(2) fire—this would kill the woody plants but not harm the perennials with their underground rhizomes and other survival parts;

(3) clay layer below surface of soil—a hard clay pan located below the surface restricts the growth of roots (the underground system of herbaceous perennials could develop above the clay layer but there would not be enough room for the root system of trees);

(4) high clay content of soils—soils with much clay shrink and swell with dry

and wet conditions creating a hostile environment that herbaceous perennials can tolerate but trees cannot;

(5) wet/drought conditions—the inhospitable conditions created by wet conditions followed by extreme drought can be tolerated by herbaceous perennials but by many woody plants;

(6) grazing animals—bison and other herbivores can seriously damage woody plants by their concentrated grazing (herbaceous plants are grazed as well but resprout much more prolifically than woody plants);

(7) wind—woody plants are often blown over by winds while the herbaceous perennials are shorter and more flexible; and

(8) mycorrhizae—these are fungus-root mutual relationships where the fungus provides water and nutrients to the roots and receives food back from the roots (mycorrhizae are important for trees and the fungus may not be present in prairie areas).

Prairies do not develop as a result of one of these factors but a combination of two or more. Fire is the one that seems to be prevalent in almost all natural prairie systems.

The big keys to prairie creation or recreation are patience and perennials. Remember that herbaceous perennials grow above ground stems yearly, die back to the ground during the Winter, and regrow the following Spring. So many people are familiar with trees and forests and treat herbaceous perennials and prairies in a similar manner. A common mistake is to think that when the above ground stems die the entire plant is dead. But, perennials survive underground and will resprout next Spring if you don't till the area or remove the sod.

Site Selection

I. Light—a full sun setting is optimal but partial shade can be tolerated by many prairie plants. A densely shaded area will not work.

II. Moisture—prairie plants can tolerate a wide range of moisture as there were wet and dry prairies. The best would be the intermediate moisture as too wet would have a tendency to develop into a marsh or bog and too dry may become a desert setting.

III. Soil—prairies can be created on just about any soil but more fertile soils tend to create problems with weed control and woody plant development.

IV. Topography—the ideal setting is flat but prairies can be created on slopes unless the slope is very extreme and erosion problems may not allow prairie development.

V. Size—prairies can be any size and the availability of the proper combination of the above four factors may dictate the size. Another factor to consider in size is the maintenance which will be covered later. If your prairie is too large to maintain, it will soon develop into a forest in most situations. The "postage stamp" prairie is a good one for the backyard in a residential situation. The front yard would also be okay but the prairie does look "weedy" or "ratty" at times of the year if you do no weed pulling. A front yard prairie is manageable with weed and height control and understanding neighbors.

VI. Prior vegetation—it is best to recreate a prairie in a site where prairie once grew but new or artificial prairies can be created just about anywhere where the above site conditions are met.

Site Preparation

All woody plants should be removed from the site and this can be accomplished mechanically or with chemicals. If possible, herbaceous plants, especially the invasive species, should also be removed. The better the site preparation, the sooner you will have a prairie. If using seeds, you have to till the site as though you were planting an agricultural crop. The better the seedbed the more seeds will germinate. Do not be concerned about annual weed seeds, some seeds will germinate the first year but very few will germinate the second and third, etc. years as long as you don't redisturb the soil. If using transplants (plugs and seedlings) to start the prairie, you do not have to prepare the area as intensely as with seeds but you will get faster results if you do.

Making the Plant List for the Prairie

Consult with a recognized expert in your area to obtain a list of plants for your prairie. A few miles can make a big difference in the choices of plants that occur (or did occur) in your area and thus should be included in your prairie. Avoid already prepared seed mixes as these are usually for a wide geographic region. It is much better to make your own list so you can more accurately recreate the prairie of your area and also be able to include your favorites, especially with small prairies. For limited budgets, space, and/or local ordinances, be sure to take into consideration the characteristics of each species; height, flower color, and time of year of flowering. Some species can tolerate a range of the environmental factors while others have a very narrow range so light tolerance, soil, and water requirements should also be considered when choosing species.

Grasses usually formed the backbone or matrix of a prairie so it is important to include a number of grasses. If you are attempting to mimic the Cajun Prairie, you would need switch grass (*Panicum virgatum*), big blue stem (*Andropogon gerardii*), little blue stem (*Schizachyrium scoparium*), slender blue stem (*Schizachyrium tenerum*), Indian grass (*Sorghastrum nutans*), and eastern gama grass (*Tripsacum dactyloides*). These species are dominants in the remaining

Cajun Prairie remnants. Other grasses that would be also important in the prairie are Florida Paspalum (*Paspalum floridanum*), brownseed paspalum (*Paspalum plicatulum*), muhly grass (*Muhlenbergia capillaris*), pineywoods dropseed (*Sporobolus junceus*), and several species of rosettegrass (*Dicanthelium* spp.).

A prairie would be very drab with grasses only so plants that produce colorful flowers should also be included in your mix. Some species or genera that would be essential are blazing stars (*Liatris* spp.), compass plant (*Silphium laciniatum*), rosin weed (*Silphium gracile*), flowering spurge (*Euphorbia corollata*), button snakeroot (*Eryngium yuccifolium*), false indigos (*Baptisia* spp.), Indian plaintain (*Arnoglossum ovatum*), grass leaved golden aster (*Pityopsis graminifolia*), snout pea (*Tephrosia onobrychoides*), phlox (*Phlox pilosa*), wine cup (*Callirhoe papaver*), sweet golden rod (*Solidago odora*), aster (*Aster* spp.), milkweeds (*Asclepias* spp.), tick seeds (*Coreopsis* spp.), wooly sunflower (*Helianthus mollis*), and Indian blanket (*Gaillardia aestivalis*).

Sources of Seeds and Transplants

Whenever possible, the seeds and/or transplants should be obtained from local sources. There are two reasons for this: (1) local plants are adapted to the local environmental conditions and (2) efforts should be made to preserve the local genetic strains and possibly prevent their extinction. It is reported that as little as 100 miles north or south can affect the time when a species begins flowering and other steps in the life cycle of a plant maybe controlled by local environmental conditions. Many prairies exist as only small remnant populations so it would behoof us to try to get those genetic strains into as many protected places as possible and not allow these genetic strains to disappear.

Seeds may be purchased from commercial producers. When purchasing seeds, use the closest seed company to you and be sure to match the species exactly and don't just plant any species in a genus. In some genera, there are some annual and some perennial species; *Gaillardia pulchella* is an annual while *Gaillardia aestivalis* is a perennial and a prairie plant. A listing of seed companies that offer prairie seeds is in Appendix B. Due to the constant flux in seed companies and the seeds that each offers, it may be necessary for you to contact the prairie organizations listed in Appendix C for more up to date data on prairie seed companies.

Seeds may also be harvested from the wild; it should be noted that the percent of germination of many wild collected seeds is often low. The gathering of seeds from perennials usually does not affect the number of plants in the area the following year because perennials regrow from underground parts. Seeds of perennials only germinate when a disturbance creates bare soil and an opening for the newly germinated seedling. Be sure to obtain permission before harvesting seeds. Seeds should be harvested as soon as possible after maturity; most seeds change from green to another color, brown, black, red, yellow, etc., with maturity.

Place the seeds in brown paper bags and allow them to dry. After drying, you may place in plastic bags but be sure to not use a plastic bag for storage until the seeds are dry. If it is time for planting, you can plant the dry seeds as is. If you are not planting right away, it is best to store the seeds in a refrigerator until time for planting. The cool storage does apply some stratification (exposure of seeds to cold temperatures to cause the embryo to mature) and also protects the seeds from insect attack. If you are hand planting, you do not have to put a lot of effort into sorting the seeds from the chaff. If you plan to use a drill or other mechanical planter, you should separate the seeds from the chaff as much as possible.

Some companies offer potted transplants for sale. A good source for transplants is the state or local Native Plant Society. As with seeds, obtain you transplant from as near as possible. Note that some local companies may be buying their plants from other areas or even other states and then growing them to the saleable size. Check into the ultimate source of the plants before purchasing. It is not recommended that you dig transplants from the wild except for rescue missions. Rescue missions should be undertaken when the plants of an area are being destroyed for construction of highways, houses, parking lots, agriculture, etc. When rescuing plants, be sure to obtain plenty of soil with the transplants, protect the transplants from wind during transport, and schedule the rescue mission during the dormant season in your area, mostly January and February. Clumps or clods of soil plus the underground parts of plants should be inside a plastic bag if you have to transport for a long distance. Many clods may be placed in the back of pickups or trailers. If traveling for a long distance, the clods should be covered to prevent drying out by the wind.

Seeding or Transplanting

The best time for seeding of prairie plants is late Fall to early Winter as would happen in nature. The plants in a prairie are staggered in flowering and seed production; there are some flowering in the Spring, some in the Summer, and many in the Fall. The seeds mature at different times but most reach the soil surface in the Fall of the year. If the soil surface is disturbed, the seed may then germinate. The general rule of thumb is that seeds should be planted as deep as the seed is thick; larger seeds deeper than smaller seeds. When planting many different seeds at one time, it is best to spread all seeds on top of the newly tilled surface and then lightly till the soil after planting the seeds. Packing the soil after the tilling will ensure that the seeds are in contact with the soil. Seeds will germinate but many will not flower the first year and it may take several years before some species flower. The important point to remember is not to till the site again as you will destroy the perennials. There are some mechanical planters available that drill the seeds into the soil. For large scale plantings, you may have to use this type equipment and possibly even consider aerial planting from airplanes.

Transplanting and rescuing work best in January and February but can be done at other seasons if watering can be used to supplement rainfall. When using transplants, dig a hole as close to the size of the ball of roots and soil of the potted plant. Remove plant from pot and disturb the soil of the root mass; some even recommend removing soil so that you have a bare root plant. Place the plant in the hole and replace the soil around the root mass. Pack the soil down around the top of the plant. You may want to cut back the stem so as to reduce water loss and increase the chances of the plant surviving. When rescuing plants, dig a hole to match the size of the clod. Place the clod in the hole and place dirt in the space between the clod and the side of the hole plus some on top of the clod.

A Recommended Twist to Planting

Most people want instant gratification in almost all aspects of life including prairies so a good way to plant prairies is to include seeds of one or a few annuals in the original seed mix that of necessity will be dominated by the seeds of perennials, and to transplant into part of the prairie site. The annuals will grow and flower the first year and a few will repeat the second year and fewer still the third year etc. The transplanted perennials will also flower the first year and start to increase in the size by growing underground parts. There will be more of these the second year and still more the third year etc. In the meantime, the perennials planted by seed will grow the first year and most will not flower but some will start to flower the second year etc. Using this method keeps the prairie from being real drab the first few years.

Maintenance of the Prairie

Remember that most prairies are in part maintained by fire so try to include fire in your plans. If fire is absolutely not an option, then use mowing once a year. The best and only time to mow would be during the dormant season. Since the herbaceous perennials are underground, the mower can be set at the lowest level possible. The mowing would simulate the action of the fire by removing the dead growth from the previous year and most importantly, cut back the woody plants. Repeated mowing should keep the woody plants under control but some woody species may require chemical or mechanical control.

Annuals will be common the first year but will decrease in numbers dramatically the second year and will steadily decrease in succeeding years. If the area is tilled or disturbed, annuals will reappear. To hasten the development of the prairie, you may use chemical or physically remove unwanted woody and especially herbaceous species. If you are patient, the native prairie species will eventually force out the other nonnative and non prairie species over a period of time.

Appendix 17
Landscaping Tips

The Cajun Prairie Habitat Preservation Society Newsletters provide a large number of tips for gardening and restoring prairie. As landscaping, prairie makes an excellent choice. Numerous books and websites address this choice for beautifying and naturalizing your habitat. My own gardens and the gardens of other members of the local society may be viewed during the entire year in order to see the variety of plants and the growth and bloom patterns. Here are general items to consider before attempting 'prairie' as a landscaping alternative.

1. decide the height of desired prairie
2. decide the size of desired prairie
3. decide the seasonal appearances
4. plan the site so as to manage for burning the site
5. match soil types to plants (although the plants will ultimately select their sites)
6. check with neighbors in order to minimize problems or disagreements that usually occur later
7. plant natives (use non-native plants sparingly and with forethought)
8. plant in winter (the plants respond better)
9. plug plants/plant seed/interseed seed—use a variety of methods to plant
10. prepare a maintenance plan for use of mowing and herbicide
11. avoid fertilizer and soil enrichment, even compost and topsoil
12. BE PATIENT

There is an old adage that Martha Harmon reminded me of several times. *First year, the plants sleep; second year, the plants creep; and third year, the plants leap.* This is the case with prairie plantings. The third year will show some obvious plants from seed or transplants, and the view will begin to appear as prairie. I have seen rapid prairies developed, for example, Bette Vidrine and Marc LeBlanc used hydromulching for seeding parts of the prairie of the Dugas Landfill. The plants literally jumped out of the soil and grew vigorously and bloomed in the first two years. The cost of hydromulching is prohibitive for large areas barring some deep pockets as a source of funds. Five years is a good length of time to allow for a typical planting of prairie to really begin to look and behave like a prairie.

Appendix 18
Benefits of Prairie Landscaping/Restoration/Preservation

1. The prairie ecosystem is extremely diverse with more than 1000 plant species.
2. 60-70% of the prairie plants are underground as a massive root/rhizome/bulb ecosystem; nearly half of it dies each winter and is converted to humus (topsoil).
3. Prairie forms deep Horizon A and becomes the major soils for agricultural production of rice, wheat, corn, soybeans, oats, rye, etc.
4. The lawn literally has no root system, requires both fertilizer and biocides, and forms a monoculture costing roughly $1000/acre/year to maintain. The lawn is anti-prairie.
5. Prairie requires neither fertilizer nor biocides, it and is a diverse system with bacteria to convert atmospheric nitrogen to usable nitrogen, mycorhizzal fungi to move phosphorus and insects and birds to pollinate plants.
6. Prairie is an insect garden with many species of butterflies—the finest example of an insect garden. The garden supports a variety of birds and other animals as well.
7. Fire is essential in prairie. It mobilizes potassium in ash. It stimulates new growth in plants and clears the previous year's debris. It stimulates prairie seeds to germinate.
8. Prairie root systems direct water from rain deep into the soil where it is filtered and percolated into the natural aquifers. Lawns runoff water with fertilizers and biocides and their residues.
9. Prairie plants remediate pollutants in air and water and soil, by sequestering these pollutants into the ground where bacteria can break them down. Prairie plants also sequester carbon, thus reducing the accumulation of carbon dioxide in the atmosphere that would lead to global warming. Cellulosic ethanol and biogasoline are potential prairie products.
10. Prairie plants and animals are used in the commercial trade. The prairie sunflowers are among the most prevalent in the cut-flower trade and in most bouquets. Other prairie plants are prominent in landscaping and butterfly gardens. Many prairie plants are edible and/or used as medicines, for example, *Echinacea*.
11. Prairie plants are native and need no watering, only rain and sunshine.
12. Prairie plants can survive on weak, worn-out soils, such as depleted agricultural soils, and thus they are used to remediate abused agricultural land. They also build new topsoil—maybe our grandchildren will eat crops grown on these rejuvenated soils.

13. Over winter, only the roots/rhizomes/bulbs survive, but they grow again in spring and produce stems, leaves, flowers and fruit. Some plants, e.g., Eastern gama grass, are being used to produce flour in polycultural agriculture, where the field mimics a prairie.
14. Prairie plants are perennial. Many are tens, possibly hundreds, of years old as individual plants—the ultimate clones.
15. The landscape of the future is the prairie/savanna—the lawn's lifetime as the predominant landscape will end in your lifetime. The lawn's list of ills is long! The prairie is beautiful with flowers blooming all spring, summer and fall and a myriad of butterflies and birds. Humans evolved in the prairies and savannas of Africa. This is home.

Appendix 19
Exotics

This short list, although informative, serves more to allow the reader to begin to appreciate the degree to which exotic species have naturalized into the Cajun Prairie ecosystem and literally become cohabitants with humans.

Exotic plants:

Chinaberry tree	*Melia azedarach*
Chinese tallow tree	*Triadica sabiferum*
Johnson grass	*Sorghum halapense*
Vasey grass	*Paspalum urveili*
Bahia grass	*Paspalum notatum*
Bermuda grass	*Cynodon dactylon*
St. Augustine grass	*Stenotaphrum secundatum*
Brazilian vervain	*Verbena brazilensis*
Southern carpetgrass	*Axonopus fissifolius*
Japanese honeysuckle	*Lonicera japonica*
Itch grass	*Rottboellia cochinchinensis*
Mimosa tree	*Albizia julibrissin*
White clover	*Trifolium repens*
Uruguayan primrosewillow	*Ludwigia uruguayensis*
Alligatorweed	*Alternathera philoxeroides*
Curly dock	*Rumex crispa*
Chocolateweed	*Melochia corchorifolia*
Purple nutsedge	*Cyperus rotundus*

Exotic animals:

Cattle
Horses
Pigs
Sheep
Asian Tiger Mosquito
Cattle Egret
Chickens
Nutria
Dogs
Cats

Mediterranean Gecko
Norway Rats
Cowbirds
Fire Ants
Formosan Termites
Eurasian Earthworms

Appendix 20
Obvious Blooming Plants and Landscaping Problems

Obvious blooms on showy plants of the Cajun Prairie are common. Some are short and showy in early season—don't discount them; some splash and bloom for a week or so—that's it—don't discount them; some are early succession and disappear in a year or three—don't discount them; some flowers are small, but there are so many of them—don't discount them; most require a burn to permit ready viewing. A diverse garden is essential with native plantings. Native plants are not genetically revved up to bloom for half a year or present a persistent garden presence or guarantee a nice look for more than a brief season. Many of the most beautiful bloomers are brief; thus a diverse garden will permit several blooming events that collectively make the garden.

The biodiversity garden, often called a natural garden, is filled with many kinds of living things that are constantly interacting. The natural garden filled with prairie plants is an insect haven. Plants that feed birds, butterfly larvae and a host of other insects are common features. Numerous insects mean numerous spiders and smaller vertebrates—predators seek out habitats with prey animals. Numerous grazers will appear, including grasshoppers and their kin. Thus natural gardening provides a variety of learning opportunities. Children from our local high school often build their insect collections in our gardens.

The horticultural aspects of prairie plants and prairie gardening are finally getting a fair amount of publicity. Numerous books and websites are now involved in communicating the values of these plants in the landscape. Publicity is not always good, as biodiversity gardens are quite different from lawns and English gardens, where plants are ordered for human eye appeal and not for the health of the life in the garden. Also problems that arise from growing prairie plants in specific neighborhoods are the major focus of some internet sites, including the site maintained by The Wild Ones—a natural landscaping group. Prairie gardening can become serious business in a neighborhood that is not prepared for the emergence of such a biodiversity garden. Times spent teaching neighbors about gardening with native plants are times well-spent.

Appendix 21
A Comparison of Native Plant Species Distribution

Plant species Common names (Search USDA PLANTSdatabase for scientific names)	Calcasieu and D'Arbonne Prairie remnants (Welsh, Iowa, Fenton, Kinder, Elton)	Mermentau Prairie remnants (Estherwood, Midland, Mermentau, Morse)	Plaquemine Prairie remnants (Frey)	Faquetaique and Mamou Prairie remnants (sEunice, wEunice, Swords, Mamou Irrigation Canal)
American aloe agave		+	+	
Aquatic milkweed	+	+	+	+
Ashy sunflower	+	+	+	
Bearded grass-pink	+			
Bergamont	+	+	+	+
Big bluestem	+	+	+	
Black-eyed Susan	+	+	+	+
Blue sage	+	+	+	
Blue-flower eryngo	+			
Butterfly-weed	+			+
Button snakeroot	+	+	+	+
Canadian louse-wort	+			
Carolina anemone	+			
Clustered mountain-mint		+		
Compass plant	+	+	+	
Dense blazing-star		+	+	
Downy phlox	+	+		+
Drumheads	+			
Eastern blue-star	+	+		+
Eastern gama grass	+	+		
Evening rain lily	+			
Fewflower milkweed		+		
Finger false dragonhead	+			+

Plant species Common names (Search USDA PLANTSdatabase for scientific names)	Calcasieu and D'Arbonne Prairie remnants (Welsh, Iowa, Fenton, Kinder, Elton)	Mermentau Prairie remnants (Estherwood, Midland, Mermentau, Morse)	Plaquemine Prairie remnants (Frey)	Faquetaique and Mamou Prairie remnants (sEunice, wEunice, Swords, Mamou Irrigation Canal)
Flowering spurge	+	+	+	
Fragrant spider-lily	+	+	+	+
Fringed sneezeweed	+			
Giant ironweed	+	+	+	+
Giant orchid	+			
Green milkweed	+	+	+	+
Green-flowered milkweed		+	+	
Golden colic-root	+	+	+	
Gulf cordgrass	+			
Hairy spiderwort	+			
Indian plantain	+	+		
Kansas gayfeather	+			
Lanceleaf blanketflower	+	+		
Lanceleaf tickseed	+	+	+	+
Largeleaf wild-indigo	+	+	+	+
Lindheimer's beeblossum	+	+	+	
Lindheimer's bergamont	+	+	+	+
Little bluestem	+	+	+	+
Long-leaf milkweed	+	+	+	
Maryland golden-aster	+		+	
Multibloom hoarypea	+	+	+	
Narrowleaf mountain-mint	+	+	+	+
Narrowleaf silkgrass	+	+		
New Jersey tea	+			
Nodding beardtongue	+	+	+	

Plant species Common names (Search USDA PLANTSdatabase for scientific names)	Calcasieu and D'Arbonne Prairie remnants (Welsh, Iowa, Fenton, Kinder, Elton)	Mermentau Prairie remnants (Estherwood, Midland, Mermentau, Morse)	Plaquemine Prairie remnants (Frey)	Faquetaique and Mamou Prairie remnants (sEunice, wEunice, Swords, Mamou Irrigation Canal)
Nodding wild-indigo	+	+	+	+
Nuttall's prairie parsley		+	+	
Nuttall's wild-indigo	+			+
Obedient plant	+	+		
Pineland milkweed	+	+	+	
Pinkscale gayfeather	+			
Purple cone flower	+			
Rayless goldenrod	+			
Rough cone flower	+			
Sampson's snakeroot	+	+	+	
Scaly blazing-star	+	+	+	
Shiny golden-rod	+	+		
Single-stem scurfpea	+		+	
Slender gayfeather	+	+	+	+
Slender bluestem	+	+	+	+
Slender rosinweed	+	+	+	+
Snowy orchid	+			
Spotted beebalm	+	+	+	+
Spring ladies' tresses	+	+	+	+
Swamp narrowleaf sunflower	+	+	+	+
Sweet goldenrod	+	+	+	
Switchgrass	+	+	+	+
Tall tickseed	+			
Talus slope penstemon	+			
Texas brown-eyed Susan	+	+		
Texas ironweed	+			

Appendix Twenty-One

Plant species Common names (Search USDA PLANTSdatabase for scientific names)	Calcasieu and D'Arbonne Prairie remnants (Welsh, Iowa, Fenton, Kinder, Elton)	Mermentau Prairie remnants (Estherwood, Midland, Mermentau, Morse)	Plaquemine Prairie remnants (Frey)	Faquetaique and Mamou Prairie remnants (sEunice, wEunice, Swords, Mamou Irrigation Canal)
Toothache grass	+			
Virginia spiderwort	+		+	+
Western horsenettle		+		
Whiteleaf mountain-mint	+			+
Wrinkled leaf goldenrod	+	+	+	
White colic-root	+			
White prairie-clover	+	+		
Whorled milkweed	+			
Woods poppymallow	+	+		
Yellow Indiangrass	+	+	+	
Yellow wild-indigo	+	+	+	+
Zigzag iris	+	+	+	+
Total	79	56	46	29

Appendix 22
Remnant Status: Morbidity and Mortality

By
Malcolm Vidrine and Tom Hillman

The prairie remnants that Charles Allen and Malcolm Vidrine had studied in the 1980's were visited on March 20, May 18 and August 17, 2008. Only the smallest piece of the Frey Prairie remnant had been burned; none of the other remnants appeared to have been burned in the last few years. Otherwise, here was the status of each of the remnants based on the visits:

Frey Prairie remnant (Plaquemine Prairie): This prairie appeared to be in poor shape in general, unburned, and with new injury. The middle segment has been made into an air-strip: however, there remained a large number of Compass plants. Other areas have a tangle of Southern dewberry and Southern blackberry (hereafter simply called briers). No burns across from the house in the area of highest diversity had been done. The southern piece has undergone a new assault. The roadside ditch was redug with a machine that tossed the material onto the remnant as far as 10 feet into the remnant and several inches deep. In several areas, dense briers were knocked down by mowing. The August trip revealed a bush-hogged prairie as most of the remnant had been mowed—only the small prairie across from the house remained unmowed.

Estherwood Prairie remnant (Mermentau Prairie): The south side of the railroad track was essentially on its last bit of life support. It was heavily overgrown with Eastern baccharis (hereafter simply called baccharis) and briers. Young Chinese tallow trees were vigorously resprouting in spring. Only a trail created by four-wheelers was traversable in the overgrown prairie, where only small areas had remnants of prairie flora. The northern side of the railroad tract had been mowed recently and had some obvious prairie plants. The western wet site, which once was nearly an acre of Eastern blue-star and Fragrant spider-lily, was reduced to small open islands within an encroaching forest. The number of blooming plants rapidly declined with each visit.

East Midland Prairie remnant (Mermentau Prairie): The eastern parts were scraped on both sections. The rest was partially overgrown with baccharis and briers. Again the northern side of the railroad tract was mowed. The May visit permitted the viewing of Western horsenettle blooming in good numbers, and a burn would have helped refurbish other plants. Again the northern side of the railroad tract had been mowed (few prairie plants were obvious even though there was no apparent evidence of mowing since our first visit).

West Midland Prairie remnant (Mermentau Prairie): The prairie on the southern side of the railroad tract appeared to have been overgrown with baccharis, briers and Chinese tallow; however, recently, the overgrowth had been scraped into 10-foot high piles for burning. The end result was bare ground. The northern side of the tract again was mowed. The scraping was approximately 30 feet wide. An additional 8-10 feet of obvious herbiciding (brown trees and prairie plants) were evident. The end result was bare ground. The northern side of the tract again was mowed (few prairie plants were evident—it appeared that the area was plowed several years ago). Few prairie plants remained. This site included a portion damaged by a 1990s train derailment.

East Mermentau Prairie remnant (Mermentau Prairie): The southern side of the railroad tract was overgrown with baccharis, briers and Chinese tallow, but there were some open prairie areas. An additional 8-10 feet of obvious herbiciding (brown trees and prairie plants) were evident. Thus it was obvious that the midland/mermentau prairies were scraped and sprayed.

West Mermentau Prairie remnant (Mermentau Prairie): The southern side of the railroad tract was overgrown with pine trees, baccharis, briers and Chinese tallow. It was so shaded as to have insufficient undergrowth to sustain a fire; however, the pine needles may be sufficient to start a fire. The northern side of the tract was partially mowed. Good numbers of Dense blazing-stars, Clustered mountain-mints, Spotted beebalm, Indian plantain, Cluster bushmint and Compass plant were evident in August.

Welsh Prairie remnant (Calcasieu Prairie): The remnant prairie south of the railroad was not been burned in several years, although it looked pretty good. Four Whorled milkweeds were in bud amongst literally thousands of White-top sedges in moist areas. Neither Purple cone flowers nor Woods poppymallows were evident, although they once were abundant in the 1990s. The remnant had deeply compacted soil down the center where one or more combines and other heavy farm equipment made an obvious trail. The remnant north of the railroad (not far from the oil fields along the Interstate 10), which was once open with hundreds of Texas brown-eyed Susans and later Indian plantain and Kansas gayfeathers was now crowded with briers, baccharis and Chinese tallow. A winding central path mowed by a bushhog every couple of years persisted, and this path contained lots of prairie plants—few were blooming. The area had not been burned in at least 5 years. Wax-myrtle was forming dense clumps all through the non-mowed region as an understory with baccharis, briers and Chinese tallow.

North Iowa Prairie remnant (Calcasieu Prairie): This prairie was mowed, except for the northern part, which was overgrown with baccharis and briers. Islands of prairie plants were apparent. The mowed regions were in bloom and looking good in May. However, most of this remnant is the overgrown part. The southern remnant was in good bloom in August, even with a small population of peppermint-scented Whiteleaf mountain-mint. Kansas gayfeathers were present in modest numbers.

South Woodlawn Prairie remnant (Calcasieu Prairie): This prairie was overgrown with briers, baccharis and Chinese tallow. Islands of prairie plants were apparent. The area had large numbers of Finger false dragonheads and Indian plantain in the 1980s; few remain. The August visit revealed that a number of the plants had bloomed in summer.

North Woodlawn Prairie remnant (Calcasieu Prairie): This prairie was now part of the four-lane highway under construction.

South-South Fenton Prairie remnant (Calcasieu Prairie): Both sides of the railroad tract (east and west) were intact. Much of it was damaged by crosstie replacement in the mid-1990's, but it seemed to be re-establishing. Baccharis and briers and some Chinese tallow were visible. A few Texas brown-eyed Susans were in bloom in May. Small islands of prairie remain evident, and Kansas gayfeathers were beginning to bloom during the August visit.

South Fenton Prairie remnant (Calcasieu Prairie): This site in the 1980s had a large seep and wet area with lots of Golden colic-root, White colic-root, Eastern blue-star, Fragrant spider-lily and Bearded grass-pinks; this was not evident. A huge circle of big bluestem was still intact. Both sides of the railroad were in moderately good shape; however, briers, baccharis and Chinese tallow were popping-up. Kansas gayfeathers were beginning to bloom in August. A number of other prairie plants were evident among the rapidly encroaching weedy species.

South Kinder Prairie remnant (D'Arbonne Prairie): Both sides of the railroad tract appeared intact; however, they were mowed. The eastern side of the railroad was heavily overgrown with briers, baccharis and Chinese tallow. Part of it had also been plowed—it appeared lost. The western side was mowed in March, but by May, the prairie bloomed well with predominately Black-eyed Susans and a few other showy plants.

Elton Prairie remnant (Calcasieu Prairie): This prairie appears to be completely forested by pines and other trees. Only the smallest islands of prairie plants were visible.

General overview: The remnants were all rapidly declining as prairie habitat. North Woodlawn was lost. Elton, Welsh, West Mermentau, West Midland, and Estherwood were in great peril and possibly already lost. The remaining prairies, including mowed sections, could be reclaimed by the reintegration of fire as a management tool. Some additional herbiciding and clearing would be necessary. In either case, there was not much time. With the price of fuel and the danger to the integrity of the railroad tracts by falling trees (trees in prairie are poorly rooted because of the clay pan and readily fall over in high winds), the use of fire was readily defensible. It is essential in the maintenance of the integrity of the prairie flora that fire be routinely placed into management schemes. The bottom line is that there are few opportunities remaining to protect the remaining remnants—restoration is not a simple option as we can testify. A small remnant east of Estherwood was in good shape as well as the two remnants on north side of the Estherwood and Mermentau prairie remnant, which had been mowed in the winter or burned more recently. It is our overall conclusion that the reintegration of fire is imperative.

Appendix 23
Characteristics of the Cajun Prairie

1. Floristically diverse for its size.
2. Two or 3 bloom peaks in a growing season.
3. Long growing season/blooming season (March-October)
4. Annual burning does not noticeably damage the prairie.
5. Ten major grasses (Poaceae) dominate.
6. Major forb families include Asteraceae, Fabaceae, Lamiaceae, Asclepidaceae, and Apiaceae (sunflowers, peas, mints, milkweeds and parsleys, respectively).
7. Numerous sedges (Cypeaceae) occur and sometimes predominate in wet sites.
8. Each named prairie is different in diversity and appearance throughout the seasons.
9. Dry and wet prairies have distinctive floras.
10. Mima mounds, *marais* and *platin* features were evident.
11. Historically divided by agriculturalists into cotton-corn prairies and rice-cattle prairies.
12. Prairie is physically taller by one foot each month of the growing season, e.g., one foot tall in March and six feet tall in August.
13. Remnant prairies are resilient, but they direly need fire; restorations are fragile and very garden-like.
14. Numerous crawfish mounds in winter and spring months.
15. Typically very wet (sometimes warm) winters and very dry (hot) summers.
16. Decidedly affected by Gulf of Mexico showers and storms.

Appendix 24
Differences between Cajun Prairie and Midwestern Prairie

The Cajun Prairie has/had:

1. a long blooming season and a long growing season (9 months minimum)
2. more rain (up to 50+ inches annually)
3. feral cattle, horses, sheep and pigs for more than 200 years
4. Chinese tallow trees, Brazilian vervain and its own variety of exotics
5. Slender bluestem and other unique native plant species
6. less Big bluestem
7. Cajun French culture
8. numerous crayfish and crayfish aquaculture
9. a clay hardpan
10. winter fire
11. intensive rice and soybean cultivation instead of corn and wheat
12. intensive cotton and sugarcane cultivation instead of corn and wheat
13. seldom ice/snow cover (white buffalo); warm, wet winters
14. intergradations with freshwater marsh and coastal marsh
15. intergradations with pine forests, including Longleaf pine forests
16. pimple (mima) mounds
17. *marais* and *platins* (maybe similar to potholes)
18. aerial spraying of biocides and fertilizer persists
19. loss Bison much earlier—records include few even in 1700s
20. recently received attention from naturalists
21. recently attempted to be restored

Appendix 25
Books / articles / films regarding the Cajun Prairie

Many books and articles have been written on prairies. The North American Prairie Conference has published 21 biannual proceedings, which are extremely useful for not only Cajun Prairie but also the entire American prairies. My favorite book on prairie is Candace Savage's (2004) *Prairie: A natural history*. My favorite beautifully illustrated books are Madson and Oberle's (1993) *Tallgrass Prairie* and Ladd and Oberle's (1995) *Tallgrass Prairie Wildflowers*—both filled with fantastic photos by Frank Oberle. M. Harris's (2003) *Botanica North America: The illustrated guide to our native plants, their botany, history, and the way they have shaped our world* contained a short introduction to major prairie plants, which is fine reading. Pat Mire's (Attakapas Productions) films, especially *Wildflowers of the Cajun Prairie* (1989) and *Anything I Catch: The Handfishing Story* (1992) provide an excellent introductions to Cajun Prairie ecology. I have attempted to cite all the natural history materials that I could find on the Cajun Prairie in the Literature Cited of this book.

There is an entire library of books dealing with the people of the prairie. Carl Brasseaux and Glenn Conrad (through the Center for Louisiana Studies and ULL Press) have edited and/or written numerous books and papers. Many other authors have privately published or published books and films through other major presses and venues throughout the south. These works provide a compelling history of the people of this region.

This book neither has the scope nor the intent to retell the stories of the peoples of the region except in the most simplistic way. With some liberties, I have tried to tell the story of the land and the life that inhabited the land—a natural history of sorts. I have taken the liberty of telling my own story by integrating my family history as part of the natural history. I have borrowed generously from all my friends and colleagues—their collective works are what we know of the natural history of the Cajun Prairie.

Appendix 26
List of Volunteers, including Those Who Helped with Prairie Restoration and with This Book

No project with the scale of this one is done in a vacuum. The science and the public information parts are universal in most organizations, but the restoration and preservation activities of the society are far beyond the usual level of volunteerism. It is common in this effort to see families involved.

Jameel Al-dujaili
Larry and Becky Allain (Andre)
Charles and Susie Allen (Dawn, Tonya and Andy)
Charles Ancona
Victor Andrews
Joan Andrus
George Archibald
Marjorie Ardoin
Wylie Barrow
Tom and Donna Barry
Patrick Bergeron
John Billodeaux
Griff Blakewood
Rebekah Bordelon
Bruno and Julie Borsari (Giovanni and Michele)
Phil Bourgeois
Jan Boydstan
John Brinkman
George Broussard
Bob Bugg
Janet Butler
Vallard Carrier
Derwin Carter
Jackie Caton
Linda Chance
Darryl and Beth Clark
Lee and Deborah Clark
Jim Cordes
Mary Lynch Courville
Dave Creech
David Daigle
Matt Dakin
Mark DeRouen
Jerry Devillier
Andy Dolan
Miranda Doucette
Jackie Duncan
Joann Hidalgo Duncan
Amy Duplechain
Arlene Dupre
Henri Escuriex
Rene Escuriex
Patricia Faulkner
George Fischer
Will Fleming
Bill and Lydia Fontenot
James Fontenot
Michael Fontenot
Paul Fontenot
Todd Fontenot
Tyrone Foreman
Sue Forrest
Jim and Paula Foret
Billy Fortier
Richard Fortier
Tommy Fortier
Margaret Frey
Jeanne Frois
Justin Fuselier
Vernon Fuselier
Connie Gardner
Celeste Gomez
Barry and Kathy Good
Gabriel Nicholas Gotte
Dave Guillory
Ginger Guillory

Harland Guillory
Robert Guillory
Jim Grace
Vicki Grafe
John Hamlin
Martha Harmon
Edius Hebert Jr.
Joel Hilburn
Tom Hillman
Jerry Hoffpauir
Mona Hollier
Bill Hudson
Wanda Huh
Jay Huner
Allan Inman
Wes Jackson
Domingo Jariel
Bryan Jobe
Stephen Johnson
Curtis Joubert
Kyla Joubert
Dirk Kavanagh
Tim Kiphart
Russel Kirk
Wayne LaFleur
Suzanne Lahaye
Garrie Landry
Madelaine Landry
Alex Lasseigne
Charlotte LaVergne
Mark LeBlanc
Lynn Lejeune
Gary Lester
David and Patricia Lewis
C. C. Lockwood
Peter and Cassandra Loos
Zoe Lynch
Tim Kiphart
Gerome (Jerry) P. Madden
Georgie Manuel
Janice Manuel
Ben Martin
David Martin
Steven Mayor
Clay Menard
Scott and Echo McClendon
Roy McLaughlin
Jan Midgley
Paul Miller
Kent and Sonie Milton
Pat Mire
RickyMonceaux
Bob Morris
Robert Murray
Chris and Jackie Naquin
Ovelia and Alinea Naquin
Bill Neiman
Anthony and Dawn Newman (Krislyn and Baylie)
Judith O'Neale
Pat O'Neil
Claude and Pat Oubre
Corinne Louise Paff
Annette Parker
Robert (Bob) Parker
Steven Parris
Cale Pastorek
Darla Pastorek
Marc and Candi Pastorek
Gerald Patout
Ken Pitre
Tim Potier
Rhonda Prejean
Janet Prudhomme
Robert D. Quebedeaux
Debra Quillman
Joann Quillman
Melanie Reed
Van Reed
William (Bill) Reese
Bobbie Richard
Randy Roach
Corinne Roberie
Renee Robichaux
Ray Robicheaux
Jim Robinson
Gary Ross
James Oakley St. Cyr
Barbara and Clarence Savoy
Mark and Anne Savoy
Rene Savoy
Charles Seale
Janet Seale
Mark Seale

Melissa Semar
Shahin Shabanian
Sara Simmonds
David Simpson
Jewel B. Slovacek
Latimore Smith and Nelwyn McGinnis
Caletta Soileau
Jane Spradling
Charles Stagg
Myriam Stanford
Thomas Strawn
Ellen Stutes
Harry Summerlin
Wayne Syron
Sara Thames
Dale Thomas
Orville Touchet
H. H. Winters
Al and Cathy Troy
Bette Vidrine
Daniel and Ashley Vidrine (Gregory)
Jacqueline O. Vidrine
John R. "Randy" Vidrine
Malcolm and Gail Vidrine (Macky, Daniel and Caroline)
Malcolm and Maureen Vidrine (Odile and Hank)
Numa and Odile Vidrine
O'Sanna Vidrine
Ryan Vidrine
Shirley Vige
Karl M. Vincent
Avery Williams
O R Willis
Ken and Becky Wilson

And the many students who transplanted and seeded the restorations.

Organizations:

City of Eunice and its City Council
Coastal Prairie Conservation Initiative
Coastal Prairie Partnership
Eunice City Museum
Imperial St. Landry Genealogical & Historical Society
International Crane Foundation
Jean Lafitte National Historic Park (Prairie Acadian Cultural Center—Eunice)
Jefferson Davis Parish Mosquito Abatement District
Lacassine National Wildlife Refuge
Lafayette Open Channel
Louisiana Department of Wildlife and Fisheries/Natural Heritage Program
Louisiana Department of Forestry
Louisiana Native Plant Society
Louisiana State University-Eunice administration, faculty and staff
Natural Resources Conservation Service
Nature Conservancy of Louisiana
Nature Conservancy of Texas
Sam Houston Rosource Conservation and Development Area, Inc.
Union Pacific Railroad
United States Fish and Wildlife Service
U. S. Forest Service
U. S. Geological Service/Wetland Center
Texas Parks and Wildlife Department

Grants from:

Jefferson Davis Parish Mosquito Abatement District and USDA Mosquito Research Lab (ARS).
Louisiana Native Plant Society
Louisiana Natural Heritage Program
Louisiana Office of the Lt. Governor to build trails (DOTD)
Undergraduate Summer Research Institute (URSI) of LSUE provided funds for students and professors working on varied topics including Cajun Prairie topics from The LSUE Foundation.
Walmart of Eunice

Endowed professorships (The LSUE Foundation):

1. Malcolm Vidrine was awarded the Opelousas General Hospital Endowed Professorship for Science in Support of Nursing (Cajun Prairie Project) in the 08-09 year through the LSUE Foundation (The Cajun Prairie: A Project Entailing the Preparation of a Book).
2. Domingo Jariel was awarded the Fritz Lang Endowed Professorship (Science in support of agriculture) in the 08-09 year (Soil Physiochemical Characteristics at Different Soil Depths in Restored and Remnant Cajun Prairies of Southwest Louisiana).
3. Jameel AL-Dujaili was awarded the BellSouth, Atmos Energy, CenturyTel, CLECO, Entergy Louisiana Endowed Professorship for Public Policy in the 09-10 year (Cajun Prairie Herbs: Their Antimicrobial Activity Against Methicillin-resistant *Staphylococcus aureus*).

Index

Abbeville irises, 59
Abbeville, LA, 2, 51, 59, 136
Acadia Parish, LA, 2, 8, 19, 25, 36, 41, 49, 51, 52, 64, 85, 96, 154, 203, 251
Advertiser (The Lafayette), 153
Africanized Killer Bees, 142
agriculture, 7, 11, 20, 29, 43, 45, 48, 49, 54, 63, 70, 79, 81, 89, 104, 105-8, 113, 115, 128, 134, 144, 146, 147, 151, 154, 157, 161, 166, 214, 251, 263, 264, 269, 273, 278
Ajilvsgi, Geyata, 60
Al-dujaili, Jameel, 132, 264, 300
alfisols, 95, 96, 98
algae, 113, 136, 159
Alligatorweed, 84, 279
Allain, Becky, 262
Allain, Larry, 68, 69, 70, 77, 85, 114, 117, 126, 136, 166, 261, 262, Plate 32
Allen, Andy, 264
Allen, Charles, 8, 20, 21, 38, 47, 61, 63, 64, 70, 72, 85, 86, 89, 96, 100, 127, 128, 130, 135, 166, 213, 261-264, 269, 287, 297, Plates 8, 23, 24, 32
Allen Parish, LA, 2, 8, 35, 39, 96, 122, 134
Allen, Susan, 262
Althans, Amy, 68
American Alligators, 153, 230
American aloe agave, 135, 219, 283
American elderberry, 132, 228
American plum, 129
amphibians, 143, 153, 215, 251; list, 257
annual rainfall, 52, 89, 90, 106, 269
Aquatic milkweed, 283
aquifers, 102, 104, 112, 138, 277
armadillos. *See* Nine-banded Armadillo
Ashy sunflower, 64, 219, 283, Plates 2, 11, 13, 14, 16
Asian Tiger Mosquito, 142, 153, 279
Asiatic Clams, 142
Atakapas. *See* Attakapas
Atchafalaya Basin, 2, 6, 26, 101, 149
Atchafalaya River, 3, 41, 48, 151, 156
Atlantic rangia, 160
atmosphere, 7, 92, 143, 149, 150, 155, 156, 161, 277
Attakapas Country (including Poste des Attakapas), 2, 8, 13, 14, 17, 23-27, 30, 31, 32, 36, 38, 41, 43, 197
Attakapas (Native Americans), 2, 13, 14, 17, 23, 27, 197
Attakapas prairie, 2, 13, 17, 28, 30, 33, 43, 213
Attwater's Prairie Chickens, 60, 148, 150, 153
Au Large, 2, 21
Avery Island, LA, 5, 8, 23, 41

Bacteria, 86, 113, 114, 118, 127, 133, 134, 138, 144, 145, 149, 166, 277
Bahia grass, 35, 44, 48, 82, 130, 228, 279
Bald cypress, 3, 93, 175, 226, 227, 229, Plate 32
Bambi (the movie), 100
Banana Spiders, 142
Baskin, Yvonne, 145
Baton Rouge, LA, 3-6, 18, 36, 93-95, 199, 202, 263
Bayou Blue, 40

Bayou Courtableu, 24, 26, 48, 197
Bayou des Cannes, 12, 17, 19, 21, 33, 45, 54, 73, 102, 103, 142, 209, 210, Plate 32
Bayou Lafourche, 13, 19, 26, 197, 199
Bayou Mallet, 159
Bayou Nezpique, 33, 37, 45, 46, 49, 73
Bayou Queue de Tortue, 17, 46, 159
Bayou Teche, 3, 13, 21, 26, 27, 32, 33, 35, 37, 38, 42, 44, 93, 103, 197, 213, 230, 251
Bayou Wikoff, 17
Bearded grass-pink, 117, 135, 219, 283, Plate 1
Beau Basin Prairie, 3, 21, 37
Beauregard Parish, LA, 39
Bereza, Daniel J., 22
Bergamont, 126, 221, 283, Plate 31
Bergerie, Maurine, 26, 35, 43
Bick, George, 54, 57, 61, 151
Bienvenu, Marcelle, 142
Big bluestem, 35, 118-120, 125, 126, 187, 213-215, 220, 283, 293, Plates 16, 19, 21-24, 26
Big Thicket, 6, 60, 101, 151
biocides, 103, 129, 134, 138, 151, 152, 153, 160, 164, 277, 293
biodiversity, 6, 10, 74, 84, 105, 107, 117-120, 134, 141, 147, 165-167, 169, 214, 217, 251, 281
biodiversity garden, 10, 169, 281
bioinformatics, 10
birds, 4, 20, 27, 29, 57, 61, 63, 78, 81, 82, 93, 99, 101, 110, 111, 113, 115, 129, 140, 143, 150, 153, 155, 157, 160, 163, 179, 180, 182, 184, 214, 215, 217, 241, 251, 252, 263, 277, 278, 281; list, 252
Bischoff family, 54
Bison (American buffalo), 35, 46, 72, 90, 105, 115, 118, 119, 139, 140, 142, 153, 169, 214-216, 231, 293
Black-eyed Susan, 225, 283, Plates 11, 14, 24, 25, 30
Black Lake, 5
blooming seasons. *See* phenology
Blue flag iris, 81
Blue-flower eryngo, 135, 225, 283
Blue sage, 126, 220, 283
Bonaparte, Napoleon, 16
Borsari, Bruno, 66-68, 78, 106, 107, 146, 261, 263, 297, Plates 14, 30
Bourgeois, Philip, 261
Bowie, Jim, 16, 197
Boydstun, Jan, 159
Brasseaux, Carl, 1, 25, 295
Brazilian vervain, 85, 228, 279, 293
Breaux Bridge, LA, 1, 2, 37
Broussard, George, 264
Brown, Clair, 6, 9, 11, 58-60, 64, 73, 89, 93, 117, 129, 223, 231, 252, 258
Browns Lake, 5
buffalo. *See* Bison.
Bullfrogs, 47, 160, 230, 257
Bull thistle, 82, 85, 228
burn. *See* fire
Bushy goldentop, 225, Plate 22
butterflies, 6, 20, 21, 54, 57, 61, 65, 67, 69, 81, 82, 86, 112, 126, 140, 142, 143, 148, 150-153, 155, 157, 160, 161, 165, 215, 217, 247, 260, 277, 278; list, 247; flight seasons, 260
Butterfly flower, 226
butterfly gardens, 82, 108, 277
Butterfly-weed, 73, 125, 126, 135, 222, 283, Plates 1, 31
Button snakeroot, 64, 125, 220, 283, Plates 2, 9, 11, 12, 26

C

Cade Farm, 251
Cajun Prairie Gardens, 12, 21, 68, 76,

80, 82-85, 99, 126, 128-130, 152, 163, 165, 169, 267, Plate 31
Cajun Prairie Habitat Preservation Society, 63, 66, 77, 79, 85, 130, 168, 185, 190, 198, 261, 275, Plate 30
Cajun Prairie Restoration Project, 12, 21, 69, 76, 77, 79, 80, 84-86, 117, 163, 165, 261, Plate 30
Calcasieu Parish, LA, 2
Calcasieu Prairie, , 3, 7, 8, 21, 25, 26, 34-36, 38-40, 42, 48, 52, 54, 59, 69, 73, 77, 96, 103, 135, 136, 156, 159, 216, 230, 283-286, 288, 289, Plates 12-26, 29
Calcasieu River, 3, 34, 35, 38, 40, 48, 54, 59, 156, 159, 190
Camel, Nancy, 142
Cameron Parish, LA, 2, 8, 40, 42
Canadian louse-wort, 135, 221, 283
canals, 4, 19, 21, 46, 51, 54, 57, 63, 103, 104, 153, 156-158, 211; canal systems list, 51
Canary Islands, 18, 19, 43, 192, 199
Carencro, LA, 37, 66, 177, 230, 263
Carolina anemone, 135, 220, 283, Plate 18
carpetgrass. *See* Southern Carpet grass
cats, 40, 119, 142, 157, 160, 217, 279
cattle, 1, 7-9, 11, 13, 19, 21, 23-28, 30, 31-39, 41, 43-48, 50, 53, 54, 61, 72, 102, 104-106, 119, 124, 125, 132, 134, 138, 139, 141, 142, 148-150, 157, 168, 197, 209, 210, 213, 216, 230, 279, 291, 293
Cattle Egret, 149, 253, 279
Center for Louisiana Studies, 295
Chaisson, Julie, 263
Chickasaw plums, 129
chickens, 1, 19, 28, 119, 134, 142, 148, 153, 157, 169, 203, 209, 210, 279
Chicken-tree. *See* Chinese tallow tree
Chicot Aquifer, 51
Chinaball tree (Chinaberry), 19, 129, 209, 228, 279
Chinese tallow tree, 55, 57, 77, 78, 83, 99, 119, 121, 134, 228, 264, 279, 287-289, 293
Choupique Prairie (French translation), 3, 230
Church Point, LA, 2, 19, 199
Civil War, 16, 35, 36, 47-49, 55, 57, 197, 204
clay, 9, 38, 51-54, 59, 80, 82, 89, 93, 95, 97, 98, 111, 126, 269, 290, 293
climate, 10, 11, 18, 20, 25, 27, 28, 39, 78, 89, 90, 92, 109, 139, 168, 216
climax community, 127
Cluster bushmint, 226, 288
Clustered mountain-mint, 79, 135, 222, 283
Coefficients of Conservatism, 7, 120, 219, 233
Comeaux, Malcolm, 26, 43, 44
Compass plant, 137, 219, 283, 288, Plates 3-5, 8-10
Conrad, Glenn, 295
Conservation International, 84
Coralbean, 46, 137, 138, 210, 222, 229, Plate 1
corn, 1, 7, 19, 27-29, 34, 38, 39, 44, 45, 47, 50-52, 54, 55, 73, 106, 167, 209, 257, 277, 291, 293
Coronado, Francisco, 23
Corpus Christi, TX, 1, 89
Cote Gelée Prairie, 2
cotton, 1, 7, 8, 19, 27, 29, 34, 38, 40, 44, 45, 46-49, 51, 54, 55, 57, 73, 91, 104, 106, 124, 158, 203, 209-211, 256, 291, 293
Coureurs du Bois, 7, 197
Coushatta, 35
Cowbirds, 280

crawfish. *See* crayfish
crayfish, 8, 47, 51, 55, 61, 102, 103, 105, 141, 143, 149, 150, 153, 156, 157, 159, 160, 241, 257, 291, 293
Crimsoneyed rose-mallow, 126, 226, Plate 31
crops, 6, 8, 9, 29, 33, 39, 42, 44, 48, 51, 52, 55, 63, 98, 101, 104-106, 110, 120, 148, 203, 277
Crowley, LA, 2, 36, 47, 49-52, 60, 96, 123, 149, 198
Crowpoison, 133, 225, Plates 2, 6, 18
Curly dock, 279
Curtis Prairie, 163
Cypremort Prairie, 2

Dakin, Matt, 155
damselflies, 151, 152, 153, 155, 244, 259; list, 244; flight seasons, 259
D'Arbonne Prairie, 3, 36, 40, 69, 73, 77, 135, 283-286, 289, Plates 26-28
De Soto, Hernando, 23
Darwin, Charles, 146
Daily World (The Opelousas), 65, 66, 69
Darby, William, 6, 16, 21, 23, 26, 31, 33, 34, 59, 73, 74, 93, 117, 197, 213
Dauterive Compact, 32, 186
DDT, 149
DeBailion, Mary S., 137
Dense blazing-star, 126, 135, 219, 283, Plates 1, 3-12
Department of Transportation and Development, LA, 129
Devillier, Jerry, 66
Diamondflowers, 220
Diamond, Jared, 161
Diboll, Neil, 117
dogs, 119, 142, 157, 160, 217, 279
Dolania americana, 155
Donaldsonville, LA, 13, 18, 19, 26, 197, 199
Dormon, Caroline, 11, 59-60
Doucette, Miranda, 264
Downy phlox, 67, 126, 221, 283, Plates 18, 24
dragonflies, 6, 21, 54, 57, 61, 65, 67, 69, 110, 143, 148, 150-153, 155-157, 160, 245, 259; list, 245; flight seasons, 259
Drumheads, 135, 223, 283
ducks, 27, 40, 253-255
Dugas Landfill prairie restoration, 68
Duncan, Jackie, 261, 264
Dung Beetles, 141
Duralde, LA, 19, 57, 68, 85, 209, 262
Duson, C. C., 50-51
Duson, LA, 2
Duson, William W., 50

Earley, Lawrence, 165
earthworms, 98, 99, 119, 138, 145-149, 156, 159, 242, 280
Eastern baccharis, 99, 228, 287
Eastern blue-eyed-grass, 225, Plate 6
Eastern blue-star, 125, 126, 223, 283, 287, 289, Plates 21, 27
Eastern box turtle, 251, Plate 32
Eastern gama grass, 76, 79, 118, 119, 125, 126, 137, 214, 220, 278, 283
Eastern yellow stargrass, 221, Plate 6
ecological niche concept, 113
Einstein/Newton garden, 87
Elton, LA, 2, 51
Elton Prairie, 36, 165, 283, 284, 285, 286, 289
erosion, 4, 10, 51, 55, 92, 93, 97, 103, 104, 108-111, 136, 144, 146, 157, 167, 270
Escherichia coli O157:H7, 132

essential oils, 132, 133
Estherwood, LA, 52, 135
Estherwood Prairie, 135, 165, 283-287, 289, 290, Plates 3-5, 29
ethanol, 167, 217, 277
Eunice Bulb and Blossom Society, 66
Eunice, LA, 2, 12, 14, 19, 21, 25, 36, 47, 50, 52, 57, 63, 66-69, 73, 76, 77, 80, 81, 84-86, 103, 108, 117, 123, 125, 146, 159, 163, 165, 198, 204, 209, 211, 214, 261, *263, 264, 299, Plates 30-31*
Eunice News (The), 66
Eurasian Earthworms. *See* earthworms
Evangeline, LA, 2, 8, 14, 18, 19, 41, 43, 45, 48, 49, 50, 55, 96, 102, 202, 209
Evening rain lily, 135, 223, 283
exotics, 55, 60, 78, 82, 85, 86, 117, 121, 122, 130, 142, 145, 159, 163, 252, 293; list, 279-280

F

Fabacher, Joseph, 49
Faquetaique Prairie, 3, 19, 21, 25, 38, 49, 69, 73, 76, 123, 135, 197, 283, 284, 285, 286
farmers, 7, 8, 27, 44, 45, 47, 48, 50, 51, 52, 54, 65, 89, 99, 103, 129, 152, 204, 214
fauna, 10, 11, 13, 21, 60, 82, 113, 119-121, 141, 142, 144, 147, 150, 151, 158, 161, 166, 217; lists, 241-258
Featherman, A., 54, 57
Fenton, LA, 2, 51
Fenton Prairie, 36, 66, 117, 122, 135, 165, 283-286, 289, Plates 18-26
fertilizer, 79, 82, 103, 109, 138, 144, 239, 275, 277, 293
Fewflower milkweed, 67, 76, 125, 126, 223, 283, Plates 1, 32
Finger false dragonhead, 60, 126, 135, 222, 283, Plates 1, 15, 17, 30
fire, 10, 11, 29, 32, 35, 52, 55, 57, 58-59, 70, 72, 77-80, 89, 92, 94, 97-100, 106, 112, 119, 127, 128, 137, 138, 140, 142, 144, 147-148, 164, 165, 167, 214, 215-217, 269, 270, 274, 280, 281, 287, 288, 290, 291, 293
Fire Ants, 119, 142, 147, 148, 217, 280
fishermen, 32, 42, 45
fishes, 42, 141, 143; list, 258
flatworms, 241
Flight records (seasons), 191, 259, 260
flora, 8, 10, 11, 13, 36, 58-60, 68, 69, 74, 93, 113, 117, 119, 120, 122-125, 134, 138, 139, 141, 144, 158, 166, 214, 217, 231, 287, 290; list, 219-229
Florence Hunting Preserve prairie, 68, 117, 122
Florida paspalum, 221, Plates 11, 22, 26
Floristic Quality Assessment (FQA) system, 120
Flowering spurge, 219, 284, Plates 7, 24, 25, 27
flukes (trematodes), 158, 241
Fontenot, Clifford Jr., 153
Fontenot, Lydia, 85, 263
Fontenot, Mary Alice, 49
Fontenot, Paul, 264
Fontenot, William R. (Bill), 14, 64, 68, 85, 142, 153, 166, 261, 263, Plate 32
Food, Inc., 106
Foreman, Tyrone, 85, 264
Forest, Sue, 130, 264
Foret, Jim, 261, 262, 264
Formosan Termites, 142, 280
Fournerat family, 54
fracking, 51, 102

Fragrant spider-lily, 125, 225, 284, 287, 289, Plates 3, 21, 31
Freeland, Paul P., 49
French and Indian War, 18, 24, 197
Frey, John, 50, 51
Frey, LA, 36, 50
Frey Prairie, 165, 242, 261, 262, 264, 283-287, 297, Plate 2
Frey, Margaret, 261, 262, 264
Fringed sneezeweed, 125, 135, 222, 284
Frois, Jeanne, 68
Fuller, Samuel L. H., 22
fungi, 86, 106, 113, 114, 118, 127, 134, 138, 144, 145, 149, 165, 166, 239, 264, 277
Fuselier, Vernon, 85

G2 (ranking), 9
Gahn, Robert, 36, 47, 48, 49, 202, 204
gas (natural), 7, 11, 41, 48, 51, 63, 64, 102, 144, 168
Gasland, 102, 177
genealogy, 11, 12, 18, 19, 163, 199, 204
geology, 11, 18, 89, 92, 120, 124, 239
Geydan, LA, 76
Giant blue iris, 76, 136
Giant ironweed, 225, 284
Giant orchid, 78, 135, 219, 284, Plate 1
Golden colic-root, 220, 284, 289
Gomez, Celeste, 264
Gomez, Gay, 143
Grace, Jim, 55, 68-70, 99, 134, 262, 298
Grafe, Vicki, 68, 85
Grand Chenière, LA, 43
Grand Lake, 5
Grand Marais, 44
Grand Prairie, 2, 3, 18
Grand Prairie Cherrami, 2
Grass shrimp, 249
Graybark grape, 227
Great Depression, 55, 57, 198
Green-flowered milkweed, 135, 221, 284
Green milkweed, 221, 227, 284
Greens Lake, 5
Gros Chevreuil Prairie, 2-3
Gueydan, LA, 2, 68, 76, 85, 98
Guillory, Firman, 264
Guillory, Harland, 67, 261
Gulf Coast Mosquito Laboratory (USDA), 154
Gulf cordgrass, 135, 225, 284
Gulf of Mexico, 1, 14, 23, 42, 54, 70, 74, 75, 90, 91, 93, 106, 291

Hackberry, 82, 229, 248
Hairyflower spiderwort, 223, Plate 21
Hansen's Disease, 148
Harman, Walter, 159
Harmon, Martha, 85, 275
Harris, Finis, 98
Hays Prairie, 3
Heal-all, 228
Hebert, Otis, 26, 35
herbicides, 29, 54, 55, 89, 108, 129, 164, 209, 217
Hickory Flats Prairie, 3, 36
Hidalgo family, 19, 199, 202-205, 297
Hidalgo, Josef, 19, 199
Hidalgo, Marie Odile. *See* Vidrine, Marie Odile Hidalgo
Hilburn, Joel, 66
Hillman, Tom, 85, 264, 287, Plates 29, 32
hillocks. *See* Mima mounds.
Hinton, Juliana, 143
Hohorst, Charlie, 142

horses, 7, 8, 9, 13, 19, 23, 25, 28, 31, 33, 34, 37, 38, 39, 41, 43, 45, 46, 54, 72, 105, 124, 134, 142, 148, 157, 168, 197, 209, 210, 213, 216, 279, 293
Huber, Zeno, 49
Huner, Jay, 160, 161, 251
hurricanes, 20, 90, 136, 139, 142, 168, 169, 210
hydrology, 11, 51, 53-55, 63, 89, 97, 98, 101-104, 106, 111, 139, 142, 162, 164, 166, 169
hydrosphere, 7, 136, 143, 149, 150, 156, 161

Iberia, LA, 1, 2, 8, 20, 21, 24, 36, 41-44, 90, 197
Illinois, 18, 24, 201
Indian blanket. *See* Lanceleaf blanketflower
Indian grass. *See* Yellow Indian grass
Indian plantain, 125, 220, 232, 284, 288, 289, Plates 2, 8, 9, 12, 13, 17, 21, 22
Indians. *See* Native Americans
indicator species, 6, 7, 112, 117-121, 125, 159
insecticides, 29, 54, 55, 104, 129, 138, 153, 155, 217
insects, 21, 28, 65, 67, 81, 82, 84, 98, 99, 106, 107, 110, 112, 113, 126, 138, 139, 141, 147-153, 155, 157, 159, 163, 215, 243, 277, 281
introduced animals. *See* exotics
Iowa, LA, 2, 50, 59
Iowa Prairie, LA, 36, 52, 59, 63, 67, 69, 135, 165, 283, 284, 285, 286, 289, Plates 15-16
Isleños, 19, 197, 199
Itch grass, 228, 279

Japanese climbing fern, 227
Japanese honeysuckle, 228, 279
Jariel, Domingo, 146, 264, 300
Jeanerette, LA, 2, 97
Jean Lafitte National Historical Park, 14, 64
Jefferson Davis Parish, LA, 2, 8, 20, 35, 36, 39, 52, 96, 149, 161, 251
Jefferson Davis Parish Mosquito Abatement District, 20, 64, 154, 299, 300
Jefferson Island, LA, 5, 8
Jefferson, Thomas, 16
Jennings, LA, 2, 20, 21, 45, 47-50, 52, 64, 198
Johnson grass, 228, 279
Johnson, Stephen, 68, 263
Johnson, William, 69
Jolliet, Louis, 23
Jones, Bill, 7, 8, 23, 26, 29, 31, 32, 34, 36, 47, 55
Joubert, Curtis, 63, 66
Justiceweed, 221, Plate 5

Kansas gayfeather, 135, 220, 284, Plates 14, 21, 22, 26, 27, 30
Kaskaskia, 18, 24, 200, 201
Kentucky blue grass, 137, 139 (in drawing)
keystone species, 6, 117, 119, 125, 142, 146, 147
Kinder, LA, 2, 36, 59
Kinder Prairie, 36, 59, 60, 135, 155, 165, 283-286, 289, Plates 26-28
Kisatchie National Forest, 59, 99

Lacassine National Wildlife Refuge, 68, 85, 299
Ladd, Doug, 295

Lafayette, LA, 1, 2, 8, 19, 21, 35-37, 41, 52, 60, 64, 66, 85, 149, 153, 156, 199, 203, 251, 262, 263, 264, 299
Lafayette Natural History Museum, 64, 263
Lafitte, Jean, 16
Lake Arthur, LA, 5, 92, 93
Lake Charles, LA, 2, 5, 35, 38, 40, 48, 60, 64, 76, 90, 159, 197
Lake Misere, 5
Lake Peigneur, 5
Lambremont, Edward, 54, 57, 61
Lanceleaf blanketflower, 219, 272, 284, Plate 16
Lance-leaf tickseed, 223, 284, Plates 2, 31
Land Ethic, 6, 20
land granting systems, 16, 17
Land Institute (The), 108
Landry, Garrie, 64
Landry, Stuart O., 26
Largeleaf wild-indigo, 224, 284, Plates 28, 31
LeBlanc, Marc, 85, 275
leeches, 242
Leopold, Aldo, 6, 10, 20, 54, 57, 84, 86, 117, 135, 141, 163
Le Page Du Pratz, 2, 128
Lewis, David, 114, 264
Lewis, Patricia, 114, 261, 264
Lindheimer's beeblossum, 226, 284, Plates 6, 8, 20, 26
Lindheimer's bergamont, 220, 284, Plate 31
Listeria monocytogenes, 132
Little bluestem, 79, 118, 119, 120, 124-126, 214, 215, 222, 284, Plates 10, 16, 23, 28, 30
Little Calcasieu Prairie, 3, 40
Lockett, Samuel, 1, 6, 8, 13, 21, 26, 36, 38, 39, 41-43, 46-48, 52, 59, 73, 74, 93, 117, 197
Lockwood, C. C., 63, 89, 143, 298
Loden, Mike, 159
Logan, William Bryant, 144
Lone Star Ticks, 147
Long-leaf milkweed, 222, 284
Longleaf pine forests, 3, 4, 74, 98, 165, 227, 293
Long Prairie, 3
Loos, Cassandra, 85, 263
Loose-head beaksedge, 221
Loos, Peter, 69, 85, 128, 166, 223, 261, 263, 264, 269, 298, Plate 32
Louisiana Conservationist, 65
Louisiana Department of Wildlife and Fisheries, 64, 299
Louisiana Irises, 81, 99, 100, 136, 224, 267, Plate 31
Louisiana Native Plant Society, 64, 66, 299
Louisiana paintbrush, 130
Louisiana Prairie Vole, 142, 153
Louisiana Purchase, 24, 26, 33, 128
Love-bugs, 110, 149, 155
Lowery, George, 150, 157
LSU Eunice, 21, 66-68, 107, 108, 130, 146, 154, 160, 264, 300
lumber. *See* timber

MacMillan, W. B., 137
Madson, John, 90, 295
Malaga, 24, 43, 197
mammals, 143, 182, 251; list, 256
Mamou (bean). *See* Coralbean
Mamou, LA, 2, 50
Mamou Prairie, 3, 14, 17, 19, 21, 33, 34, 36-38, 45, 46, 48, 51, 57, 73, 97, 110, 123, 135, 138, 141, 186, 209, 210, 222, 229, 283-286
Manuel, Georgie, 52
marais, 4, 8, 39, 41, 52, 55, 59, 68, 72,

73, 80-83, 94, 96, 97, 101, 106, 125, 229, 291, 293
Marksville, LA, 215
Marquette, Jacques, 23
marsh, 1, 3, 4, 6, 7, 9, 10, 20, 22, 31, 38-43, 45, 48, 52, 54, 58, 59, 68, 71, 72, 74-76, 80, 90, 93, 94, 96, 97, 99, 100, 101, 122-124, 130, 136-137, 139, 142-143, 153, 157, 160, 162, 166, 169, 215, 216, 227, 230, 244, 267, 270, 293
Maryland golden-aster, 135, 221, 284
mayflies, 243
Maypop. *See* Purple passionflower
McLaughlin, Roy, 154
Mediterranean Gecko, 142, 190, 280
melons, 29, 39, 54, 106, 209
Mermentau, LA, 2, 36
Mermentau Prairie, 2, 3, 36, 38, 69, 73, 77, 135, 156, 158, 165, 216, 283-290, Plates 3-12, 29
Mermentau River, 3, 17, 33, 41, 42, 46, 54, 103, 156, 158, 197, 213
Meyer, Harry, 143
micro-relief, 68, 73, 74, 80, 81, 101, 166
midges, 150, 155, 243
Midland, LA, 64
Midland Prairie, 64, 66, 97, 122, 135, 165, 283-289, Plates 6-8
Miller, Brian, 142
Miller, Doug, 261, 264
Miller's Lake, 48, 49, 51
Milton, Kent, 264
Milton, Sonie, 264
mima mounds, 4, 34 (hillocks), 55, 59, 68, 71, 73, 80, 81, 96-98, 216
Mimosa tree, 279
Mire, Pat, 54, 66, 142, 263, 295, 298, Plate 32
Mirex, 147
Mississippi River, 7, 14, 15, 23, 24, 26, 93, 157, 199, 201
mites, 98, 104, 105, 143, 145, 156-159, 249
Monarch, 67, 215, 248, 260, Plate 32
Moncharvaux, Elizabeth Tisserand de, 18, 201
Morales, Isabel Sambrana, 19, 199
Morse, LA, 135
Morse Prairie, 135, 283-286
mosquitoes, 21, 27, 43, 64, 82, 83, 110, 129, 142, 143, 150, 153-155, 157, 159-161, 230, 244, 279, 299, 300
moss animals, 241
moths, 152, 241, 247
mowing, 20, 38, 59, 61, 83, 102, 108, 111, 112, 134, 139, 164, 239, 274, 275, 287, 288
Multibloom-hoarypea, 220, Plates 14, 27
music, 15, 53, 55, 164
musicians, 53
mussels, 21, 22, 54, 67, 90, 102-105, 119, 136, 142, 143, 148, 156-159, 161, 215, 242
mycorrhizae, 89, 98, 108, 127, 138, 139, 239, 270

Nabhan, Gary, 10, 84, 89, 108, 126, 165
Naquin, Chris, 264
Nardi, James B., 144
Narrowleaf mountain-mint, 222, 284, Plate 24
Narrowleaf silkgrass, 220, 284
Narrow-leaf springbeauty, 223, Plate 6
Natchez, MS, 24
Natchitoches, LA, 8, 24, 31, 197
Native Americans, 2, 5, 7-9, 13-20, 23, 24, 29, 31, 35, 41, 45, 46, 51-53, 67, 72, 98, 99, 101, 106, 128-132, 153,

160, 163, 167, 197, 204, 214-216, 229, 230
Natural Heritage Program (Louisiana), 9, 63-65, 68, 96, 128, 299, 300
Nelson, Ira, 137
New Iberia, LA, 1, 2, 20, 21, 24, 36, 42, 43, 44, 90, 197
New Jersey tea, 135, 223, 231, 284
Newman, Dawn Allen, 264
New Orleans, LA, 13, 24, 26, 27, 28, 31, 32, 33, 34, 48, 49, 136, 197, 199, 213, 263
Nine-banded Armadilloes, 20, 148, 157
nitrogen fixation, 134
Nodding beardtongue, 221, 284
Nodding wild-indigo, 224, 285, Plate 18
No Man's Land, 31, 48
North American Prairie Conference, 63, 69, 295
Norway Rats, 280
Nova Scotia, Canada, 14, 25, 32, 197, 203
nutrias, 141, 142, 257, 279
Nuttall's prairie parsley, 135, 219, 285
Nuttall's wild-indigo, 135, 224, 285

Obedient plant, 60, 222, 224, 285
Oberle, Frank, 295
oil (petroleum), 7, 11, 35, 40, 41, 45, 46, 48, 51, 55, 63, 102, 136, 147, 198, 288
Old field toadflax, 228
Old Spanish Trail, 36, 46, 51
Opelousas (Native Americans), 2, 14, 23, 24, 27, 197
Opelousas Prairie, 2, 13, 23, 27, 28, 33, 38, 197, 213
Opelousas General Hospital, 300
Opelousas (including Poste des Opelousas), LA, 1, 2, 8, 13, 14, 19, 21, 24-28, 31, 32, 34, 36-38, 41, 47, 52, 60, 61, 123, 125, 197, 203
orchids, 125, Plate 1
Ornate Box Turtles, 153
Oubre, Claude, 13, 197

Paille jaune, 61, 120, 215, 230
Painted coreopsis, 228
Pale coneflower, *See* Purple coneflower
Partridge pea, 20, 227
Passenger Pigeons, 142
Pastorek, Candi, 263
Pastorek, Marc, 69, 76, 85, 111, 166, 261, 263, 264, 269, Plate 32
Pecan, 43, 229
Pecan Island, LA, 43
Petit habitants, 7, 30, 45, 47, 230
phenology, 6, 70, 113, 123, 126, 127, 216, 234-237, 239
phosphate mobilization, 134
pigs, 1, 8, 19, 23, 28, 31, 40, 46, 54, 72, 102, 106, 134, 142, 157, 159, 209, 210, 213, 216, 279, 293
pimple mounds. *See* mima mounds
Pineland milkweed, 78, 222, 285
Pine Prairie, LA, 2, 47, 96
Pinkscale gayfeather, 135, 219, 285
Pitre, Kenneth, 261
PLANTSdatabase, 46, 115, 117, 122, 219, 283-286
Plaquemine Prairie, 3, 17, 19, 25, 36, 38, 69, 73, 77, 135, 283-287, Plate 2
platins, 4, 8, 44, 46, 52, 55, 72, 80-83, 94, 97, 101, 106, 125, 291, 293
pock-pock plants, 132
pollen, 92
pollinators, 55, 73, 84, 106, 108, 109,

112, 118, 126, 127, 129, 138, 139, 153, 165
polychetes, 242
Post, Lauren, 26, 44
pot pourri, 129
prairie chickens, 119, 142, 169
Prairie Faquetaique. See Faquetaique Prairie.
Prairie Fire Pink, 59
Prairienymph, 78, 220
prairies, Midwestern, 1, 2, 8, 35, 72, 91, 94, 101, 122, 125, 129, 239, 293
Prairie Swallow, 3, 40
prairie voles. See Louisiana Prairie Vole
Purple cone flower, 125, 135, 219, 285, Plates 1, 14, 24, 25
Purple Martins, 160, 217, 255
Purple passionflower, 20, 226

Quillman, Joann, Plate 12

Raoul grass. *See* Itch grass.
Rayless goldenrod, 135, 220, 285
Rayne, LA, 2, 52, 198
Red River, 48, 103, 128, 157, 176, 177, 179, 190
Red Wolves, 119, 142, 153
Reese, William (Bill), 64
reptiles, 115, 143, 153, 157, 176, 177, 179, 215, 251; list, 257
resource partitioning, 91, 155, 156
restoration procedure for prairies, 269-274
restoration projects, 84-85, 113
rhizosphere, 7, 98, 138, 143, 144, 149, 161
rice, 1, 7, 8, 9, 11, 19, 21, 34, 36, 38-40, 44-55, 57, 61, 64, 65, 68, 73, 80, 101-103, 105, 106, 124, 134, 142, 149, 150, 153, 154, 156-160, 197, 198, 209-211, 228, 256, 277, 291, 293
Ritchie (formerly Fabacher), LA, 49, 50
Roanoke, LA, 52
Roberie, Corrine, 264
Robert's Cove, LA, 3, 49, 198
Robichaux, Renee, 264
Robin, C. C., 13, 16, 21, 23, 26, 29, 30, 72, 74, 117, 197, 252
Ross, Gary, 61
rotifers, 142, 143, 157, 159, 160, 241
Rough coneflower, 135, 221, 285, Plates 17, 19, 30
Rough green snake, 251, Plate 32
railroads, 2, 6, 8, 9, 12, 21, 36, 45, 46, 50-53, 60, 63, 64, 66, 69, 70, 72, 75-77, 89, 99, 122, 125, 129, 164, 197, 198, 214, 269, 287-290

S1 (ranking), 9
Sabine Lake, 5, 42
Sabine Prairie, 3, 5, 6, 8, 24, 25, 31, 34, 38, 40, 42, 48, 51, 70, 73, 76, 103, 136
Sabine River, 3, 6, 24, 25, 38, 40, 70, 136
salt (salt domes), 6, 8, 16, 28, 42, 48, 51, 93, 101, 121, 244
Sampson's snakeroot, 221, 285
Savage, Candace, 295
Savoy, Barbara, 261
Scaly blazing-star, 219, 285, Plate 5
Scarlet Indian paintbrush, 130
Seale, Janet, 261
sedimentation, 102-104, 142, 152, 157
Sensitive brier, 221, Plates 24-25
shallow groundwater discharge, 103, 104, 157

Shannon Index, 118
sheep, 23, 28, 31, 39, 40, 46, 72, 134, 142, 157, 210, 213, 251, 279, 293
Shiny golden-rod, 222, 285, Plates 5, 9, 26
Simmonds, Sara, 264
Simpson's Index, 118
Single-stem scurfpea, 135, 221, 285
Slender bluestem, 79, 219, 285, 293
Slender gayfeather, 221, 285, Plate 17
Slender rosinweed, 220, 285, Plates 6, 7, 12, 13, 20, 21, 27, 30, 31
Smith, Latimore, 96
Smokey the Bear, 58, 100
snails, 143, 157, 159, 242
snakes, 132, 161
Snowy orchid, 67, 135, 219, 285, Plate 1
soils, 1, 2, 6, 7, 11, 27, 30, 34, 38, 39, 44, 52, 53, 55, 58-60, 63, 67, 70, 71, 73-75, 77, 79, 80, 82, 83, 87, 89, 90, 92-99, 101-104, 106-113, 120, 124, 137, 138, 140, 143-148, 156, 157, 163, 164, 168, 169, 214, 215, 239, 264, 269, 270-275, 277, 288
soil sickness, 73
Southern blackberry, 228, 287
Southern carpet grass, 44, 228, 279
Southern dewberry, 228, 287
soybeans, 8, 48, 51, 55, 124, 277
Spanish Lake, 5, 42
species evenness, 6, 117, 118, 127, 214
species richness, 6, 112, 117-119, 121, 127, 138, 214, 215, 251
spiders, 110, 241, 249, 281
sponges, 241
Spotted beebalm, 129, 227, 285, 288
Spring ladies'-tresses, 223, 285, Plate 1
St. Augustine grass, 130, 146, 279
St. Landry Parish, LA, 2, 8, 14, 24, 35, 37, 39, 41, 47, 49, 59, 73, 74, 123, 197
St. Martin Parish, LA, 2, 8, 14, 24, 41, 251
St. Martinville, LA, 2, 24, 26, 32, 37, 44
Stanford, Myriam, 129, 264
Stephen F. Austin State University, 263
Stewart, Amy, 146
Strawberry, 129
Strecker, John K., 148
succession (ecological), 40, 78, 79, 81, 85, 86, 113, 121, 126-128, 161, 239, 281
Sugarcane, 44, 55, 61, 293
Sulphur, LA, 35
Swamp narrowleaf sunflower, 225, 285
Sweet goldenrod, 129, 219, 285, Plates 16, 22
Sweet Lake, 5
Switchgrass, 118, 119, 125, 167, 214, 215, 223, 285, Plates 8, 10, 16, 22, 23
Syron, Wayne, 85

Taber, Stephen W., 147
Tall tickseed, 135, 222, 285
Talus slope penstemon, 135, 220, 285
tardigrades, 143
Texas brown-eyed Susan, 220, 285, Plates 12-13
Texas ironweed, 135, 223, 285
Thames, Sara, 85, 261, 264
Three Sisters (agricultural system), 7, 29, 52, 128, 214
tick fever, 34
timber, 16, 19, 34, 35, 38-40, 117
Toothache grass, 78, 135, 221, 286
topsoil, 33, 44, 51, 82, 97, 98, 107, 113,

138, 140, 164, 275, 277
Touchet, Orville, 96
traiteurs, 131

Union Pacific Railroad, 66, 77, 299
University of Louisiana-Lafayette, 64, 85, 149, 251, 264
University of Louisiana-Monroe, 66
Urbatsch, Lowell, 262
Uruguayan primrosewillow, 279

Vachers, 7, 8, 29, 99, 230
Vacheries, 7, 33, 230
Valvilov, Nikolay, 74, 84
Varichaetadrilus harmani, 159, 242
Vasey grass, 82, 84, 85, 112, 228, 279
Védrines, Jean Baptiste Lapaise, 18, 197, 200
vegetables, 39, 46, 51, 81, 84, 106, 132, 146, 203, 209, 210
Vermilion Bay, 5
Vermilion Parish, LA, 2
Vermilion Prairie, 3, 8, 25, 33, 37, 38, 41-43, 71, 73, 77, 96, 103, 117, 136, 156, 178, 213, 216, 256
vertisols, 94-97
Vidrine, Bette, 85, 275
Vidrine, Caroline, 80, 169, 267, 299
Vidrine, Daniel, 78, 80, 267, 299, Plate 12
Vidrine, Euzebe, 55
Vidrine family, 200-205
Vidrine, Gail, 68, 80, 130, 137, 267, 299, Plate 12
Vidrine, Jacqueline O., 18, 200, 204
Vidrine, Jean Baptiste Pierre. *See* Jean Pierre Baptiste Lapaise Védrines
Vidrine, John R. (Randy), 201, 205
Vidrine, LA, 49, 204
Vidrine, Malcolm F., 199, 261, 269, Plate 32
Vidrine, Marie Odile Hidalgo, 199, 203, 205
Vidrine, Numa Lee, 199, 204
Vidrine silt loam, 96
Vige, Shirley Jr., 261
Ville Platte, LA, 1, 2, 89, 90, 202, 204
Villeré, Sidney Louis, 26, 43, 205
Vinton, LA, 2, 76
Virginia spiderwort, 135, 223, 286, Plate 2
Virginia strawberry, 129

Wade, Michael, 44
Washington, LA, 2, 14, 18, 24, 26, 45, 48, 197, 202, 207
Water hemlock, 85, 227
Wax-myrtle, 228, 288
Weeks Island, LA, 5, 8
Welsh, LA, 2, 51
Welsh Prairie, 36, 45, 47, 52, 129, 135, 165, 283-289, Plates 12-14, 29
Western horsenettle, 135, 224, 286, 287
Wetland Classification, 126
wheat, 1, 29, 55, 277, 293
White clover, 279
White colic-root, 135, 220, 286, 289
White Lake, 5
Whiteleaf mountain-mint, 135, 224, 286, 289
White prairie-clover, 220, 286
White-tailed Deer, 139, 153, 257
Whooping Crane, 60, 150, 153
Whorled milkweed, 125, 135, 222, 286
Wild Ones (The), 281
Wild Turkeys, 40
Williams, Avery, 251

Willow Lake, 5
Wilson, Becky, 264
Wilson, Edward O., 114, 147
Wilson, Ken, 261, 262, 264
wind, 4, 37, 38, 89, 90, 91, 96, 97, 106, 111, 114, 270, 273
Woodlawn Prairie, LA, 165, 289, Plate 17
Woods poppymallow, 67, 138, 221, 286, Plates 1, 14, 24, 25
Wooly sunflower, 219, 221, 228
World War II, 36, 48, 55, 198
Wrinkled leaf goldenrod, 219, 286, Plate 16

Yellow Indiangrass, 67, 72, 98, 118, 119, 124-126, 214, 219, 271, 286
Yellow wild-indigo, 224, 286, Plates 7, 30, 31

Zebra Mussels, 142
Zigzag iris, 125, 224, 286

About the Author

Malcolm F. Vidrine is a biologist, who earned a B. S. and M. S. in Zoology from Louisiana State University in 1970 and 1974, and a Ph.D. in biology from University of Louisiana-Lafayette (then University of Southwestern Louisiana) in 1980. He has worked with Gulf South Research Institute on armadillos and Hansen's disease, with the Jefferson Davis Parish Mosquito Abatement District on mosquito control and biology/ecology of rice fields and with LSU-Eunice as professor. He has published more than 100 scientific papers and 10 popular articles. This is his 10th book. He was with Charles Allen when Charles rediscovered remnant prairies in southwest Louisiana, and he has visited, photographed and studied each of the remnants that were found along railroad rights-of-way in southwestern Louisiana. He also directly participated in the reconstruction of 3 prairies (the Cajun Prairie Restoration Project in Eunice, the LSU-Eunice Prairie Project and his own Cajun Prairie Gardens).